D. Schomburg · M. Salzmann (Eds.)

GBF – Gesellschaft für Biotechnologische Forschung

# Enzyme Handbook
# 4

## Class 3: Hydrolases

Springer-Verlag
Berlin Heidelberg GmbH

Professor Dr. Dietmar Schomburg
Margit Salzmann
GBF – Gesellschaft für Biotechnologische Forschung mbH
Mascheroder Weg 1
W-3300 Braunschweig, FRG

This collection of datasheets was generated from the database „BRENDA"

ISBN 978-3-642-48986-0        ISBN 978-3-642-84437-9 (eBook)
DOI 10.1007/978-3-642-84437-9

Media conversion: Brühlsche Universitätsdruckerei, Giessen
Production of the plasticfiles: Lux-Plastik oHG, Murnau
51/3130-543210 – Printed on acid-free paper

# Preface

Recent progress in enzyme immobilisation, enzyme production, coenzyme regeneration and enzyme engineering has opened up fascinating new fields for the potential application of enzymes in a large range of different areas. As more progress in research and application of enzymes has been made the more apparent has become the lack of an up-to-date overview of enzyme molecular properties. The need for such a data bank was also expressed by the EC-task force "Biotechnology and Information". Therefore we started the development of an enzyme data information system as part of protein-design activities at GBF. The present book "Enzyme Handbook" represents the printed version of this data bank. In future it is also planned to make a computer searchable version available.

The enzymes in the Handbook are arranged according to the 1984 Enzyme Commission list of enzymes and later supplements. Some 3000 "different" enzymes are covered. Frequently very different enzymes are included under the same E. C. number. Although we intended to give a representative overview on the molecular variability of each enzyme, the Handbook is not a compendium. The reader will have to go to the primary literature for more detailed information. Naturally it is not possible to cover all numerous, up to 40 000, literature references for each enzyme if data representation is to be concise as is intended.

The authors are grateful to the following bioligists and chemists for invaluable help in the compilation of data, expecially Cornelia Munaretto, Dr. Ida Schomburg, Dr. Sabine Vogel-Ziebolz, Uwe Hirschgänger, Inka Siegmund and Roland Vogt. Mrs. C. Munaretto and Dr. I. Schomburg are also thanked for the correction of the final manuscript.

Braunschweig
June, 1990

Margit Salzmann
Dietmar Schomburg

# BRENDA – Enzyme Data for Research and Production

Enzymes are used in all parts of the living world for catalysis of innumerable biochemical reactions. It has been known for some time that they represent potentially highly interesting catalysts for chemical research and production because of their high efficiency, stereospecifity, and their regio- and enantioselectivity. Enormous progress has been made in recent years in enzyme immobilisation, stabilisation, coenzyme regeneration etc., while gene technology has made possible the production of large quantities of otherwise inaccessible enzymes. Enzyme-design methods and recent work on enzyme behaviour in organic solvents have opened further possibilities for their use in new application areas. In addition to the chemical industry their use in the food industry and environmental technology is worthy of mention.

A number of problems still have to be overcome before enzymes take their place besides the other commonly used catalysts in the chemical laboratory, productions and in the awareness of the chemist and the general public. So the current industrial use of enzymes is more or less limited to proteases and carbohydrases, i. e. hydrolytic enzymes. Other synthetically important reactions such as the forming of C-C bonds are rarely achieved enzymatically at present. In addition to the real problems (price, too high selectivity etc.) reservations on the part of the chemists prohibit decisions about their potential use.

This is mainly caused by the undeniable fact that information about enzymes is not as easily obtained as that about other synthetic means and catalysts. Results of enzyme research are often published in journals which are rarely read by most chemists and are often not available in organic-chemistry libraries. The published data on molecular weight, stability etc. can be contradictory, the use of a number of different names for the same enzyme is common, and the systematic classification of the enzyme is usually unsatisfactory.

The apparently simple question:

*"Is there an enzyme, that catalyzes the enantioselective replacement of an hydrogen atom in an α-position to an aromatic ring and is stable in a water solution with a certain pH-value and a given temperature?"*

can often only be answered after intensive work in a library or the use literature data bases. A rational choice between several potential enzyme catalyzed reaction paths is imposible in a reasonable time. When planning research projects in the area of enzyme design, information about potential enzymes and use of the correct initial enzyme is also essential.

In addition to the above types of chemical problem, systematic biochemical investigations are central to the wide interest in enzymes, therefore a great deal of information can be gained from establishment of a comprehensive collection of enzyme data.

In the GBF research programme, enzyme technology has always played a significant role. This is documented by innovative contributions in the field of cofactor regeneration, enzyme production by genetic engineering and downstream processing after fermentation.

The area has gained new significance from the activities in protein design where the central theme is protein structure determination and biocomputing.

It soon became evident that the next logical step was the development of an enzyme data bank which could be undertaken by the group for molecular structure research. We hope that publication of this comprehensive, critical literature evaluation will be of wide interest to external users and, in this way, we hope the GBF has provided a further important contribution to the information infrastructure in biotechnology.

Braunschweig                                             Joachim Klein
June, 1990                                    GBF, Scientific Director

# List of Abbreviations

| | | | |
|---|---|---|---|
| A | adenosine | *E. coli* | *Escherichia coli* |
| Ac | acetyl | EDTA | ethylene diaminetetraacetate |
| ADP | adenosine 5'-diphosphate | EGTA | ethylene glycol bis ($\beta$-aminoethyl ether) tetraacetate |
| Ala | alanine | | |
| All | allose | | |
| Alt | altrose | ER | endoplasmic reticulum |
| AMP | adenosine 5'-monophosphate | Et | ethyl |
| | | EXAFS | extended x-ray absorption fine structure |
| Ara | arabinose | | |
| Arg | arginine | | |
| Asn | asparagine | FAD | flavin-adenine dinucleotide |
| Asp | aspartic acid | | |
| ATP | adenosine 5'-triphosphate | FMN | flavin mononucleotide (riboflavin 5'-monophosphate) |
| c | cytidine | | |
| cal | calories | Fru | fructose |
| CDP | cytidine 5'-diphosphate | Fuc | fucose |
| CDTA | trans-1,2-diaminocyclohexane-N,N,N,N-tetra-acetic acid | G | guanosine |
| | | Gal | galactose |
| | | GDP | guanosine 5'-diphosphate |
| CMP | cytidine 5'-monophosphate | Glc | glucose |
| | | GlcN | glucosamine |
| CoA | coenzyme A | GlcNAc | N-acetylglucosamine |
| CTP | cytidine 5'-triphosphate | Gln | glutamine |
| Cys | cysteine | Glu | glutamic acid |
| d | deoxy- | Gly | glycine |
| D- and L- | prefixes indicating configuration | GMP | guanosine 5'-monophosphate |
| DFP | diisopropyl fluorophosphate | GSH | glutathione |
| | | GTP | guanosine 5'-triphosphate |
| DNA | deoxyribonucleic acid | | |
| DPN | diphosphopyridinium nucleotide (now NAD) | Gul | gulose |
| | | $H_4$ | tetrahydro |
| DTNB | 5,5'-dithiobis (2-nitrobenzoate) | His | histidine |
| | | HPLC | high pressure liquid chromatography |
| DTT | dithiothreitol (i.e. Cleland's reagent) | Hyl | hydroxylysine |
| EC | number of enzyme in Enzyme Commission's system | Hyp | hydroxyproline |
| | | IAA | iodoacetamide |
| | | Ig | immunoglobulin |

| | |
|---|---|
| Ile | isoleucine |
| Ido | idose |
| IDP | inosine 5'-diphosphate |
| IMP | inosine 5'-monophosphate |
| ITP | inosine 5'-triphosphate |
| $K_m$ | concentration of substrate giving half maximum velocity (Michaelis constant) |
| L- | see D- |
| Leu | leucine |
| Lys | lysine |
| Lyx | lyxose |
| M | gm molecule (1 mole) per litre |
| m- | meta- |
| Man | mannose |
| MES | 2-(N-morpholino)ethane sulfonate |
| Met | methionine |
| mM | $10^{-3}$ Mole |
| Mur | muramic acid |
| MW | molecular weight |
| NAD | nicotinamide-adenine dinucleotide (state of oxidation unspecified) |
| $NAD^+$ | nicotinamide-adenine dinucleotide (oxidized form) |
| NADH | reduced NAD |
| NADP | NAD phosphate (state of oxidation unspecified) |
| $NADP^+$ | NADP (oxidized form) |
| $NAD(P)^+$ | indicates either $NAD^+$ or $NADP^+$ |
| NADPH | reduced NADP |
| NAD(P)H | indicates either NADH or NADPH |
| NDP | nucleoside 5'-diphosphate |
| NEM | N-ethylmaleimide |
| Neu | Neuraminic acid |
| nm | nanometre ($10^{-9}$ metre) |
| NMN | nicotinamide mononucleotide |
| NMP | nucleoside 5'-monophosphate |
| NTP | nucleoside 5'-triphosphate |
| o- | ortho- |
| Orn | ornithine |
| p- | para- |
| PCMB | p-chloro-mercuribenzoate |
| PEP | phosphoenolpyruvate |
| pH | $-\log_{10}$ [H] |
| Ph | phenyl |
| Phe | phenylalanine |
| PIXE | proton-induced X-ray emission |
| PMSF | phenylmethane-sulfonylfluoride |
| Pro | proline |
| $Q_{10}$ | temperature coefficient for a reaction |
| Rha | rhamnose |
| Rib | ribose |
| RNA | ribonucleic acid |
| mRNA | messenger RNA |
| rRNA | ribosomal RNA |
| tRNA | transfer RNA |
| Sar | N-methylglycine (sarcosine) |
| SDS-PAGE | sodium dodecyl sulphate (= sodium lauryl sulphate)- polyacrylamide gel electrophoresis |
| Ser | serine |
| T | ribosyl-thymine |
| $t_{1/2}$ | time for half-completion of reaction |
| Tal | talose |
| TDP | ribosylthymine 5'-diphosphate |

| | | | |
|---|---|---|---|
| Thr | threonine | u | uridine |
| TMP | ribosylthymine 5'-monophosphate | U/mg | $\mu$mol/(mg*min) |
| | | UDP | uridine 5'-diphosphate |
| Tos- | tosyl- (p-toluenesulfonyl-) | UMP | uridine 5'-monophosphate |
| TPN | triphosphopyridinium nucleotide (now NADP) | UTP | uridine 5'-triphosphate |
| | | Val | valine |
| | | Xaa | symbol for an amino acid of unknown con- stitution in peptide formula |
| Tris | tris(hydroxymethyl)- aminomethane | | |
| Trp | tryptophan | | |
| TTP | ribosylthymine 5'-triphosphate | XAS | X-ray absorption spectroscopy |
| Tyr | tyrosine | Xyl | xylose |

# Index

(Alphabetical order of Enzyme names)

## 1 NOMENCLATURE

**EC number**
   3.2.2.5

**Systematic name**
   NAD$^+$ glycohydrolase

**Recommended name**
   NAD$^+$ nucleosidase

**Synonymes**
   Nucleosidase, nicotinamide adenine dinucleotide
   NAD hydrolase
   Diphosphopyridine nucleosidase
   Nicotinamide adenine dinucleotide nucleosidase
   NADase
   Diphosphopyridine nucleotidase
   DPNase
   NAD glycohydrolase
   DPN hydrolase
   NAD nucleosidase
   Nicotinamide adenine dinucleotide glycohydrolase

**CAS Reg. No.**
   9032-65-9

## 2 REACTION AND SPECIFICITY

**Catalysed reaction**
   NAD$^+$ + H$_2$O →
   → nicotinamide + ADPribose (enzyme from some animal sources also
   catalyzes transfer of ADPribose residues)

**Reaction type**
   N-Glycosyl bond hydrolysis

**Natural substrates**
   NAD$^+$ + H$_2$O (possible participation in the mechanism of Ca$^{2+}$ release
   from rat liver mitochondria) [28]

## Substrate spectrum

1 NAD$^+$ + H$_2$O (Bacillus subtilis: irreversible [3]) [1–30]
2 NADP$^+$ + H$_2$O (43% of the rate of NAD hydrolysis [1], microsomal enzyme [25], chromatin enzyme not [25]) [1, 2, 3, 5, 15, 18, 21, 25, 26]
3 Thio NAD$^+$ + H$_2$O (28% the rate of NAD hydrolysis) [1]
4 Nicotinamide hypoxanthine dinucleotide + H$_2$O (90% the rate of NAD hydrolysis) [2]
5 Nicotinamide ethenoadenine dinucleotide + H$_2$O [15, 27]
6 Nicotinamide hypoxanthine dinucleotide + H$_2$O [15]
7 More (transglycosidase activity [14, 16, 27], catalyzes alcoholysis of NAD$^+$ to form O-alkyl-ADP-ribosides and an ADP-ribosylation of imidazole derivatives [4], neglegible activity towards NAD$^+$ analogues [5], mechanism [7], self-inactivation [7], rat erythrocyte: no pyridine base exchange reaction [15], no phosphatase, no phosphodiesterase activity [28], not: deamino-NAD$^+$, deamino-NMN, nicotinamide ribonucleoside [1], NADH, NADPH [2]) [1, 2, 4, 5, 7, 15, 16, 27, 29, 28]

## Product spectrum

1 Nicotinamide + ADP ribose
2 ?
3 Thionicotinamide + ADP ribose
4 ?
5 Nicotinamide + ethenoadenine ribose
6 Nicotinamide + hypoxanthine ribose
7 ?

## Inhibitor(s)

N$^1$-Alkylnicotinamide chlorides [2, 24]; NADH [2]; HgCl$_2$ [21]; Nicotinamide (only at very high concentrations [1], microbial enzyme not inhibited [3]) [1, 2, 9, 12, 21, 22, 26, 29]; NADP$^+$ [1, 9]; Thio-NAD$^+$ [1]; Deamino-NAD$^+$ [1]; NMN [1]; Alpha-NAD$^+$ [1]; ADP-ribose [2, 4, 12, 14, 15, 24, 26, 29]; Adenosine [2, 12, 15]; AMP [2, 4, 12, 14, 24]; ADP [2, 4, 12, 14, 15, 24]; n-Alkylphosphates [2]; Aliphatic carboxylic acids [2]; 3-Acetylpyridine [2, 3]; Pyridine dinucleotides [2]; Thionicotinamide [3]; Anionic phospholipids [26]; Nicotinylhydroxamic acid derivatives (of NAD$^+$) [3]; Pyridine bases [4]; Pyridine nucleotides [4]; Thionicotinamide NAD$^+$ [5]; HgCl$_2$ (only at high concentration) [5]; p-Chloromercuribenzoate (only at high concentration) [5]; Iodoacetamide (only at high concentration) [5]; EDTA [6]; Benzoic acid [9]; NAD$^+$ (membrane-bound: no inhibition, soluble: inhibition by excess of substrate [12], substrate inhibition disappeares at high ionic strength and low pH, below pH 5: substrate activation [17]) [12, 17]; cAMP [12]; Salicylic acid [12, 14];

NADH [12]; 4-Carboxyhydrazide-pyridine-adenine dinucleotide [12]; 2,
3-Butanedione [14]; Arginyl-modifiying reagent [14]; Woodward's reagent K
[14]; dAMP [14]; ATP [14]; 2-Hydroxy-5-iodo-benzoic acid [14]; 2-Hydroxy-3,
5-diiodo-benzoic acid [14]; Adenylic acid [15]; Pyridine derivatives [15, 25];
NAD$^+$ analogs [15]; Alpha-iodoacetamide [20]

## Cofactor(s)/prostethic group(s)
More (high-molecular weight component required for full activity) [8]

## Metal compounds/salts
Cu$^{2+}$ (Agkistrodon acutus, snake venom: 1 mol enzyme contains 1 mol
Cu$^{2+}$, essential for activity) [6]

## Turnover number (min$^{-1}$)
21500 (NAD$^+$) [3]; More [30]

## Specific activity (U/mg)
850 [3]; 996 [5]; 7.0 [17]; More [2, 4, 6, 16, 19, 20, 28, 30]

## K$_m$-value (mM)
3.2 (pyridine base exchange reaction, 3-aminopyridine) [4]; 7.9 (pyridine
base exchange reaction, 3-pyridyl carbinol) [4]; 1.4 (pyridine base exchange reaction, pyridine) [4]; 20.0 (pyridine base exchange reaction,
3-acetylpyridine) [4]; 6.1 (pyridine base exchange reaction,
3-pyridylacetonitrile) [4]; 6.5 (pyridine base exchange reaction, 3-methyl-
pyridine) [4]; 0.50 (beta-NAD$^+$) [6]; 0.13 (NADP$^+$) [6]; 0.16 (beta-nicotine
guanine dinucleotide) [6]; 0.020 (NADP$^+$) [15]; 0.014 (epsilon-NAD$^+$) [15];
0.0285 (NAD$^+$, chromatin enzyme) [1]; 0.060 (NAD$^+$, nuclear envelope enzyme) [1]; 0.1 (NAD$^+$, 25°C) [2]; 0.32 (NAD$^+$, 38°C) [2]; More (of membrane
bound and soluble form [12], effect of ionic strength on K$_m$ [12]) [2, 3, 4, 5, 7
, 12, 13, 14, 15, 16, 17, 18, 20, 25]

## pH-optimum
7.5 [2, 4, 6, 29]; 5–6 [5]; 7.2 (broad) [6]; 6.5–7.0 (chromatin enzyme [1]) [1,
18]; 6.5 (nuclear envelope enzyme) [1, 16]; 8.0–8.5 [15]; 7.3 [20]; 6.0 [21];
6.3–7.0 (nucleus) [22]; 6.4 (microsomal) [25]; 7.6–8.0 (chromatin) [25]; More
[7]

## pH-range
3–9 (3: about 75% of activity maximum, 9: about 70% of activity maximum)
[5]; 6.0–9.0 (6.0: about 10% of activity maximum, 9.0 about 80% of activity
maximum) [15]; 5.0–9.0 (5.0: about 10% of activity maximum, 9.0: about
15% of activity maximum) [16]; 4.5–8.0 (4.5: about 65% of activity maximum, 8.0: about 80% of activity maximum) [21]

## Temperature optimum (°C)
37 (assay at) [5]; 40 [6, 20]; More [12, 15]

Enzyme Handbook © Springer-Verlag Berlin Heidelberg 1991
Duplication, reproduction and storage in data banks are only
allowed with the prior permission of the publishers

**Temperature range** (°C)
35–45 (active between) [20]; More [12]

---

## 3 ENZYME STRUCTURE

**Molecular weight**
90000 (calf, permeation chromatography) [10]
29000 (Vibrio cholera, inner membrane, SDS-PAGE) [11]
22000 (Vibrio cholera, outer membrane, SDS-PAGE) [11]
83000 (rat, sucrose density gradient sedimentation) [13]
24000 (calf, gel filtration [17], Bacillus subtilis, gel filtration [19]) [17, 19]
26200 (Bacillus subtilis, ultracentrifugation) [19]
36500 (Neurospora crassa, ultracentrifugation) [19]
31000 (Neurospora crassa, gel filtration) [19]
55000 (Streptococcus pyogenes, gel filtration, SDS-PAGE) [20]
36300 (bull, high speed sedimentation equilibrium) [2, 23]
120000 (bovine, gel filtration) [24]
62000 (rat, SDS-PAGE) [28]
33000 (Neurospora crassa, SDS-PAGE) [30]
26200 (Bacillus subtilis, ultracentrifugation) [3]
27000 (Fusarium nivale, gel filtration) [5]
98000 (Agkistrodon acutus, gel filtration) [6]
More [8]

**Subunits**
Dimer (Bungarus fasciatus, 2 × 62000) [4, 29]
Dimer (Agkistrodon acutus, 2 × 50000) [6]
More [7]

**Glycoprotein/Lipoprotein**
Glycoprotein (10.4% carbohydrate [2], 52.8% carbohydrate [3],
Neurospora crassa: 55% carbohydrate, Bacillus subtilis: 76% carbohydrate
[19]) [2, 3, 4, 6, 13, 19, 29, 30]

---

## 4 ISOLATION/PREPARATION

**Source organism**
Rat [1, 13, 16, 18, 25, 28]; Bovine [2, 15, 23, 24, 26]; Bacillus subtilis [3];
Bungarus fasciatus (snake venom [4]) [4, 27, 29]; Fusarium nivale [5]; Agkistrodon acutus (snake [6]) [6]; Rabbit [21]; Mouse (Ehrlich ascites tumor
cells) [22]; Streptococcus pyogenes [8, 20]; Mycobacterium phlei [9]; Calf
[10, 12, 14, 17]; Vibrio cholera [13]; Bacillus subtilis [19]; Neurospora crassa
[19, 30]; More (widely distributed among procaryotes and eucaryotes [7],
widely distributed in mammals, found in some microorganisms, rarely
reported in higher plants [17]) [7, 17]

## Source tissue

Liver [1, 13, 16, 18, 21, 25, 28]; Seminal plasma [2]; Snake venom [4, 6, 27, 29]; Cells [19]; Semen [23, 24]; Spleen [10, 12, 14, 17]; Erythrocytes [15]; Ehrlich ascites tumor cells [22]; Thyroid tissue [26]; Conidia [30]

## Localisation in source

Plasma membrane [7, 13, 26]; Extracellular [7, 8]; Microsomes (membrane [13]) [7, 13, 25]; Nucleus [7, 18, 22]; Membrane (firmly bound) [10–12]; Lysosomes [21]; More (microbial enzyme: soluble, mammalian enzyme: except that from bull semen: membrane bound) [17]

## Purification

Streptococcus pyogenes [8, 20]; Mycobacterium phlei [9]; Bacillus subtilis [19]; Neurospora crassa [19, 30]; Rat [1, 13]; Bovine [2, 26]; Bacillus subtilis [3]; Fusarium nivale [4]; Agkistrodon acutus (snake) [6]; Calf [17]; Bungarus fasciatus [29]; More [7]

## Crystallization

–

## Cloned

–

## Renaturated

–

---

## 5 STABILITY

## pH

More (more labile at higher pH than at lower pH) [17]

## Temperature (°C)

60 (pH 4.5, 5 minutes, 45% loss of activity) [1]; 100 (5 minutes, complete loss of activity) [2]; 50 (1 hour, at or below 50°C, no loss of activity) [2]; 52 (5 minutes stable) [16]; 37 (pH 7.4, membrane-bound, no loss of activity after 6 hours [12], soluble: $t_{1/2}$ is 25 minutes [17]) [12, 17]; 63 (50% loss of activity after 5 minutes) [16]; 45 (pH 7, 1 minute, chromatin enzyme: 75% loss of activity, microsomal enzyme: stable) [25]; More [2, 22]

## Oxidation

## Organic solvent

## General stability information

Immobilization (on concanavalin A-Sepharose, 70% retention of catalytic activity) [4]

**Storage**

–20°C (stable for several weeks) or –80°C (stable for several months) [17];
4°C (50% loss of activity) [17]; 4 °C, pH 7.4, 0.01 M potassium phosphate
buffer, 4 weeks (40 % loss of activity) [2]; –20°C (stable for over 1 year) [3];
4°C (stable for several months) [3]; Lyophilized, –20 °C (stable for several
weeks) or –90°C (stable for several months) [5]; –20°C (stable for at least 1
month) [13]

## 6 CROSSREFERENCES TO STRUCTURE DATABANKS

**PIR/MIPS code**

**Brookhaven code**

## 7 LITERATURE REFERENCES

[1] Okayama, H., Ueda, K., Hayaishi, O.: Methods Enzymol., 66, 151–154 (1980)
(Review)

[2] Anderson, B.M., Yuan, J.H.: Methods Enzymol., 66, 144–150 (1980) (Review)

[3] Everse, K.E., Everse, J., Simeral, L.S.: Methods Enzymol., 66, 137–144 (1980) (Review)

[4] Anderson, B.M., Yost, D.A., Anderson, C.D.: Methods Enzymol., 122, 173–181 (1986)
(Review)

[5] Stathakos, D., Isaakidou, I., Thomou, H.: Biochim. Biophys. Acta, 302, 80–89 (1973)

[6] Huang, W.-Z., Wang, C., Luo, L.-Q., Lu, Z.-X.: Toxicon, 26, 535–542 (1988)

[7] Price, S.R., Pekala, P.H. in "Coenzymes Cofactors", 2, 513–548 (Review)

[8] Zahradnik, F.J.: Folia Microbiol., 25, 40–49 (1980)

[9] Davis, W.B.: Antimicrob. Agents Chemother., 17, 663–668 (1980)

[10] Schuber, F., Muller, H., Schenherr, I.: FEBS Lett., 109, 247–251 (1980)

[11] Fernandes, P.B., Welsh, K.M., Bayer, M.E.: J. Biol. Chem., 254, 9254–9261 (1979)

[12] Travo, P., Muller, H., Schuber, F.: Eur. J. Biochem., 96, 141–149 (1979)

[13] Diaugustine, R.P., Abe, T., Voytek, P.: Biochim. Biophys. Acta, 526, 518–530 (1978)

[14] Schuber, F., Pascal, M., Travo, P.: Eur. J. Biochem., 83, 205–214 (1978)

[15] Pekala, P.H., Anderson, B.M.: J. Biol. Chem., 253, 7453–7459 (1978)

[16] Fukushima, M., Okayama, H., Takahashi, Y., Hayaishi, O.: J. Biochem., 80, 167–176
(1976)

[17] Schuber, F., Travo, P.: Eur. J. Biochem., 65, 247–255 (1976)

[18] Ueda, K., Fukushima, M., Okayama, H., Hayaishi, O.: J. Biol. Chem., 250, 7541–7546
(1975)

[19] Everse, J., Everse, K.E., Kaplan, N.O.: Arch. Biochem. Biophys., 169, 702–713 (1975)

[20] Grushoff, P.S., Shany, S., Bernheimer, A.W.: J. Bacteriol., 122, 599–605 (1975)

[21] Mellors, A.M., Lun, A.K.L., Peled, O.N.: Can. J. Biochem., 53, 143–148 (1974)

[22] Green, S., Dobrjansky, A.: Biochemistry, 11, 4108–4113 (1972)

[23] Yuan, J.H., Barnett, L.B., Anderson, B.M.: J. Biol. Chem., 247, 511–514 (1972)

[24] Yuan, J.H., Anderson, B.M.: J. Biol. Chem., 247, 515–520 (1972)

[25] Nakazawa, K., Ueda, K., Honjo, T., Yoshihara, K., Nishizuka, Y., Hayaishi, O.:
Biochem. Biophys. Res. Commun., 32, 143–149 (1968)

[26] De Wolf, M.J.S., Van Dessel, G.A.F., Lagrou, A.R., Hilderson, H.J.J., Dierick, W.S.H.: Biochem. J., 226, 415–427 (1985)

[27] Anderson, B.M., Anderson, C.D.: Anal. Biochem., 140, 250–255 (1984)

[28] Moser, B., Winterhalter, K.H., Richter, C.: Arch. Biochem. Biophys., 224, 358–364 (1983)

[29] Yost, D.A., Anderson, B.M.: J. Biol. Chem., 256, 3647–3653 (1981)

[30] Menegus, F., Pace, M.: Eur. J. Biochem., 113, 485–490 (1981)

## 1 NOMENCLATURE

**EC number**
3.2.2.6

**Systematic name**
NAD(P)$^+$ glycohydrolase

**Recommended name**
NAD(P)$^+$ nucleosidase

**Synonymes**
Nucleosidase, nicotinamide adenine dinucleotide (phosphate)
Nicotinamide adenine dinucleotide (phosphate) nucleosidase
Triphosphopyridine nucleotidase
NAD(P) nucleosidase
NAD(P)ase
Nicotinamide adenine dinucleotide (phosphate) glycohydrolase
NAD(P)-glycohydrolase

**CAS Reg. No.**
9025-46-1

---

## 2 REACTION AND SPECIFICITY

**Catalysed reaction**
NAD(P)$^+$ H$_2$O →
→ nicotinamide + ADPribose(P) (also catalyses transfer of
ADPribose(P) residues)

**Reaction type**
N-Glycosyl bond hydrolysis

**Natural substrates**
NAD(P)$^+$ + H$_2$O

**Substrate spectrum**
1 NAD(P)$^+$ + H$_2$O [1, 2]
2 NAD$^+$ + H$_2$O [1, 2]
3 Thio-NAD$^+$ + H$_2$O [2]
4 More (at low rate: acetylpyrimidine-adenine dinucleotide,
nicotinamide-hypoxanthine dinucleotide, nicotinamide-hypoxanthine
dinucleotide phosphate) [2]

**Product spectrum**
1  Nicotinamide + ADPribose(P)
2  Nicotinamide + ADPribose
3  ?
4  ?

**Inhibitor(s)**
$Hg^{2+}$ [1]; Heat-labile inhibitor [2]

**Cofactor(s)/prostethic group(s)**

**Metal compounds/salts**

**Turnover number** (min$^{-1}$)

**Specific activity** (U/mg)
409 [1]; More [2, 3]

**$K_m$-value** (mM)
0.52 (NAD(P)ase I, NAD$^+$) [1]; 0.66 (NADP$^+$) [2]; 0.85 (NAD$^+$) [2]; 0.25
(thio-NAD$^+$) [2]; 1.33 (NAD$^+$) [3]; 0.65 (NADP$^+$) [3]; 1.44
(nicotinamide-hypoxanthine dinucleotide) [2]; 0.51
(nicotinamide-hypoxanthine dinucleotide phosphate) [2]; More (effect of
temperatures on $K_m$) [2]

**pH-optimum**
5.5 [1]

**pH-range**
5–9 [3]; 5.5–8.0 [3]

**Temperature optimum** (°C)
50 [1]; 56 [3]

**Temperature range** (°C)
More [3]

# 3 ENZYME STRUCTURE

**Molecular weight**
50000 (2 forms: I (50000) and II (45000), Saccharomyces cerevisiae, gel
filtration) [1]
45000 (2 forms: I (50000) and II (45000), Saccharomyces cerevisiae, gel
filtration) [1]
23500 (Pseudomonas putida, gel filtration) [2]
160000 (gel filtration, Streptomyces griseus) [3]

Subunits

Glycoprotein/Lipoprotein

–

---

## 4 ISOLATION/PREPARATION

Source organism
   Human (Burkitt's lymphoma lines, nonmalignant lymphoblastoid cell lines)
   [5]; Neurospora crassa [4]; Saccharomyces cerevisiae (2 forms: I, II) [1];
   Pseudomonas putida [2]; Streptomyces griseus [3]; Rabbit [6]

Source tissue
   Cell [1, 2]; Erythrocytes [6]

Localisation in source

Purification
   Saccharomyces cerevisiae [1]; Pseudomonas putida [2]; Streptomyces
   griseus (partial) [3]

Crystallization
   –

Cloned
   –

Renaturated
   –

---

## 5 STABILITY

pH
   3–11 [1]

Temperature (°C)
   37 (5 minutes, 20% loss of activity) [1]; 50 (5 minutes, 100% loss of activity)
   [1]; 60 (5 minutes, 20% loss of activity) [3]; 100 (60 minutes, 30% loss of activity [2], 1 minute, 100% loss of activity [3]) [2, 3]

Oxidation

Organic solvent

General stability information
   Mineral acids (stable) [3]

**Storage**

–20°C, 0.1 M potassium phosphate buffer, pH 7.0, several weeks [2]; 4°C, pH 7 (50% loss of activity) [2]

## 6 CROSSREFERENCES TO STRUCTURE DATABANKS

**PIR/MIPS code**

**Brookhaven code**

## 7 LITERATURE REFERENCES

[1] Yamasaki, N., Mori, I., Takakuwa, M.: J. Ferment. Technol., 60, 131–137 (1982)
[2] Mather, I.H., Knight, M.: Biochem. J., 129, 141–152 (1972)
[3] Bröcker, M., Schindelmeiser, J., Pape, H.: FEMS Microbiol. Lett., 6, 245–247 (1979)
[4] Jorge, J.A., Terenzi, H.F.: J. Gen. Microbiol., 130, 1563–1568 (1984)
[5] Skala, H., Lenoir, G.M., Pichard, A.L., Vuillaume, M., Dreyfus, J.C.: Blood, 60, 912–917 (1982)
[6] Artman, M., Frankl, G.: Can. J. Microbiol., 28, 696–702 (1982)

# 1 NOMENCLATURE

**EC number**
   3.2.2.7

**Systematic name**
   Adenosine ribohydrolase

**Recommended name**
   Adenosine nucleosidase

**Synonymes**
   Adenosinase
   Nucleosidase, adenosine
   N-Ribosyladenine ribohydrolase
   Adenosine hydrolase
   ANase [1]

**CAS Reg. No.**
   9075-41-6

# 2 REACTION AND SPECIFICITY

**Catalysed reaction**
   Adenosine + $H_2O$ →
   → adenine + D-ribose (also acts on adenosine N-oxide)

**Reaction type**
   N-Glycosyl bond hydrolysis

**Natural substrates**
   Adenosine + $H_2O$ (adenine salvage metabolism [4], nucleoside recycling
   [5]) [4, 5]

**Substrate spectrum**
   1  Adenosine + $H_2O$ (ir [4, 12]) [1–13]
   2  Adenosine $N^1$-oxide + $H_2O$ [11, 13]
   3  7-Methyladenosine + $H_2O$ [1]
   4  2'-Deoxyadenosine + $H_2O$ [3, 9, 11, 12]
   5  $N^6$-(Delta-isopentenyl)adenosine + $H_2O$ [7]
   6  Purine ribosides + $H_2O$ [11]
   7  Purine ribonucleosides + $H_2O$ (low rate) [13]

## Product spectrum

1 Adenine + D-ribose (ir [4, 12]) [4]
2 Adenine-$N^1$-oxide + D-ribose
3 7-Methyladenine + D-ribose
4 2'-Deoxyadenine + D-ribose
5 $N^6$-(Delta-isopentenyl)adenine + D-ribose
6 Purine + D-ribose
7 ?

## Inhibitor(s)

Arsenite [13]; $Mn^{2+}$ [13]; $N^6$-Methyladenine (6-methylaminopurine) [8];
Dithiothreitol [11]; Adenosine [7]; Cytokinin nucleosides [7]; Adenine (and
derivatives [11]) [8, 11, 12, 13]; Cytokinin ribosides [2];
$N^6$-Isopentenyladenosine [2]; Zeatin riboside [2]; $N^6$-Benzyladenosine [2];
p-Chloromercuribenzoate (no effect [7]) [3]; $Zn^{2+}$ [3]; $Cu^{2+}$ [3]; $Hg^{2+}$ [3];
Cytokinin [11]

## Cofactor(s)/prostethic group(s)

## Metal compounds/salts

More (no metal ion requirement for optimal activity) [3]

## Turnover number (min$^{-1}$)

## Specific activity (U/mg)

0.736 [7]; 30.50 [11]; 7.97 [8]; 0.62 [13]; More [12]

## $K_m$-value (mM)

2.4 (adenosine) [13]; 0.025 (adenosine, R1) [2]; 0.009 (adenosine, R2) [2];
0.006 (adenosine, Lf) [2]; 0.00238 (cytokinin nucleoside) [7]; 0.120
(deoxyadenosine) [11]; More [2, 3, 5, 11, 12]

## pH-optimum

5.0 (R1 [2]) [2, 3]; 6.0 (R2, Lf) [2]; 4.5 (II) [8]; 4.0 (I/II) [8]; 4.0 [12]; 4.7 (citrate
buffer) [11]; 5. 4 (cytokinin nucleoside) [7]; 3.5–4.5 [13]; More [5]

## pH-range

3.5–7.0 (3.5: about 40% of maximal activity, 7.0: about 55 % of maximal activity) [12]; 2–8 [8]; 3–8 (R1 /pH 3 and 8: about 30% of maximal activity) [2];
4–9 (R2/Lf) [2]; 3.5–7.5 (3.5: about 55% of maximal activity, 7.5: about 30%
of maximal activity) [3]; 4.0–5.5 (4.0: about 50% of maximal activity, 5.5:
about 60% of maximal activity) [7]

## Temperature optimum (°C)

50 (II/III) [8]; 58 [11]; 60 [3]; 45 (I) [8]

## Temperature range (°C)

30–70 (30–40% of maximal activity) [3]

# 3 ENZYME STRUCTURE

## Molecular weight
 59000 (gel filtration, Triticum sativum) [7]
 68000 (gel filtration, Lycopersicon esculentum, R2) [2]
 120000 (gel filtration, barley) [3]
 68000 (gel filtration, Camellia sinensis) [9]
 66000 (gel filtration, barley) [11]
 62400 (sucrose density gradient sedimentation, Solanum tuberosum) [12]

## Subunits
 Dimer (2 × alpha: 33000, SDS-PAGE, barley) [11]

## Glycoprotein/Lipoprotein
 –

---

# 4 ISOLATION/PREPARATION

## Source organism
 Triticum sativum [7]; Trypanosoma cruzi [8]; Leishmania braziliensis [8];
 Leishmania donovani [8]; Leishmania tarentolae [8]; Leishmania mexicana
 [8]; Crithidia fasciculata [8]; Lupinus luteus [10]; Solanum tuberosum [12];
 Brussels sprouts [13]; Camellia sinensis (3 forms: I, II, III [9]) [1, 9]; Lycoper-
 sicon esculentum (root: 2 distinct forms, R1, R2, leaves: one form, Lf) [2];
 Helianthus tuberosus [5]; Acacia dealbata [4]; Barley (malted [3]) [3, 11];
 Fraxinus excelsior [4]; Tilia cordata [4]; Ulmus carpinifolia [4]; More [4]

## Source tissue
 Leaves [1, 2, 4, 9, 11, 12]; Roots [2]; Germ [7]; Cotyledons [10]; Seeds [10]

## Localisation in source

## Purification
 Barley (malted [3]) [3, 11]; Lycopersicon esculentum (partial, root: 2 distinct
 forms, R1, R2, leaves: one form Lf) [2]; Camellia sinensis (3 forms: I, II, III [9])
 [9]; Triticum sativum (partial) [7]; Solanum tuberosum [12]; Brussels
 sprouts [13]

## Crystallization
 –

## Cloned
 –

## Renaturated
 –

## 5 STABILITY

### pH
4 (unstable in acidic media) [8]; 6.0–8.0 [8]; More (adenine and adenosine protect against heat inactivation) [11]

### Temperature (°C)
60 (15 minutes, complete inactivation) [5]; 100 (enzyme stabilized in Sephadex, 5% loss of activity after 1 hour) [5]; 40 (15 minutes, rapid loss of activity above 40°C) [8]; 10 (15 days) [13]; 100 (3 minutes, complete loss of activity) [13]; 22 (8 days, 50% loss of activity) [13]

### Oxidation

### Organic solvent

### General stability information
Adenosine (protects against heat inactivation) [11]; Adenine (protects against heat inactivation) [11]; Freezing and thawing (stable) [11, 13]; Dilution (complete loss of activity when frozen in diluted salt solution) [11]; Enzyme entrapped in Sephadex G-50 and dried at above 0°C, 4.5 days at room temperature, stable [6]

### Storage
–20°C, 1 months (stable), 4°C (gradual loss of activity) [8]; –10°C, 0.01 M $NaHCO_3$, 4 months (5% loss of activity) [13]

---

## 6 CROSSREFERENCES TO STRUCTURE DATABANKS

### PIR/MIPS code

### Brookhaven code

---

## 7 LITERATURE REFERENCES

[1] Negishi, O., Ozawa, H., Imagawa, H.: Agric. Biol. Chem., 52, 169–175 (1988)
[2] Burch, L.R., Stuchbury, T.: J. Plant Physiol., 125, 267–273 (1986)
[3] Lee, W.L., Pyler, R.E.: J. Am. Soc. Brew. Chem., 44, 86–90 (1986)
[4] Leszczynska, D., Schneider, Z., Tomaszewski, M., Mackowiak, M.: Ann. Bot., 54, 847–849 (1984)
[5] Le Floc'h, F., Lafleuriel, J.: Phytochemistry, 20, 2127–2129 (1981)
[6] Schneider, Z., Friedmann, H.C.: J. Appl. Biochem., 3, 135–146 (1981)
[7] Chong-Maw Chen, Kristopeit, S.M.: Plant Physiol., 68, 1020–1023 (1981)
[8] Nolan, L.L., Kidder, G.W.: Antimicrob. Agents Chemother., 17, 567–571 (1980)
[9] Imagawa, H., Yamano, H., Inoue, K., Takino, Y.: Agric. Biol. Chem., 43, 2337–2342 (1979)

[10] Guranowski, A., Pawelkiewicz, J.: Planta, 139, 245–247 (1978)
[11] Guranowski, A., Schneider, Z.: Biochim. Biophys. Acta, 482, 145–158 (1977)
[12] Clark, M.C., Page, O.T., Fisher, M.G.: Phytochemistry, 11 , 3413–3419 (1972)
[13] Mazelis, M., Creveling, R.K.: J. Biol. Chem., 238, 3358–3361 (1963)

## 1 NOMENCLATURE

**EC number**
3.2.2.8

**Systematic name**
Nucleoside ribohydrolase

**Recommended name**
Ribosylpyrimidine nucleosidase

**Synonymes**
Nucleosidase, pyrimidine
N-Ribosylpyrimidine nucleosidase
Pyrimidine nucleosidase
N-Ribosylpyrimidine ribohydrolase [4]

**CAS Reg. No.**
37288-60-1

## 2 REACTION AND SPECIFICITY

**Catalysed reaction**
An N-D-ribosylpyrimidine + $H_2O$ →
→ a pyrimidine + D-ribose (also hydrolyzes purine D-ribonucleosides,
more slowly)

**Reaction type**
N-Glycosyl bond hydrolysis

**Natural substrates**
N-D-Ribosylpyrimidine + $H_2O$

**Substrate spectrum**
1 N-D-Ribosylpyrimidine + $H_2O$ [1–4]
2 Purine D-ribonucleosides + $H_2O$ (slowly) [3]
3 Cytidine + $H_2O$ [1, 3]
4 Uridine + $H_2O$ [1, 3]
5 Adenosine + $H_2O$ (not [3]) [4]
6 Xanthosine + $H_2O$ [3]
7 Inosine + $H_2O$ [3]
8 More (not: adenosin, guanosine) [3]

## Product spectrum

1 Pyrimidine + D-ribose
2 Purine + D-ribose
3 Cytosine + D-ribose
4 Uracil + D-ribose
5 Adenine + D-ribose
6 Xanthine + D-ribose
7 Hypoxanthine + D-ribose
8 ?

## Inhibitor(s)

4-Hydroxyl-1-beta-D-ribofuranosylpyrazolo-(3, 4-d) pyrimidine [3];
4-Thio-1-beta-D-ribofuranosylpyrazolo-(3, 4-d)pyrimidine [3];
4-Methylthio-1-beta-D-ribofuranosylpyrazolo-(3, 4-d)pyrimidine [3];
Guanosine [3]; 4, 6-Dihydroxy-1-beta-D-ribofuranosylpyrazolo-(3, 4-d)
pyrimidine [3]; Inosine [4]; Cytidine [4]; Adenosine [3, 4]; $MnSO_4$ [4]; $FeCl_3$
[4]; $HgCl_2$ [4]; $CoCl_2$ [4]; $CdCl_2$ [4]; $ZnCl_2$ [4]; EDTA [4]; More (poor inhibitors [4]) [3, 4]

## Cofactor(s)/prostethic group(s)

## Metal compounds/salts

More (no metal ion requirement) [4]

## Turnover number (min$^{-1}$)

## Specific activity (U/mg)

8.5 [3]; 50 [4]

## $K_m$-value (mM)

More [3]; 0.8 (uridine) [4]; 1.0 (cytidine) [4]; 0.5 (adenosine) [4]; 0.2 (inosine)
[4]; 2.5 (5-bromouridine) [4]; 7.5 (1-beta-D-ribofuranosylthymidine) [4]; 0.8
(uridine) [4]; 0.6 (guanosine) [4]; 6.5 (xanthosine) [4]

## pH-optimum

6.2 [1]; 6.5 [3]; 8.5 (uridine, cytidine, adenosine) [4]; 7.0 (xanthosine) [4]; 6.0
(inosine) [4]; 6.5 (guanosine) [4]; 8.0 (1-beta-D-ribofuranosylthymidine) [4]

## pH-range

## Temperature optimum (°C)

40 [1]

## Temperature range (°C)

# 3 ENZYME STRUCTURE

**Molecular weight**
   180000 (Leishmania donovani, gel filtration) [3]

**Subunits**

**Glycoprotein/Lipoprotein**
   –

---

# 4 ISOLATION/PREPARATION

**Source organism**
   Penicillium chrysogenum [1]; Penicillium oxalicum [1]; Leishmania
   donovani [2, 3]; Fusarium moniliforme [1]; Pseudomonas fluorescens [4]

**Source tissue**

**Localisation in source**

**Purification**
   Leishmania donovani [3]; Pseudomoas fluorescens [4]

**Crystallization**
   –

**Cloned**
   –

**Renaturated**
   –

---

# 5 STABILITY

**pH**

**Temperature (°C)**
   45 (10 minutes, pH 7.0–9.2, less than 10% loss of activity) [4]; 70 (5 minutes,
   pH 8.5, complete inactivation) [4]

**Oxidation**

**Organic solvent**

**General stability information**

**Storage**
   –15°C [4]; More [4]

# 6 CROSSREFERENCES TO STRUCTURE DATABANKS

PIR/MIPS code

Brookhaven code

# 7 LITERATURE REFERENCES

[1] Allam, A.M., Hassan, M.M., Ghanem, B.S., Elzainy, T.A. : Biochem. Syst. Ecol., 15, 515–517 (1987)
[2] Krenitsky, T.A., Koszalka, G.W., Tuttle, J.V., Adamczyk, D.L., Elion, G.B., Marr, J.J.: Adv. Exp. Med. Biol., 122, 51–56 (1980)
[3] Koszalka, G.W., Krenitsky, T.A.: J. Biol. Chem., 254, 8185–8193 (1979)
[4] Terada, M., Tatibana, M., Hayaishi, O.: J. Biol. Chem. , 242, 5578–5585 (1967)

## 1 NOMENCLATURE

**EC number**
   3.2.2.9

**Systematic name**
   S-Adenosyl-L-homocysteine homocysteinylribohydrolase

**Recommended name**
   Adenosylhomocysteine nucleosidase

**Synonymes**
   Nucleosidase, adenosylhomocysteine
   S-Adenosylhomocysteine hydrolase
   S-Adenosylhomocysteine nucleosidase
   5'-Methyladenosine nucleosidase [2]
   S-Adenosylhomocysteine/5'-methylthioadenosine nucleosidase [5]
   AdoHcy/MTA nucleosidase [5]

**CAS Reg. No.**
   9055-10-1

## 2 REACTION AND SPECIFICITY

**Catalysed reaction**
   S-Adenosyl-L-homocysteine + $H_2O$ →
   → adenine + S-D-ribosyl-L-homocysteine (also acts on 5'-
   methylthioadenosine to give adenine and 5-methylthioribose)

**Reaction type**
   N-Glycosyl bond hydrolysis

**Natural substrates**
   S-Adenosyl-L-homocysteine + $H_2O$ (catabolism) [1]
   5'-Methylthioadenosine + $H_2O$ (catabolism) [1]
   More (purine metabolism) [6]

**Substrate spectrum**
   1 S-Adenosyl-L-homocysteine + $H_2O$ [1–3]
   2 5'-Methylthioadenosine + $H_2O$ (not: enzyme from Alcaligenes faecalis
     and Sarcina lutea [4]) [1–3]
   3 5'-Ethylthioadenosine + $H_2O$ [2]
   4 5'-n-Propylthioadenosine + $H_2O$ [2]

**Product spectrum**
1 Adenine + S-D-ribosyl-L-homocysteine [1]
2 Adenine + 5-methylthioribose [1]
3 Adenine + 5-ethylthioribose
4 Adenine + 5-n-propylthioribose

**Inhibitor(s)**
5'-Methylthioinosine [2]; L-Methionine [1]; AMP [1]; 5'-Ethylthioadenosine
[2]; 5'-n-Propylthioadenosine [2]; S-Adenosylhomocysteine [2];
N-2-Hydroxyethylpiperazine-N'-2-ethanesulfonic acid [2]; Tris [2]; Adenine
[2]; 5'-Methylthioformycin [5]; 5'-Chloroformycin [5];
S-Formicinylhomocysteine [5]; 5'-Methylthiotubercidin [5];
S-Tubercidinylhomocysteine [5]; S-8-Aza-Adenosylhomocysteine [5]; More
(poor inhibitors) [5]

**Cofactor(s)/prostethic group(s)**
More (no cofactor required) [1]

**Metal compounds/salts**

---

**Turnover number** (min$^{-1}$)

**Specific activity** (U/mg)
9.93 [2]; 3.75 [1]; 3.6 [3]; 373 [5]

**$K_m$-value** (mM)
1.8 (5'-methylthioadenosine) [3]; 3.0 (S-adenosylhomocysteine) [3]; 0.0004
(5'-methylthioadenosine) [5]; 0.0043 (S-adenosylhomocysteine) [5]; 0.00031
(5'-methylthioadenosine) [2]; 1.8 (5'-methylthio-adenosine) [1]; 3.0
(S-adenosylhomocysteine) [1]

**pH-optimum**
6.5 [1, 3]

**pH-range**
5.7–7.1 (little variation) [1, 3]; 4.0–8.0 (low activity at) [1]

**Temperature optimum** (°C)
37 (assay at) [1, 2]

**Temperature range** (°C)

---

## 3 ENZYME STRUCTURE

**Molecular weight**
31000 (gel filtration, E. coli) [2]

**Subunits**

**Glycoprotein/Lipoprotein**

–

---

## 4 ISOLATION/PREPARATION

**Source organism**
E. coli [1–5]; Bacillus cereus [4]; Staphylococcus aureus [4]; Lycopersicon esculentum [6]; Bacteria [4]; Aerobacter aerogenes [1, 3]; Salmonella typhimurium [1, 3]; Alcaligenes faecalis [4]; Sarcina lutea [4]; Citrobacter freundii [4]; Enterobacter aerogenes [4]; Proteus vulgaris [4]; More (not: Candida utilis, Saccharomyces cerevisiae, rat /spleen/liver/kidney [3], not: eucaryotes, yeast, plants, birds, mammals [4]) [3, 4]

**Source tissue**
Cell [1, 3, 6]

**Localisation in source**
Cytoplasm [6]

**Purification**
E. coli [1, 2, 3, 5]

**Crystallization**
–

**Cloned**
–

**Renaturated**
–

---

## 5 STABILITY

**pH**

**Temperature (°C)**
60 (15 minutes, rapidly inactivated) [5]

**Oxidation**

**Organic solvent**

**General stability information**
Dilution (extremly sensitive to dilution) [2]; Glycerol (20%, stabilizes) [2]; Low ionic strength (unstable) [1, 3]; Dialysis (unstable) [1, 3]; 2-Mercaptoethanol (stabilizes) [3]; EDTA (stabilizes) [3]; Freezing and thawing (stable) [2]

**Storage**

−20°C, several weeks (enzyme after Sephadex treatment: stable, enzyme from DEAE-Sephadex column: unstable) [1]; 4 °C (50% loss of activity after 3 weeks) [2]; −20°C, 5.0 mg/ml, repeatedly frozen and thawed, 9 months (no loss of activity) [2]; 0°C, 0.1M phosphate buffer, 3 mM 2-mercaptoethanol, 0.5 mM EDTA (enzyme from DEAE-Sephadex, stable for 2 weeks) [2]; −20°C, 2 months [5]

# 6 CROSSREFERENCES TO STRUCTURE DATABANKS

**PIR/MIPS code**

**Brookhaven code**

# 7 LITERATURE REFERENCES

[1] Duerre, J.A.: J. Biol. Chem., 237, 3737–3741 (1962)
[2] Ferro, A.J., Barrett, A., Shapiro, S.K.: Biochim. Biophys. Acta, 438, 487–494 (1976)
[3] Duerre, J.A.: Methods Enzymol., 17B, 411–415 (1971)
[4] Walker, R.D., Duerre, J.A.: Can. J. Biochem., 53, 312–319 (1975)
[5] Della Ragione, F., Porcelli, M., Carteni-Farina, M., Zappia, V.: Biochem. J., 232, 335–341 (1985)
[6] Wasternack, C., Guranowski, A., Glund, K., Tewes, A., Walther, R.: J. Plant Physiol., 120, 19–28 (1985)

# 1 NOMENCLATURE

**EC number**
3.2.2.10

**Systematic name**
Pyrimidine-5'-nucleotide phosphoribo(deoxyribo)hydrolase

**Recommended name**
Pyrimidine-5'-nucleotide nucleosidase

**Synonymes**
Pyrimidine nucleotide N-ribosidase

**CAS Reg. No.**
9023-31-8

---

# 2 REACTION AND SPECIFICITY

**Catalysed reaction**
A pyrimidine 5'-nucleotide + $H_2O$ →
→ a pyrimidine + D-ribose 5-phosphate

**Reaction type**
N-Glycosyl bond hydrolysis

**Natural substrates**
Pyrimidine 5'-nucleotides + $H_2O$ [5]

**Substrate spectrum**
1 Pyrimidine 5'-nucleotides + $H_2O$ (ir) [5]

**Product spectrum**
1 Pyrimidines + pentose 5-phosphate [5]

---

**Inhibitor(s)**
p-Chloromercuribenzoate [5]; EDTA [5]; KCN [5]; $Ag^+$ [5]; $Hg^{2+}$ [5]; $Cu^{2+}$ [5]; $Zn^{2+}$ [5]; $Cd^{2+}$ [5]; $Cr^{3+}$ [5]

**Cofactor(s)/prostethic group(s)**

**Metal compounds/salts**
$Ca^{2+}$ [5]

---

**Turnover number** (min$^{-1}$)

**Specific activity** (U/mg)
  2.5 [5]

**K$_m$-value** (mM)
  0.2 (ribosylthymine) [4]; 6.95 (5'-UMP) [5]

**pH-optimum**
  8.0 (5'-UMP) [1] 7.5 (ribosylthymine) [4]; 7.0 (pyrimidine nucleotides) [5];
  6.0–7.0 (5'-UMP) [6]

**pH-range**

**Temperature optimum** (°C)
  55 (5'-UMP) [6]

**Temperature range** (°C)

---

## 3 ENZYME STRUCTURE

**Molecular weight**

**Subunits**

**Glycoprotein/Lipoprotein**
  –

---

## 4 ISOLATION/PREPARATION

**Source organism**
  Neisseria meningitidis [1]; Mammals [2]; Pseudomonas [3]; Alcaligenes
  faecalis [3]; Neurospora crassa [4]; Streptomyces [5, 6]

**Source tissue**
  Mammalian erythrocytes [2]

**Localisation in source**

**Purification**
  Streptomyces virginiae [5]

**Crystallization**
  –

**Cloned**
  –

**Renaturated**
  –

## 5 STABILITY

pH
  5.5–8.8 [5]

Temperature (°C)

Oxidation

Organic solvent

General stability information

Storage
  2 months, –4°C [5]

## 6 CROSSREFERENCES TO STRUCTURE DATABANKS

PIR/MIPS code

Brookhaven code

## 7 LITERATURE REFERENCES

[1] Jyssum, S.: APMIS, 97, 343–346 (1989)
[2] Lestas, A.N., Nicolaides, K.H., Rodeck, C.H., Bellingham, A.J.: Br. J. Haematol., 63, 471–476 (1986)
[3] Sakai, T., Yu, T., Omata, S.: Agric. Biol. Chem., 40, 1893–1895 (1976)
[4] Bankel, L., Lindstedt, G., Lindstedt, S.: FEBS Lett., 71, 147–153 (1976)
[5] Imada, A.: J. Gen. Appl. Microbiol., 13, 267–278 (1967)
[6] Imada, A., Kuno, M., Igarasi, S.: J. Gen. Appl. Microbiol., 13, 255–265 (1967)

# 1 NOMENCLATURE

**EC number**
3.2.2.11

**Systematic name**
1-Beta-aspartyl-N-acetyl-D-glucosaminylamine L-asparaginohydrolase

**Recommended name**
Beta-aspartyl-N-acetylglucosaminidase

**Synonymes**
Acetylglucosaminidase, .beta.-aspartyl
.beta.-Aspartylacetylglucosaminidase

**CAS Reg. No.**
9027-31-0

---

# 2 REACTION AND SPECIFICITY

**Catalysed reaction**
1-Beta-aspartyl-N-acetyl-D-glucosamine + $H_2O$ →
→ N-acetyl-D-glucosamine + L-asparagine

**Reaction type**
N-Glycosyl bond hydrolysis

**Natural substrates**
1-Beta-aspartyl-N-acetyl-D-glucosamine + $H_2O$
1-Beta-aspartyl-2-acetamido-1, 2-dideoxy-D-glucosamine + $H_2O$ (cleavage
of beta-aspartylglycosylamine linkage in ovalbumin glycopeptide)

**Substrate spectrum**
1  1-Beta-aspartyl-N-acetyl-D-glucosamine + $H_2O$ (amide bond in
   glycopeptides only when NH2- and $CO_2$-group of aspartic acid are free)
   [2]
2  1-Beta-aspartyl-2-acetamido-1, 2-dideoxy-D-glucosylamine + $H_2O$ [1]
3  More (no release of asparagine from glycopeptides obtained from al-
   pha$_1$-glycoprotein, fetuin or ovalbumin after pronase treatment)

**Product spectrum**
1  N-Acetyl-D-glucosamine + L-asparagine
2  2-Acetamido-1, 2-dideoxy-D-glucosamine + L-asparagine
3  ?

**Inhibitor(s)**

**Cofactor(s)/prostethic group(s)**

**Metal compounds/salts**

---

**Turnover number** (min$^{-1}$)

**Specific activity** (U/mg)
0.105 [1]

**K$_m$-value** (mM)

**pH-optimum**
7.7 (2 similar enzymes with different pH-optima: 7.7 and 7.5) [2]; 7.5 (2 similar enzymes with different pH-optima: 7.7 and 7.5) [2]

**pH-range**
6–8 (nearly maximal activity from 6–8) [1]; 4.4 (inactive at pH 4.4) [1]

**Temperature optimum** (°C)
37 (assay at) [1]

**Temperature range** (°C)

---

## 3 ENZYME STRUCTURE

**Molecular weight**

**Subunits**

**Glycoprotein/Lipoprotein**
–

---

## 4 ISOLATION/PREPARATION

**Source organism**
Sheep [1]; Limnea stagnalis (2 similar enzymes with different pH-optima) [2]

**Source tissue**
Epididymis [1]

**Localisation in source**

**Purification**
Sheep [1]

**Crystallization**
–

Cloned

–

Renaturated

–

## 5 STABILITY

pH

Temperature (°C)

Oxidation

Organic solvent

**General stability information**
Lyophilization (stable) [1]; Freezing (after butanol extraction, stable) [1]

**Storage**
–20°C, several months [1]

## 6 CROSSREFERENCES TO STRUCTURE DATABANKS

PIR/MIPS code

Brookhaven code

## 7 LITERATURE REFERENCES

[1] Eylar, E.H., Murakami, M.: Methods Enzymol., 8, 597–600 (1966)
[2] Kaverzneva, E.D., Tschuchrova, A.I., Kisseleva, V.V.: Liebigs Ann. Chem., 738, 130–135 (1970)

## 1 NOMENCLATURE

**EC number**
3.2.2.12

**Systematic name**
5'-Inosinate phosphoribohydrolase

**Recommended name**
Inosinate nucleosidase

**Synonymes**
Nucleosidase, inosinate

**CAS Reg. No.**
37288-61-2

## 2 REACTION AND SPECIFICITY

**Catalysed reaction**
5'-Inosinate + $H_2O$ →
→ hypoxanthine + D-ribose 5-phosphate

**Reaction type**
N-Glycosyl bond hydrolysis

**Natural substrates**
5'-Inosinate + $H_2O$

**Substrate spectrum**
1 5'-Inosinate + $H_2O$ [1]

**Product spectrum**
1 Hypoxanthine + D-ribose 5-phosphate

**Inhibitor(s)**

**Cofactor(s)/prostethic group(s)**

**Metal compounds/salts**

**Turnover number** (min$^{-1}$)

**Specific activity** (U/mg)

**$K_m$-value** (mM)

pH-optimum

pH-range

Temperature optimum (°C)

Temperature range (°C)

---

## 3 ENZYME STRUCTURE

Molecular weight

Subunits

Glycoprotein/Lipoprotein

—

---

## 4 ISOLATION/PREPARATION

Source organism

Source tissue

Localisation in source

Purification

Crystallization

—

Cloned

—

Renaturated

—

---

## 5 STABILITY

pH

Temperature (°C)

Oxidation

Organic solvent

General stability information

Storage

2

# 6 CROSSREFERENCES TO STRUCTURE DATABANKS

**PIR/MIPS code**

**Brookhaven code**

# 7 LITERATURE REFERENCES

[1] Kuninaka, A.: Koso Kagaku Shinpojiumu, 12, 65 (1957)

## 1 NOMENCLATURE

**EC number**
3.2.2.13

**Systematic name**
1-Methyladenosine ribohydrolase

**Recommended name**
1-Methyladenosine nucleosidase

**Synonymes**
Nucleosidase, 1-methyladenosine
1-Methyladenosine hydrolase [2]

**CAS Reg. No.**
37367-71-8

## 2 REACTION AND SPECIFICITY

**Catalysed reaction**
1-Methyladenosine + $H_2O$ →
→ 1-methyladenine + D-ribose

**Reaction type**
N-Glycosyl bond hydrolysis

**Natural substrates**
1-Methyladenosine + $H_2O$ (enzyme is related to oocyte maturation and spawning in starfish) [3]

**Substrate spectrum**
1 1-Methyladenosine + $H_2O$

**Product spectrum**
1 1-Methyladenine + D-ribose

**Inhibitor(s)**
SH-blocking agents [1]; N-Carboxyphenylmaleimide [1];
p-Chloromercuribenzoate [1]; N-Ethylmaleimide [1]; o-Iodosobenzoate [1];
1-Methylionosine [2]; 1-Methylguanosine [2]

**Cofactor(s)/prostethic group(s)**

**Metal compounds/salts**

**Turnover number** (min$^{-1}$)

**Specific activity** (U/mg)
  More [2]

**K$_m$-value** (mM)
  0.665–0.715 (1-methyladenosine) [2]

**pH-optimum**
  7.8–8.0 [1]; 7.5 [3]

**pH-range**
  6.2–9.5 (half-maximum activities) [2]; More [3]

**Temperature optimum** (°C)
  25 (assay at) [2, 3]

**Temperature range** (°C)

---

## 3 ENZYME STRUCTURE

**Molecular weight**
  96000 (gel filtration, Asterina pectinifera) [3]

**Subunits**

**Glycoprotein/Lipoprotein**
  –

---

## 4 ISOLATION/PREPARATION

**Source organism**
  Asterina pectinifera [1, 3]; Pisaster ochraceous [2]; Asterias amurensis [3];
  Patiria miniata [3]

**Source tissue**
  Follicle cells (isolated) [1]; Testes [2]; Ovaries [2, 3]; Stomach [2]; Pyloric
  caeca [2]

**Localisation in source**

**Purification**
  Asterina pectinifera (partial) [1]; Pisaster ochraceous [2]

**Crystallization**
  –

**Cloned**
  –

**Renaturated**
–

---

## 5 STABILITY

**pH**

**Temperature (°C)**
40 (21% loss of activity after 30 minutes) [2]

**Oxidation**

**Organic solvent**

**General stability information**
Freezing and thawing (inactivates) [2]

**Storage**

---

## 6 CROSSREFERENCES TO STRUCTURE DATABANKS

**PIR/MIPS code**

**Brookhaven code**

---

## 7 LITERATURE REFERENCES

[1] Mita, M.: Dev. Growth Differ., 28, 67–74 (1986)
[2] Tarr, H.L.A.: J. Fish. Res. Board Can., 30, 1861–1866 (1973)
[3] Shirai, H., Kanatani, H.: Exp. Cell Res., 75, 79–88 (1972)

# 1 NOMENCLATURE

**EC number**
3.2.2.14

**Systematic name**
Nicotinamide-nucleotide phosphoribohydrolase

**Recommended name**
NMN nucleosidase

**Synonymes**
NMNase
Nucleosidase, nicotinamide mononucleotide
Nicotinamide mononucleotidase
NMN glycohydrolase
NMNGhase [3]

**CAS Reg. No.**
37237-49-3

---

# 2 REACTION AND SPECIFICITY

**Catalysed reaction**
Nicotinamide D-ribonucleotide + $H_2O \rightarrow$
$\rightarrow$ nicotinamide + D-ribose 5-phosphate

**Reaction type**
N-Glycosyl bond hydrolysis

**Natural substrates**
NMN + $H_2O$
More (enzyme is important regulatory component of pyridine nucleotide cycle metabolism [1], pyridine nucleotide cycle) [4, 5]

**Substrate spectrum**
1 Nicotinamide D-ribonucleotide + $H_2O$ (ir [1, 6], highly specific for NMN [6, 7]) [1–7]

**Product spectrum**
1 Nicotinamide + D-ribose 5-phosphate (ir [1]) [1–7]

## Inhibitor(s)

p-Chloromercuribenzoate [6]; N-Ethylmaleimide [6]; Nicotinamide [6]; Guanosine [5]; D-Ribose-5-phosphate [5 , 6]; ATP [1]; NAD$^+$ [5]; CTP [1]; 7-Methyl GTP [1]; NADP (less effective) [5]; GMP (in presence of GTP [6]) [5, 6]; dGMP (in presence of GTP) [6]; dCMP (in presence of GTP) [6]; GTP [5]; Inorganic polyphosphates [1]; Cd$^{2+}$ [6]; Cu$^{2+}$ [6]; Zn$^{2+}$ [6]; EDTA [6]

## Cofactor(s)/prostethic group(s)

More (activators [1, 2], GTP activates [1, 2, 6], obligatory functional requirement for GTP [1, 2, 6], regulatory enzyme with absolute requirement for an intracellular and high-molecular-weight component: RNA [2], absolute dependence on guanylic acid derivatives [6]) [1, 2, 6]

## Metal compounds/salts

More (metal ion not required for full activity) [6]

## Turnover number (min$^{-1}$)

## Specific activity (U/mg)

0.317 [6]; More [5, 6]

## K$_m$-value (mM)

0.220 (NMN) [4]; 4.5 (NMN) [6]

## pH-optimum

8.5–9.0 (Tris-HCl buffer) [6]

## pH-range

7.5–10 (7.5: less than 10% of maximal activity, 10: 50% of maximal activity) [6]

## Temperature optimum (°C)

30 (assay) [4]; 39 [6]

## Temperature range (°C)

25–45 (25°C: about 45% of maximal activity at, 45°C: about 95% of maximal activity) [6]

# 3 ENZYME STRUCTURE

## Molecular weight

67000 (gel filtration, Salmonella typhimurium) [5]
213000 (gel filtration, Azotobacter vinelandii) [6]
240000 (gel filtration, Azotobacter vinelandii) [6]

## Subunits

**Glycoprotein/Lipoprotein**

  –

---

## 4 ISOLATION/PREPARATION

**Source organism**
  Azotobacter vinelandii [1, 2, 6]; Nicotiana tabacum [3, 4]; Salmonella
  typhimurium [5]; E. coli [7]

**Source tissue**
  Callus (callus cultures from roots of seedlings) [3, 4]; Cell [1, 2, 5, 7]; Root
  [4]

**Localisation in source**
  Membrane (87%) [7]; Cytoplasm (13%) [7]

**Purification**
  Azotobacter vinelandii [6]

**Crystallization**

  –

**Cloned**

  –

**Renaturated**

  –

---

## 5 STABILITY

**pH**

**Temperature (°C)**

**Oxidation**

**Organic solvent**

**General stability information**

**Storage**
  1 week at 4°C or 1 month at 0°C or –16°C, 0.025 M, Tris-HCl, pH 7.5, 1 mM
  reduced glutathione [6]

## 6 CROSSREFERENCES TO STRUCTURE DATABANKS

PIR/MIPS code

Brookhaven code

## 7 LITERATURE REFERENCES

[1] Imai, T.: J. Biochem., 101, 163–173 (1987)
[2] Imai, T.: J. Biochem., 101, 153–161 (1987)
[3] Wagner, R., Feth, F., Wagner, K.G.: Planta, 168, 408–413 (1986)
[4] Wagner, R., Feth, F., Wagner, K.G.: Planta, 167, 226–232 (1986)
[5] Foster, J.W.: J. Bacteriol., 145, 1002–1009 (1981)
[6] Imai, T.: J. Biochem., 85, 887–899 (1979)
[7] Andreoli, A.J., Okita, T.W., Bloom, R., Grover, T.A.: Biochem. Biophys. Res. Commun., 49, 264–269 (1972)

## 1 NOMENCLATURE

**EC number**
3.2.2.15

**Systematic name**
DNA-deoxyinosine deoxyribohydrolase

**Recommended name**
DNA-deoxyinosine glycosidase

**Synonymes**
DNA(hypoxanthine)glycohydrolase
Deoxyribonucleic acid glycosylase
Hypoxanthine-DNA glycosylase [1, 2]

**CAS Reg. No.**

## 2 REACTION AND SPECIFICITY

**Catalysed reaction**
Hydrolyses DNA and polynucleotides, releasing free hypoxanthine

**Reaction type**
N-Glycosyl bond hydrolysis

**Natural substrates**
DNA + $H_2O$ (DNA repair by preventing deaminated adenine residues from being expressed)

**Substrate spectrum**
1 DNA + $H_2O$ [1, 2]
2 Polyribonucleotides + $H_2O$ (double-stranded having either thymidine or cytosine residues in the complementary strand, single stranded polymers slowly attacked [1]) [1, 2]

**Product spectrum**
1 DNA (hydrolyzed) + hypoxanthine
2 Polyribonucleotides (hydrolyzed) + hypoxanthine

**Inhibitor(s)**
$Mn^{2+}$ (high concentration) [1]; $Ca^{2+}$ (high concentration) [1]; NaCl [2]

**Cofactor(s)/prostethic group(s)**
More (no cofactors required) [1, 2]

**Metal compounds/salts**
More (divalent cations not required) [2]

**Turnover number** (min$^{-1}$)

**Specific activity** (U/mg)

**$K_m$-value** (mM)
0.0009 (dIMP residues) [1]

**pH-optimum**
7.2–7.8 [1]

**pH-range**

**Temperature optimum** (°C)
37 (assay at) [1]

**Temperature range** (°C)

## 3 ENZYME STRUCTURE

**Molecular weight**
31000 (calf, sedimentation coefficient) [1]
30000 (E. coli, gel filtration) [2]

**Subunits**

**Glycoprotein/Lipoprotein**
–

## 4 ISOLATION/PREPARATION

**Source organism**
Calf [1]; E. coli [2]

**Source tissue**
Thymus [1]; Cell [2]

**Localisation in source**

**Purification**
Calf [1]; E. coli (partial) [2]

**Crystallization**
–

**Cloned**
–

**Renaturated**

–

---

## 5 STABILITY

**pH**

**Temperature (°C)**
56 (half-life: 3 minutes) [2]

**Oxidation**

**Organic solvent**

**General stability information**

**Storage**

---

## 6 CROSSREFERENCES TO STRUCTURE DATABANKS

**PIR/MIPS code**

**Brookhaven code**

---

## 7 LITERATURE REFERENCES

[1] Karran, P., Lindahl, T.: Biochemistry, 19, 6005–6011 (1980)
[2] Karran, P., Lindahl, T.: J. Biol. Chem., 253, 5877–5879 (1978)

## 1 NOMENCLATURE

**EC number**
3.2.2.16

**Systematic name**
Methylthioadenosine methylthioribohydrolase

**Recommended name**
Methylthioadenosine nucleosidase

**Synonymes**
Nucleosidase, methylthioadenosine
Methylthioadenosine nucleosidase
5'-Methylthioadenosine nucleosidase
MTA nucleosidase [2]
MeSAdo nucleosidase [3]

**CAS Reg. No.**
50812-28-7

## 2 REACTION AND SPECIFICITY

**Catalysed reaction**
Methylthioadenosine + $H_2O$ →
→ adenine + 5'-methylthio-D-ribose

**Reaction type**
N-Glycosyl bond hydrolysis

**Natural substrates**
Methylthioadenosine + $H_2O$ (purine metabolism in plants [4]) [1–8]

**Substrate spectrum**
1 Methylthioadenosine + $H_2O$ (ir [4], also thioether analogs [3, 4]) [1–8]
2 Methylthioinosine + $H_2O$ [3]

**Product spectrum**
1 Adenine + 5'-methylthio-D-ribose (ir [4]) [1–8]
2 Hypoxanthine + 5-methylthio-D-ribose

## Inhibitor(s)

$F^-$ [5]; $CuSO_4$ [5]; EDTA [5]; Thiol reagents [6]; p-Chloromercuribenzoate [6]; Iodoacetamide (slight) [6] Sinefungin [3, 4]; Decoyinine [3, 4]; Adenosine 5'-carboxamide [3]; 9-Erythro-(2-hydroxy 3-nonyl)adenine [3, 4]; 5'-Ethylthioadenosine [2]; 5'-Chloroformycin [2]; 5'-Chloroadenosine [2]; 5'-Isobutylthioadenosine [2]; 5'-Isopropylthioadenosine [2]; (S)-9-(2, 3-Dihydroxypropyl)adenine [3, 4]; 3-Deazaadenine [3, 4]; Adenosine [5]; S-Adenosylmethionine [5]; S-Adenosylhomocysteine [5]; Adenine [3–5]; 5'-Methylthioribose (slight) [3, 4]

## Cofactor(s)/prostethic group(s)

More (reducing agents required, phosphate ion required for activity) [6]

## Metal compounds/salts

$Mg^{2+}$ (activity depends on) [5]; $Mn^{2+}$ (activity depends on) [5]

---

## Turnover number (min$^{-1}$)

## Specific activity (U/mg)

27.0 [4]; 0.276 [6]

## $K_m$-value (mM)

0.00031 (5'-methylthioadenosine) [8]; 0.00041 (methylthioadenosine) [3, 4]; 0.55 (methylthioinosine) [3]; 0.00725 (5'-methylthioadenosine) [5]; 0.0103 (methylthioadenosine) [8]

## pH-optimum

8–8.5 [4]; 8.6 [5]

## pH-range

6–9.5 (6: about 30% of maximal activity, 9.5: about 75% of maximal activity) [4]; 4.5–10 (4.5: about 20% of maximal activity, 10.0: about 30% of maximal activity) [5]

## Temperature optimum (°C)

45 [5]; 67 [6]

## Temperature range (°C)

20–60 (less than 50% of maximal activity at 20°C and 60°C) [5]; 37–75 (37°C: about 30% of maximal activity, 75°C: about 5% of maximal activity) [6]

---

## 3 ENZYME STRUCTURE

## Molecular weight

95000 (human, gel filtration) [6]
62000 (gel filtration, Lupinus luteus) [4]
46000 (gel filtration, Acetabularia mediteranea) [5]

## Subunits
Dimer (2 × 31000, SDS-PAGE, Lupinus luteus) [4]

## Glycoprotein/Lipoprotein
–

# 4 ISOLATION/PREPARATION

## Source organism
Persea americana [1]; Human [6]; Ochromonas malhamensis [7]; Vinca
rosea [8]; Lycopersicon esculentum [2]; Lupinus luteus [3, 4]; Acetabularia
mediteranea [5]; Dasycladaceae [5]; Acetabularia ryukyuensis [5];
Acetabularia major [5]; Acetabularia exigura [5]; Acetabularia clavata [5];
Acetabularia cliftonii [5]

## Source tissue
Fruit (mesocarp tissue) [1, 2]; Placenta [6]; Seeds [3, 4]

## Localisation in source
More (supernatant) [5]

## Purification
Lupinus luteus [3, 4]; Human [6]; E. coli [8]; Vinca rosea [8]

## Crystallization
–

## Cloned
–

## Renaturated
–

# 5 STABILITY

## pH

## Temperature (°C)
100 (3 minutes, complete inactivation) [5]; 30–50 [6]; 70 (15 minutes, complete loss of activity) [6]

## Oxidation
Reducing agent required [6]

## Organic solvent

## General stability information

2-Mercaptoethanol (stabilizes) [5]; Glutathione (reduced, stabilizes) [5]; L-Cystein (stabilizes) [5]

## Storage

−18°C (less than 10% loss of activity after 2 weeks) [5]

## 6 CROSSREFERENCES TO STRUCTURE DATABANKS

PIR/MIPS code

Brookhaven code

## 7 LITERATURE REFERENCES

[1] Kushad, M.M., Yelenosky, G., Knight, R.: Plant Physiol., 87, 463–467 (1988)
[2] Kushad, M.M., Richardson, D.G., Ferro, A.J.: Plant Physiol., 79, 525–529 (1985)
[3] Guranowski, A.B., Chiang, P.K., Cantoni, G.L.: Methods Enzymol., 94, 365–369 (1983)
[4] Guranowski, A.B., Chiang, P.K., Cantoni, G.L.: Eur. J. Biochem., 114, 293–299 (1981)
[5] Yamakawa, M., Schweiger, H.-G.: Dev. Cell Biol. (Dev. Biol. Acetabularia) 3, 241–253 (1979)
[6] Cacciapuoti, G., Oliva, A., Zappia, V.: Int. J. Biochem., 9, 35–41 (1978)
[7] Sugimoto, Y., Toraya, T., Fukui, S.: Arch. Microbiol., 108, 175–182 (1976)
[8] Baxter, C., Coscia, C.J.: Biochem. Biophys. Res. Commun., 54, 147–154 (1973)

## 1 NOMENCLATURE

**EC number**
3.2.2.17

**Systematic name**
Deoxy-D-ribocyclobutadipyridine polynucleotidodeoxyribohydrolase

**Recommended name**
Deoxyribodipyrimidine endonucleosidase

**Synonymes**
Endonuclease V (2 distinct domains in a single enzyme: pyrimidine dimer-DNA glycosylase and apurinic/apyrimidinic endonuclease activity) [7, 10]
Glycosidase, deoxyribonucleate pyrimidine dimer
Pyrimidine dimer DNA glycosylase
Pyrimidine dimer DNA-glycosylase
Pyridine dimer-DNA glycosylase [1]
PD-DNA glycosylase [1]
T4-induced UV endonuclease (a single protein contains both pyrimidine dimer-DNA glycosylase and apyrimidinic endonuclease) [3]
More (a single protein contains both pyrimidine dimer-DNA glycosylase and apyrimidinic endonuclease [3, 4, 8], glycosylase has an associated apyrimidinic/apurinic (AP)endonuclease [5]) [3–5, 8]

**CAS Reg. No.**
75302-33-9

---

## 2 REACTION AND SPECIFICITY

**Catalysed reaction**
Cleaves the N-glycosidic bond between the 5'-pyrimidine residue in cyclobutadipyrimidine (in DNA) and the corresponding deoxy-D-ribose residue (mechanism) [11]

**Reaction type**
N-Glycosyl bond hydrolysis

**Natural substrates**
DNA + $H_2O$ (repair of pyrimidine dimer-containig DNA) [5, 6]

## Substrate spectrum
1 DNA + $H_2O$ (pyrimidine dimers in double-stranded DNA preferred to those in single-stranded DNA [5, 9], under conditions of substrate excess, dimers containing 5'-thymidine are preferred to those with a 5'-cytosine residue) [5]

## Product spectrum
1 Hydrolyzed DNA

## Inhibitor(s)
2-N-(Deoxyguanosin-8-yl)acetyl aminofluorene [13]

## Cofactor(s)/prostethic group(s)

## Metal compounds/salts

## Turnover number (min⁻¹)

## Specific activity (U/mg)
More [5, 7]

## $K_m$-value (mM)
0.000001 (dipyrimidine in DNA) [4]; 0.000015 (thymidine-labeled DNA) [5]; 0.000280 (cytosine-labeled DNA) [5]

## pH-optimum
7.5 (assay at) [8]

## pH-range

## Temperature optimum (°C)
37 (assay at) [3, 8]

## Temperature range (°C)

## 3 ENZYME STRUCTURE

## Molecular weight
16000 (bacteriophage T4, SDS-PAGE) [7, 8]
26000 (gel filtration, E. coli infected with bacteriophage T4) [3]
17500 (Micrococcus luteus, SDS-PAGE) [5]

## Subunits
Monomer (Micrococcus luteus, SDS-PAGE) [5]

## Glycoprotein/Lipoprotein
–

## 4 ISOLATION/PREPARATION

**Source organism**
   Bacteriophage T4 (E. coli infected with bacteriophage T4 [3, 6]) [1, 3, 4, 6, 7, 8, 9, 10]; Micrococcus luteus [2, 5, 9]; E. coli (infected with bacteriophage T4) [3, 6, 10]

**Source tissue**

**Localisation in source**

**Purification**
   Micrococcus luteus [2, 5]; Bacteriophage T4 (partial) [7]

**Crystallization**
   –

**Cloned**
   (bacteriophage T4, endonuclease V gene) [12]

**Renaturated**
   –

## 5 STABILITY

**pH**

**Temperature (°C)**
   40 (about 55% loss of activity after 8 minutes) [3]; 42 (loss of activity, linear rate: 4% per minute) [8]

**Oxidation**

**Organic solvent**

**General stability information**

**Storage**

## 6 CROSSREFERENCES TO STRUCTURE DATABANKS

**PIR/MIPS code**

**Brookhaven code**

# 7 LITERATURE REFERENCES

[1] Radany, E.H., Friedberg, E.C.: Nature, 286, 182–185 (1980)
[2] Haseltine, W.A., Gordon, L.K., Lindan, C.P., Grafstrom, R.H., Shaper, N.L., Grossman, L.: Nature, 285, 634–641 (1980)
[3] Warner, H.R., Christensen, L.M., Persson, M.-L.: J. Virol., 40, 204–210 (1981)
[4] McMillan, S., Edenberg, H.J., Radany, E.H., Friedberg, R.C., Friedberg, E.C.: J. Virol., 40, 211–223 (1981)
[5] Grafstrom, R.H., Park, L., Grossman, L.: J. Biol. Chem., 257, 13465–13474 (1982)
[6] Radany, E.H., Friedberg, E.C.: J. Virol., 41, 88–96 (1982)
[7] Nakabeppu, Y., Yamashita, K., Sekiguchi, M.: J. Biol. Chem., 257, 2556–2562 (1982)
[8] Nakabeppu, Y., Sekiguchi, M.: Proc. Natl. Acad. Sci. USA, 78, 2742–2746 (1981)
[9] Gordon, L.K., Haseltine, W.A.: J. Biol. Chem., 255, 12047–12050 (1980)
[10] Seawell, P.C., Smith, C.A., Ganesan, A.K.: J. Virol., 35, 790–797 (1980)
[11] Bailly, V., Sente, B., Verly, W.G.: Biochem. J., 259, 751–759 (1989)
[12] Recinos III, A., Lloyd, R.S.: Biochemistry, 27, 1832–1838 (1988)
[13] Duker, N.J., Merkel, G.W.: Biochemistry, 24, 408–412 (1985)

## 1 NOMENCLATURE

**EC number**
3.2.2.19

**Systematic name**
$N^2$-(ADP-D-ribosyl)-L-arginine ADPribosylhydrolase

**Recommended name**
ADPribosylarginine hydrolase

**Synonymes**
ADP-ribose-L-arginine cleavage enzyme
ADP-ribosylarginine hydrolase [1]

**CAS Reg. No.**

---

## 2 REACTION AND SPECIFICITY

**Catalysed reaction**
$N^2$-(ADP-D-ribosyl)-L-arginine + $H_2O$ →
→ L-arginine + ADPribose

**Reaction type**
N-Glycosyl bond hydrolysis

**Natural substrates**
$N^2$-(ADP-D-ribosyl)-L-arginine + $H_2O$ (ADP-ribosylation cycle) [1, 2]

**Substrate spectrum**
1 $N^2$-(ADP-D-ribosyl)-L-arginine + $H_2O$ [1–3]
2 ADP-ribosylguanidine + $H_2O$ [2]
3 (2'-Phospho-ADP-ribosyl)arginine + $H_2O$ [2]
4 More (poor substrates: (phosphoribosyl)arginine ribosylarginine) [2]

**Product spectrum**
1 L-Arginine + ADPribose [1–3]
2 Guanidine + ADPribose [2]
3 Arginine + 2'-phospho-ADPribose [2]
4 ?

---

**Inhibitor(s)**
Dithiothreitol [1, 2]; More (inactivation by $NAD^+$: arginine
ADP-ribosyltransferase in presence of $NAD^+$, $Mg^{2+}$ and $Mg^{2+}$ plus
dithiothreitol protect) [1]; ADP-ribose [2]

## Cofactor(s)/prostethic group(s)

**Metal compounds/salts**
$Mg^{2+}$ (and thiol required for activity) [1–3]

## Turnover number (min⁻¹)

**Specific activity** (U/mg)
More [1, 2]

**$K_m$-value** (mM)
0.065 (ADP-ribosylarginine) [2]; 0.047 ((2'-phospho-ADP-ribosyl)arginine) [2]; 0.027 (ADP-ribosylguanidine) [2]

**pH-optimum**
6.5–7.5 [2]

**pH-range**
5.5–9.0 (5.5: 10–15% of maximal activity, 9.0: about 10 % of maximal activity) [2]

**Temperature optimum** (°C)
30 (assay at) [2]

**Temperature range** (°C)

## 3 ENZYME STRUCTURE

**Molecular weight**
39000 (turkey, SDS-PAGE under reducing conditions, gel permeation) [1]

**Subunits**
Monomer (turkey, SDS-PAGE under reducing conditions, gel permeation) [1]

**Glycoprotein/Lipoprotein**
–

## 4 ISOLATION/PREPARATION

**Source organism**
Turkey (HS: thiol sensitive hydrolase, HR: thiol resistant hydrolase [1]) [1–3]

**Source tissue**
Erythrocytes [1–3]

**Localisation in source**
Soluble [1]

## Purification
Turkey [1, 2]

## Crystallization
–

## Cloned
–

## Renaturated
–

---

## 5 STABILITY

### pH

### Temperature (°C)
More ($Mg^{2+}$ stabilizes against thermal inactivation in absence and presence of thiol) [1, 2]

### Oxidation

### Organic solvent

### General stability information
$Mg^{2+}$ (stabilizes against thermal inactivation in absence and presence of thiol) [1, 2]; Thiols (unstable) [2]

### Storage

---

## 6 CROSSREFERENCES TO STRUCTURE DATABANKS

### PIR/MIPS code

### Brookhaven code

---

## 7 LITERATURE REFERENCES

[1] Moss, J., Tsai, S.-C., Adamik, R., Chen, H.-C., Stanley, S.J.: Biochemistry, 27, 5819–5823 (1988)
[2] Moss, J., Oppenheimer, N.J., West, R.E., Stanley, S.J. : Biochemistry, 25, 5408–5414 (1986)
[3] Moss, J., Jacobson, M.K., Stanley, S.J.: Proc. Natl. Acad. Sci. USA, 82, 5603–5607 (1985)

# 1 NOMENCLATURE

**EC number**
3.2.3.1

**Systematic name**
Thioglucoside glucohydrolase

**Recommended name**
Thioglucosidase

**Synonymes**
Glucosidase, thio-
Beta-Thioglucosidase [1]
Glucosinolase
Beta-Thioglucoside glucohydrolase [2]
Myrosin
Myrosinase
Sinigrinase
Sinigrase

**CAS Reg. No.**
9025-38-1

# 2 REACTION AND SPECIFICITY

**Catalysed reaction**
A thioglucoside + $H_2O$ →
→ a thiol + a sugar

**Reaction type**
S-Glycosyl bond hydrolysis
More (O-glycosidase activity [2], sulfatase activity [32], beta-glucosidase
activity [33], Aspergillus sydowi: thioglucosidase) [2, 32, 33]

**Natural substrates**
Thioglucoside + $H_2O$
Glucosinolates + $H_2O$ (degradation upon tissue disruption) [2]

**Substrate spectrum**
1  Thioglucoside + $H_2O$
2  Sinigrin (2-propenylglucosinolate) + $H_2O$ [1, 2, 5, 6 , 20]
3  p-Nitrophenyl-beta-D-glucoside + $H_2O$ [2, 20, 33]
4  2-Hydroxybut-3-enylglucosinolate + $H_2O$ [4, 8, 9]
5  Allylglucosinolate (+ epithiospecifier protein) + $H_2O$ [10, 12]
6  More (Brassica napus cv Zephyr substrate [22], stereoselectivity [8], has
   a wide specifity for thioglycosides) [8, 22]

**Product spectrum**
1  Thiol + sugar
2  ? + glucose [1, 2]
3  ?
4  1-Cyano-2-hydroxy-but-3-ene + 5-vinyloxyazolidine-2-thione + glucose
   [4, 9]
5  1-Cyano-2, 3-epithiopropane + $H_2SO_4$ + D-glucose [10, 12]
6  More (aglucone product can either undergo spontaneous Lossen rear-
   rangement to give a isothiocyanate or can form a nitrile [14], $Fe^{2+}$ has
   effect on product formation [4], effect of pH on product formation [9]) [4,
   9, 14]

---

**Inhibitor(s)**
2-Methoxy 5-nitrotropone [17, 18]; 5,5'-Dithiobis-(2-nitrobenzoic acid) [17,
27]; 2-Hydroxy-5-nitrobenzyl bromide (slight inhibition) [17]; L-Ascorbic
acid (Enterobacter cloacae: inhibition, plants: activation) [20]; Glucosides
[20, 23, 35]; Sugar (Aspergillus niger not [23]) [20, 35];
Delta-gluconolactone [20, 23, 34]; Salicin [20, 34, 35]; Amygdalin [20, 34];
Arbutin [20, 34, 35]; p-Nitrophenyl beta-glucoside [20, 34];
Beta-phenylglucoside [20, 34]; $Co^{2+}$ (slight inhibition) [20]; $Ca^{2+}$ (slight inhibition) [20]; $Mg^{2+}$ (slight inhibition) [20]; $Sr^{2+}$ (slight inhibition) [20]; $Ni^{2+}$
(slight inhibition) [20]; $Fe^{3+}$ (slight inhibition) [20]; Alpha-methylglucoside
[34]; Beta-methylglucoside [34]; Glucose [20, 34, 35]; Xylose [35]; Fructose
[20, 35]; Galactose [20, 35]; Mannose [10]; Sucrose [20]; Sorbitol [20];
$NaNO_3$ (effect on ascorbate-activated enzyme, no effect on non-activated
enzyme) [19]; NaBr (effect on ascorbate-activated enzyme, no effect on
non-activated enzyme) [19]; $KNO_3$ (effect on ascorbate-activated enzyme,
no effect on non-activated enzyme) [19]; p-Chloromercuribenzoate [23];
Diisopropylfluorophosphate (slight inhibition) [23]; Iodoacetate (slight inhibition) [23]; Fluorodinitrobenzene [27];
Monochlorotrifluoro-p-benzoquinone [27]; p-Diazobenzenesulfonic acid
[27]; NaCl [34]; $Cu^{2+}$ [20]; $Zn^{2+}$ [20]; $Sn^{2+}$ [23]; $Fe^{2+}$ [34];
Castanospermine [2]; $Fe^{2+}$ (slight inhibition [4]) [4, 20]; $Cu^{2+}$ [4, 17, 20, 34];
$Hg^{2+}$ [17, 20, 23, 34]; $Fe^{3+}$ (slight inhibition) [17]; EDTA [17, 20];
o-Phenanthroline (not: Enterobacter cloacae [20]) [17]; Trinitrobenzenesul-
fonic acid [17, 27]; p-Substituted mercuribenzoate [17, 18, 27]; More [19, 27]

## Cofactor(s)/prostethic group(s)

More (plant enzyme: ascorbic acid activates [4, 13, 15, 18, 22 , 30],
Enterobacter cloacae: enzyme inhibited by L-ascorbic acid [20], Aspergillus
niger, Aspergillus sydowi: no effect [34, 34]) [4, 13, 15, 18, 20, 22, 30]

## Metal compounds/salts

$Fe^{2+}$ (effect on course of reaction [4, 6], thiol compounds greatly accelerate the formation of nitriles from glucosinolate in the presence of
$Fe^{2+}$) [5]; $Cu^{2+}$ (effect on the course of reaction [4], stimulates [23]) [4, 23];
$Mn^{2+}$ (stimulates) [17, 23]; $Ca^{2+}$ (stimulates) [17]; $Sn^{2+}$ (stimulates) [17];
$Co^{2+}$ (stimulates) [17, 23, 34]; $Ni^{2+}$ (slight stimulation) [17]; $Sr^{2+}$ (slight
stimulation) [17]; $Mg^{2+}$ (slight stimulation [17], stimulates [34]) [17, 34];
$Zn^{2+}$ (slight stimulation [17], stimulation [34]) [7, 34]; $K^+$ (slight stimulation)
[17]; $Na^+$ (slight stimulation) [17]; $Li^{2+}$ (slight stimulation) [20]

## Turnover number (min⁻¹)

## Specific activity (U/mg)

32.11 [1]; 78.5 [2]; 11.5 [20]; 1.916 [23]; 10.17 [32]; 111 [33]; More [16 , 17, 23,
25, 33]

## $K_m$-value (mM)

0.070 (sinigrin) [1]; 0.115 (sinigrin) [2]; 0.71
(p-nitrophenyl-beta-D-glucoside) [20]; 2.0 (p-nitrophenyl-beta-D-glucoside)
[2]; 0.156 (sinigrin) [13]; 1.5 (p-nitrophenyl-beta-D-glucoside) [23]; More ($K_m$
increases 3–7 times in the presence of 0.4 ascorbic acid [1]) [1, 17, 19, 20,
23, 32]

## pH-optimum

6.8 [20]; 6.2 [22]; 7 (about) [32]; 5.2–5.5 [1]; 5.5 (sinigrin) [2]; 6.5
(p-nitrophenyl-beta-D-glucoside) [2]; 4.5–7.8 [13]; 6.5–7.0 [17]; More [30]

## pH-range

6.0–9.0 (6.0: about 25% of activity maximum, 9.0: about 50 % of activity
maximum) [23]; 5.3–8.0 (5.3, 8.0: about 20% of activity maximum) [20]; 3–8
(3: about 15% of activity maximum, 8: about 25% of activity maximum) [1];
4–10 (4: about 40% of activity maximum, 10: about 25% of activity maximum) [17]; More (effect on relative amounts of products of spontaneous
degradation of aglucone) [14]

## Temperature optimum (°C)

70–75 [1]; 37 [17]; 60 [24]; 34 [23]; 55 (without L-ascorbic acid) [17]; 35
(with ascorbic acid) [17]

## Temperature range (°C)

30–55 (30°C: about 40% of activity maximum, 55°C: about 5% of activity
maximum) [17]; More [18]

## 3 ENZYME STRUCTURE

### Molecular weight
150000 (Sinapis alba, gel filtration) [31]
154000 (Brassica napus, native polyacrylamide electrophoresis) [1]
130000 (Lepidium sativum, FPLC-gel filtration) [2]
580000 (Wasabia japonica, gel filtration) [17]
61000 (Enterobacter cloacae, gel filtration) [20]
135000 (Brassica napus, gel filtration) [25]
More [29]

### Subunits
Dimer (molecular sieving on Sephadex in 6 M guanidine-HCl) [25]
Tetramer (SDS-PAGE, mustard powder, at least 4 subunits, about 40000)
[29]
Dimer (Sinapis alba, chromatography of reduced and alkylated enzyme on
Sepharose in guanidine-HCl, 2 × 62000) [31]
Dimer (2 × 77000, SDS-PAGE, Brassica napus [1], 1 × 62000, 1 × 65000,
Lepidium sativum, SDS-PAGE [2]) [1, 2]
Dodecamer (about 12 subunits, 12 × 45000–47000, SDS-PAGE, Wasabia
japonica) [17]

### Glycoprotein/Lipoprotein
Glycoprotein (9.6–18.9% carbohydrate [1], 14% carbohydrate [25]) [1, 2,
25, 29, 31]

## 4 ISOLATION/PREPARATION

### Source organism
Brassica napus L. (isoenzymes [7, 9, 24, 25], 3 forms: Ca, Cb, Cc [1]) [1, 4, 5,
6, 7, 9, 16, 22, 24, 25, 30]; Brassica nigra (isoenzymes [7]) [7, 30]; Lepidium
sativum [2, 11, 15, 16]; Sinapis alba (isoenzymes [7, 24], immobilized enzyme [3]) [3, 7, 12, 16, 22, 24, 30, 31]; Brassica juncea [5, 18, 30]; Aspergillus
sydowi [10]; Brassica chinensis [11, 16]; Brassica oleraceae [11, 16];
Raphanus sativus [11, 16]; Brassica [12]; Crambe [12]; Amoracia [12];
Sinapis [12]; Brassicaceae (several species) [16]; Iberis amara [16];
Wasabia japonica [17]; Enterobacter cloacae [20, 21]; Aspergillus niger [23,
26]; Carica papaya [28]; Brassica campestris [30]; Crambe abyssinica [30];
Aspergillus sydowi [32, 33, 34]

### Source tissue
Seeds [1, 3, 9, 15, 19, 22, 24, 28, 31]; Seedlings [2]; Callus culture [16]; Com-
mercial mustard powder [14, 29]; Fruit [28]; Root [16]; Steem [16]; Leaf [16];
Culture broth [32, 33]

## Localisation in source
Protoplast (specific activity of protoplast is less than that of intact root tissue [11]) [11, 16]; Intracellular [23, 26]; Sarcotestae [28]; More (not: in papaya latex and endosperm) [28]

## Purification
Brassica napus [1, 7, 9, 25]; Lepidium sativum [2]; Sinapis alba [7, 31]; Brassica nigra [7]; Wasabia japonica [17]; Enterobacter cloacae [20]; Aspergillus sydowi [32]

## Crystallization
–

## Cloned
–

## Renaturated
–

---

## 5 STABILITY

### pH
5.0–7.0 (24 hours, below 40°C, stable) [20]; 5.5–8.5 (below 45°C, stable) [3]; 7.0 (about 20 minutes, stable) [17]; 7.6–8.0 (5°C, 24 hours, stable) [23]; More [24]

### Temperature (°C)
24 (1 month, immobilized enzyme, 20% loss of activity) [3]; 30 (20 minutes, stable below [17], 30 minutes, stable below [20]) [17, 20]; 40 (stable up to) [20]

### Oxidation
Photooxidation (by Methylene Blue, sensitive to) [27]

### Organic solvent

### General stability information
Freezing (completely destroys purified enzyme) [24]; Purified enzyme is more stable than crude preparation [24]; 2-Mercaptoethanol (enzyme unstable, but stabilized by coexistence with 2-mercaptoethanol and ascorbic acid) [26]; Ascorbic acid (enzyme unstable, but stabilized by coexistence with 2-mercaptoethanol and ascorbic acid) [26]

### Storage
24°C, 1 month (20% loss of activity, immobilized enzyme) [3]; 4°C, pH 6.0, imidazole buffer (best way of storing) [24]

---

# 6 CROSSREFERENCES TO STRUCTURE DATABANKS

PIR/MIPS code

Brookhaven code

# 7 LITERATURE REFERENCES

[1] Bones, A., Slupphaug, G.: J. Plant Physiol., 134, 722–729 (1989)

[2] Durham, P.L., Poulton, J.E.: Plant Physiol., 90, 48–52 (1989)

[3] Iori, R., Leoni, O., Palmieri, S.: Biotechnol. Lett., 10, 575–578 (1988)

[4] MacLeod, A.J., Rossiter, J.T.: Phytochemistry, 26, 669–673 (1987)

[5] Uda, Y., Kurata, T., Arakawa, N.: Agric. Biol. Chem., 50, 2741–2746 (1986)

[6] Uda, Y., Kurata, T., Arakawa, N.: Agric. Biol. Chem., 50, 2735–2740 (1986)

[7] Buchwaldt, L., Larsen, L.M., Plöger, A., Sorensen, H.: J. Chromatogr., 363, 71–80 (1986)

[8] Petroski, R.J.: Plant Sci., 44, 85–88 (1986)

[9] MacLeod, A.J., Rossiter, J.T.: Phytochemistry, 25, 1047–1051 (1986)

[10] Petroski, R.J., Kwolek, W.F.: Phytochemistry, 24, 213–216 (1985)

[11] Iversen, T.-H., Myhre, S., Evjen, K., Baggerud, C.: Z. Pflanzenphysiol., 112, 391–401 (1983)

[12] Petroski, R.J., Tookey, H.L.: Phytochemistry, 21, 1903–1905 (1982)

[13] Palmieri, S., Leoni, O., Iori, R.: Anal. Biochem., 123, 320–324 (1982)

[14] Gil, V., MacLeod, A.J.: Phytochemistry, 19, 2547–2551 (1980)

[15] Gil, V., MacLeod, A.J.: Phytochemistry, 19, 2071–2076 (1980)

[16] Iversen, T.-H., Baggerud, C.: Z. Pflanzenphysiol., 97, 399–407 (1980)

[17] Ohtsuru, M., Kawatani, H.: Agric. Biol. Chem., 43, 2249–2255 (1979)

[18] Ohtsuru, M., Hata, T.: Biochim. Biophys. Acta, 567, 384–391 (1979)

[19] Tsuruo, I., Hata, T.: Agric. Biol. Chem., 32, 479–483 (1968)

[20] Tani, N., Ohtsuru, M., Hata, T.: Agric. Biol. Chem., 38, 1623–1630 (1974)

[21] Tani, N., Ohtsuru, M., Hata, T.: Agric. Biol. Chem., 38, 1617–1622 (1974)

[22] Srivastava, V.K., Hill, D.C.: Phytochemistry, 13, 1043–1046 (1974)

[23] Ohtsuru, M., Hata, T.: Agric. Biol. Chem., 37, 2543–2548 (1973)

[24] Björkman, R., Lönnerdal, B.: Biochim. Biophys. Acta, 327, 121–131 (1973)

[25] Lönnerdal, B., Janson, J.-C.: Biochim. Biophys. Acta, 315, 421–429 (1973)

[26] Ohtsuru, M., Tsuruo, I., Hata, T.: Agric. Biol. Chem. , 37, 967–971 (1973)

[27] Ohtsuru, M., Hata, T.: Agric. Biol. Chem., 37, 269–275 (1973)

[28] Tang, C.-S.: Phytochemistry, 12, 769–773 (1973)

[29] Ohtsuru, M., Hata, T.: Agric. Biol. Chem., 36, 2495–2503 (1972)

[30] Henderson, H.M., McEwen, T.J.: Phytochemistry, 11, 3127–3133 (1972)

[31] Björkman, R., Janson, J.-C.: Biochim. Biophys. Acta, 276, 508–518 (1972)

[32] Ohtsuru, M., Tsuruo, I., Hata, T.: Agric. Biol. Chem. , 33, 1309–1314 (1969)

[33] Ohtsuru, M., Tsuruo, I., Hata, T.: Agric. Biol. Chem. , 33, 1320–1325 (1969)

[34] Ohtsuru, M., Tsuruo, I., Hata, T.: Agric. Biol. Chem. , 33, 1315–1319 (1969)

[35] Tsuruo, I., Hata, T.: Agric. Biol. Chem., 32, 1420–1424 (1968)

## 1 NOMENCLATURE

**EC number**
3.5.1.1

**Systematic name**
L-Asparagine amidohydrolase

**Recommended name**
Asparaginase

**Synonymes**
L-Asparaginase
Asparaginase II
Colaspase
Elspar
Leunase
Crasnitin
Alpha-asparaginase
Elspar [1]

**CAS Reg. No.**
9015-68-3

## 2 REACTION AND SPECIFICITY

**Catalysed reaction**
L-Asparagine + $H_2O$ →
→ L-aspartate + $NH_3$

**Reaction type**
Carboxylic acid amide hydrolysis

**Natural substrates**
L-Asparagine + $H_2O$

**Substrate spectrum**
1 L-Asparagine + $H_2O$
2 DL-Alanyl-DL-asparagine + $H_2O$ [11]
3 Glycyl-L-asparagine + $H_2O$ [11]
4 Glycyl-D-asparagine + $H_2O$ [11]
5 D-Asparagine + $H_2O$ [1 , 5, 8, 11, 14, 15, 16]
6 L-Asparagine + hydroxylamine [8, 15]

    7  5-Diazo-4-oxo-L-norvaline + $H_2O$ [1, 6, 11]
    8  Beta-aspartyl-hydroxamic acid + $H_2O$ [1, 2, 8, 11, 14, 15]
    9  L-Glutamine + $H_2O$ [1, 5, 7, 11, 14]
  10  L-Phenylalanineamide + $H_2O$ [15]
  11  L-Leucinamide + $H_2O$ [15]
  12  L-Tyrosinamide + $H_2O$ [15]
  13  Succinamic acid + $H_2O$ [11]
  14  D-Glutamine + $H_2O$ [5, 7]
  15  Beta-cyano-L-alanine + $H_2O$ [1, 8, 11]
  16  More (asparagine derivatives) [3, 5, 14]

## Product spectrum

    1  L-Aspartate + $NH_3$
    2  Alanyl-aspartic acid (75%) + alanyl asparagine (25%) [11]
    3  Glycyl-L-aspartic acid [11]
    4  Glycyl-D-aspartic acid [11]
    5  D-Aspartic acid [11]
    6  L-Aspartic acid-beta-hydroxamate [8]
    7  ?
    8  ?
    9  L-Glutamate + $NH_3$
  10  L-Phe + $NH_3$
  11  L-Leu + $NH_3$
  12  L-Tyr + $NH_3$
  13  ?
  14  D-Glutamate + $NH_3$
  15  ?
  16  ?

## Inhibitor(s)

Glycine [2]; D-Asparagine [2, 16]; 5-Bromo-4-oxo-L-norvaline [3, 6, 11]; DL-Aspartylhydroxamate [2, 3]; L-Aspartate [2, 16]; D-Aspartate [2]; 3-Cyano-L-alanine [2, 3]; Glycylglycine [2]; Glycine methylester [2]; Cysteine [2]; 2-Amino-5-chloro-4-oxo-pentanoic acid [3]; N-Ethylmaleimide [4]; Carbobenzoxy-L-asparagine [5]; L-Leucine amide [5]; Succinamide [5]; p-Substituted mercuribenzoate [15]; p-Chlorobenzoate [7, 12, 16]; p-Hydroxymercuribenzoate [4]; Iodoacetate [4]; Beta-mercaptoethanol [7, 12]; $NH_3$ (at pH 8.5) [15]; Bromide [3]; Dichromate [3]; $Hg^{2+}$ [4, 5, 7, 12, 15]; $Cu^{2+}$ [3, 4, 7, 12]; $Ni^{2+}$ [3, 7, 12]; More (other divalent metal ions) [3]

## Cofactor(s)/prostethic group(s)

## Metal compounds/salts

$K^+$ (activates) [16, 17]

**Turnover number** (min$^{-1}$)

**Specific activity** (U/mg)
1.65–2.97 (Lupinus) [2, 3]; 10.9 (Bacillus coagulans) [16]; 32 (Acinetobacter calcoaceticus, Asparaginase A) [12]; 70 (Acinetobacter calcoaceticus, Asparaginase B) [7]; 74 (Azotobacter vinelandii) [4]; 202 (Vibrio succinogenes) [8]; 226–400 (Enterobacteria) [9, 10, 13, 15]

**$K_m$-value** (mM)
6 ($N^4$-methoxy-L-asparagine) [3]; 0.43 (D-asparagine) [14]; 5.0–6.25 (L-glutamine) [11, 14]; More [1–3, 11, 15–16]

**pH-optimum**
6.0 [5]; 8.0 [2, 15]; 7.3 [12]; 8.5 [10, 2]; 8.6 [4]; 7.0–8. 0 [14]; 7.5–8.5 [15]; More [1, 12, 13, 16]

**pH-range**
4–10 [14]; 6.5–8.8 [8]; 6.0–9.6 [3]; 7.3–10 [16]

**Temperature optimum** (°C)
55 [16]; 57 [14]; 41 [13]; 48 [4]

**Temperature range** (°C)
50 (up to) [13]; 60 (up to) [14]

---

## 3 ENZYME STRUCTURE

**Molecular weight**
72000–75000 (gel filtration, sucrose density gradient, Lupinus sp.) [2, 3]
84000 (PAGE, Azotobacter vinelandii) [4]
97000 (gel filtration, Acinetobacter glutaminasificans) [11]
130000 (sedimentation equilibrium, Acinetobacter glutaminasificans) [11]
130000–150000 (gel filtration, Serratia marcescens) [1, 10, 11]
133000–138000 (gel filtration, guinea pig) [1, 15, 11]
More [7, 12, 11, 8, 1]

**Subunits**
Tetramer (4 × 32000–37000, SDS-PAGE, sedimentation equilibrium with urea, Vibrio succinogenes [8], Serratia marcescens [10], overview [11]) [8, 10, 11]
Tetramer (2 × alpha, 2 × beta, dansyl experiments [10], 2 × 11000–14000, 2 × 13000–155000, SDS-PAGE, Lupinus sp. [2]) [2, 10]

**Glycoprotein/Lipoprotein**
–

## 4 ISOLATION/PREPARATION

### Source organism

Plants [2]; Bacteria [1]; Marine algae [1]; Fungi [1]; Yeast [1]; Mammalia [11]; Birds [11]; Guinea pig [15]; Lupinus sp. [2]; Vibrio succinogenes [1]; Azotobacter vinelandii [4]; Candida utilis [5]; Acinetobacter calcoaceticus [7, 12]; Proteus vulgaris [9, 14]; Serratia marcescens [10]; Erwinia aroideae [13]; E. coli [15]; Bacillus coagulans [16]; E. coli [18]; Proteus vulgaris [9, 14]; Acinetobacter glutaminasificans [11]

### Source tissue

Serum [11, 15]; Liver [11]; Kidney [11]; Culture medium [5]

### Localisation in source

Extracellular [5]; Cytoplasm [3]; Vacuole [3]; Soluble [4]

### Purification

Guinea pig [15]; Lupinus sp. [2]; Vibrio succinogenes [1]; Azotobacter vinelandii [4]; Candida utilis [5]; Acinetobacter calcoaceticus [7, 12]; Proteus vulgaris [9, 14]; Serratia marcescens [10]; Erwinia aroideae [13]; E. coli [15]; Bacillus coagulans [16]; E. coli [18]; Proteus vulgaris [9, 14]

### Crystallization

–

### Cloned

–

### Renaturated

[11]

## 5 STABILITY

### pH

4–10 (37°C) [5]; 5–7 (60°C) [5]; 8.4 (above) [15]; 11.2 (up to) [11]

### Temperature (°C)

40 (up to) [11]; 50 (up to) [4]; 60 (up to) [5]

### Oxidation

### Organic solvent

Acetone (stable) [4]; Butanol (stable) [4]; Ethanol (stable) [14]; More [6]

### General stability information

### Storage

At –20°C for 3 months without loss of activity [3, 4, 8]; 1 week at –20°C [14]; More [15, 16, 7, 14]

## 6 CROSSREFERENCES TO STRUCTURE DATABANKS

**PIR/MIPS code**

XDEC (Escherichia coli); A35132 (precursor, Escherichia coli); JU0047
(Escherichia coli K-12); A26054 (Erwinia chrysanthemi); S03681 (precursor,
Erwinia chrysanthemi)

**Brookhaven code**

## 7 LITERATURE REFERENCES

[1] Wriston, J.C.: Methods Enzymol., 113, 608–618 (1985) (Review)

[2] Chang, K.S., Farnden, K.J.F.: Arch. Biochem. Biophys., 208, 49–58 (1981)

[3] Lea, P.J., Fowden, L., Miflin, B.J.: Phytochemistry, 17, 217–222 (1978)

[4] Gaffar, S.A., Shetna, Y.I.: Appl. Environ. Microbiol., 33, 508–514 (1977)

[5] Sakamoto, T., Araki, C., Beppu, T., Arima, K.: Agric. Biol. Chem., 41, 1359–1364
(1977)

[6] Handschumacher, R.E.: Methods Enzymol., 46, 432–435 (1977)

[7] Joner, P.E.: Biochim. Biophys. Acta, 438, 287–295 (1976)

[8] Distasio, J.A., Niederman, R.A., Kafkewitz, D., Goodman, D.: J. Biol. Chem., 251,
6929–6933 (1976)

[9] Chibata, I., Tosa, T., Sato, T., Sano, R., Yamamoto, K., Matuo, Y.: Methods Enzymol.,
34, Pt. B, 405–411 (1974)

[10] Whelan, H.A., Wriston, J.C.: Biochim. Biophys. Acta, 365, 212–222 (1974)

[11] Wriston, J.C., Yellin, T.O.: Adv. Enzymol. Relat. Areas Mol. Biol., 39, 185–248 (1973)
(Review)

[12] Joner, P.E., Kristiansen, T., Einasson, M.: Biochim. Biophys. Acta, 327, 146–156
(1973)

[13] Liu, F.S., Zajic, J.E.: Can. J. Microbiol., 18, 1953–1957 (1972)

[14] Tosa, T., Sano, R., Yamamoto, K., Nakamura, M., Chibata, I.: Biochemistry, 11,
217–222 (1972)

[15] Wriston, J.C. in "The Enzymes", 3rd Ed. (Boyer, P.D.) , 4, 101–121 (1971)

[16] Law, A.S., Wriston, J.C.: Arch. Biochem. Biophys., 147, 744–752 (1971)

[17] Sodek, L., Lea, P., Mifflin, B.J.: Plant Physiol., 65, 22–26 (1980)

[18] Ho, P.P.K., Frank, B.H., Burch, P.J.: Science, 165, 510 (1969)

# 1 NOMENCLATURE

**EC number**
   3.5.1.2

**Systematic name**
   L-Glutamine amidohydrolase

**Recommended name**
   Glutaminase

**Synonymes**
   Glutaminase I
   L-Glutaminase
   Glutamine aminohydrolase

**CAS Reg. No.**
   9001-47-2

---

# 2 REACTION AND SPECIFICITY

**Catalysed reaction**
   L-Glutamine + $H_2O$ →
   → L-glutamate + $NH_3$

**Reaction type**
   Carboxylic acid amide hydrolysis

**Natural substrates**
   L-Glutamine + $H_2O$ [16, 17]

**Substrate spectrum**
   1 L-Glutamine + $H_2O$ [16, 17]
   2 L-Glutamine + hydroxylamine [3, 16]
   3 L-Glutamate + methanol [16]
   4 D-Glutamine + $H_2O$ [3, 12]
   5 5-Diazo-4-oxo-L-norvaline + $H_2O$ [3]
   6 L-Asparagine + $H_2O$ [3, 12]
   7 D-Asparagine + $H_2O$ [3, 12]
   8 Beta-cyanoalanine + $H_2O$ [3]
   9 L-Theamine + $H_2O$ [12]
   10 Glutamylmethylamide + $H_2O$ [16]
   11 L-Isoglutamine + $H_2O$ [17]
   12 Alpha-methyl-DL-glutamine [17]
   13 More [12]

## Product spectrum

1 L-Glutamate + $NH_3$
2 Glutamic acid gamma-monohydroxamate [3, 16]
3 Methylglutamate [16]
4 ?
5 ?
6 ?
7 ?
8 ?
9 ?
10 ?
11 ?
12 ?
13 ?

## Inhibitor(s)

L-Glutamate [2, 4, 7, 11, 17]; cAMP [2]; cGMP [2]; Palmityl CoA [2]; Stearyl CoA [2]; Ammonia [2, 7, 17]; N-Ethylmaleimide [2, 7, 10, 17]; p-Substituted mercuribenzoate [16, 17]; 6-Diazo-5-oxo-L-norleucine [3, 7, 8, 16]; Methylene Blue (light) [3]; Rose Bengal (light) [3]; 2-Oxoglutarate [7]; Citrate [7]; Cresol Green [16, 17]; $Cl^-$ [11, 17]; Borate [13]; $Hg^{2+}$ [16, 17]; $Ag^+$ [16]; $Pb^{2+}$ [16]; $Cu^{2+}$ [16]; More [17]

## Cofactor(s)/prostethic group(s)

## Metal compounds/salts

## Turnover number (min$^{-1}$)

75900 (L-glutamine) [16]; 38700 (methyl glutamate) [16]; 480 (glutamyl methylamide) [16]; 304800 (L-glutamic acid) [16]; More [16]

## Specific activity (U/mg)

31.6 [1]; 121.1 [2]; 130 [2]; 245 [4]; 377 [13]; 36 [15]; 28 [15]

## $K_m$-value (mM)

0.01 (L-asparagin) [3]; 0.4–1.2 ($NH_4^+$) [5]; 4 ($HCO_3^-$) [9]; 1 (ATP) [9]; 4.5–6.1 (phosphate) [6]; More [3, 4, 7, 11, 6, 12, 16]

## pH-optimum

5.0 (E. coli) [13]; 6.0 (Acinetobacter) [3]; 7.7–8.6 (mammalian liver) [5, 6]; 7.9–8.8 (mammalian kidney) [6]; 7.1–9.0 (E. coli) [13]; 6.0–8.0 (Pseudomonas) [15]; 8.3–9.2 (E. coli) [13]

## pH-range

7.0–10.0 [7]; 5.5–10.5 [13]

**Temperature optimum (°C)**
40 [17]

**Temperature range (°C)**

## 3 ENZYME STRUCTURE

**Molecular weight**
138000–141000 (glutaminase-asparaginase) [3, 12]
118000–122000 (sedimentation equilibrium, thin layer chromatography,
Pseudomonas) [15]
120000–160000 (sucrose density gradient, mammalia) [1, 2, 4]
250000–320000 (sedimentation equilibrium, gel chromatography) [4, 5, 8]
90000 (gel filtration, SDS-PAGE, E. coli, glutaminase B) [13]
100000 (gel filtration, E. coli, glutaminase A) [16]
2000000–10000000 (large polymers) [1, 2, 4, 5, 11]

**Subunits**
Tetramer (4 × 33000–36400) [12, 14]
Dimer (2 × 50000, E. coli, glutaminase A) [16]
Dimer (2 × 65000, rat) [2, 4]

**Glycoprotein/Lipoprotein**
–

## 4 ISOLATION/PREPARATION

**Source organism**
Rat [1, 2, 4, 5, 7, 8, 9, 10, 11]; Pig [2, 17]; Man [2, 6]; Monkey [2]; Cow [2];
Rabbit [2]; Mouse [2]; Hen [2]; Acinetobacter glutaminasificans [3]; Pseu-
domonas aeruginosa [12, 14, 15]; E. coli [3, 13, 16]; More [3]

**Source tissue**
Liver [1, 5, 6, 9]; Brain [2, 4, 10]; Kidney [2, 4, 8, 11]

**Localisation in source**
Mitochondria (loosely associated with inner membrane [1, 9] , outer face of
inner membrane [2, 10]) [1, 2, 9, 10]

**Purification**
Pseudomonas aeruginosa (glutaminase A and B) [12]; E. coli (glutaminase
A [16], glutaminase B [13]) [13, 16]; Acinetobacter glutaminasificans [3]

**Crystallization**
(glutaminase-asparaginase) [15, 18]

## Cloned
–

## Renaturated
(after guanidine hydrochloride or urea treatement) [12]

## 5 STABILITY

### pH

### Temperature (°C)

### Oxidation
Photooxidation in presence of Methylene Blue and Rose Bengal [3]

### Organic solvent

### General stability information
Enzyme aggregation, protection against heat inactivation [11]; EDTA stabilizes at 37°C for 15 hours [14]; Avoid freezing/thawing [3, 13]; More [1]

### Storage
Lyophilized powder at 5°C indefinitely [3]; At least 1 year at –20°C [2]; Sterile solution at least 10 days [3]; At 4°C several months , pH 8.9 [4]; More [13]

## 6 CROSSREFERENCES TO STRUCTURE DATABANKS

### PIR/MIPS code
A35444 (hepatic, rat, fragment)

### Brookhaven code

## 7 LITERATURE REFERENCES

[1] Heini, H.G., Gebhard, R., Brecht, A., Mecke, D.: Eur. J. Biochem., 162, 541–546 (1987)
[2] Kvamme, E., Torgner, I.A.A., Svenneby, G.: Methods Enzymol., 113, 241–256 (1985) (Review)
[3] Holcenberg, J.S.: Methods Enzymol., 113, 257–263 (1985)
[4] Haser, W.G., Shaphiro, R.A., Curthoys, N.P.: Biochem. J., 229, 399–408 (1985) (Review)
[5] Patel, M., McGivan, J.D.: Biochem. J., 220, 583–590 (1984)
[6] Snodgrass, P.J., Lund, P.: Biochim. Biophys. Acta, 798 , 21–27 (1984)
[7] Ardawi, M.S., Newsholme, E.A.: Biochem. J., 217, 289–296 (1984)
[8] Morehouse, R.F., Curthoys, N.P.: Biochem. J., 193, 709–716 (1981)
[9] McGivan, J.D., Lacey, J.H., Joseph, S.K.: Biochem. J., 192, 537–542 (1980)

[10] Kvamme, E., Olsen, B.E.: FEBS Lett., 107, 33–36 (1979)

[11] Kovacervic, Z., Breberina, M., Pavlovic, M., Bajin, K.: Biochim. Biophys. Acta, 567, 216–224 (1979)

[12] Oshima, M., Yamamoto, T., Soda, K.: Agric. Biol. Chem., 40, 2251–2256 (1976)

[13] Prusiner, S., Davis, J.N., Stadtman, E.R.: J. Biol. Chem., 251, 3447–3456 (1976)

[14] Abe, T., Takenaka, O., Inada, Y.: Biochim. Biophys. Acta, 358, 113–116 (1974)

[15] Katsumata, H., Katsumata, R., Abe, T., Takenaka, O., Inada, Y.: Biochim. Biophys. Acta, 289, 405–409 (1972)

[16] Hartman, S.C. in "The Enzymes", 3rd Ed. (Boyer, P.D., Ed.) 4, 79–100 (1971) (Review)

[17] Robers, E. in "The Enzymes", 2nd Ed. (Boyer, P.D., Ed.) 4, 285–300 (1960) (Review)

[18] Ammon, H.L., Murphy, K.C., Sjolin, L., Wlodawer, A., Holcenberg, J.S., Roberts, J.: Acta Crystallogr. Sect. B Struct. Sci., 39, 250 (1983)

# 1 NOMENCLATURE

**EC number**
3.5.1.3

**Systematic name**
Omega-amidodicarboxylate amidohydrolase

**Recommended name**
Omega-amidase

**Synonymes**
Amidase, .omega.
Alpha-keto acid-omega-amidase

**CAS Reg. No.**
9025-19-8

# 2 REACTION AND SPECIFICITY

**Catalysed reaction**
A monoamide of dicarboxylic acid + $H_2O$ →
→ a dicarboxylate + $NH_3$

**Reaction type**
Carboxylic acid amide hydrolysis

**Natural substrates**
Alpha-ketoglutaramate [1, 2]

**Substrate spectrum**
1 Alpha-ketoglutaramate + $H_2O$ [1, 6]
2 Alpha-ketosuccinamate + $H_2O$ [1, 6]
3 Hydroxylamine + $H_2O$ [1, 3, 6]
4 Succinamate + $H_2O$ [1, 3, 6]
5 Glutaramate + $H_2O$ [1, 6]
6 Succinate + $H_2O$ [3]
7 Succinamide + $H_2O$ [3]
8 Succinimide + $H_2O$ [3]
9 Asparagine + $H_2O$ [3]
10 Malate + $H_2O$ [3]
11 Fumarate + $H_2O$ [3]
12 Alpha-keto-glutarate + $H_2O$ [5]
13 Valeramide + $H_2O$ [6]

14 Caproamide + $H_2O$ [6]
15 L-Isoglutamine + $H_2O$ [6]
16 L-Isoasparagine + $H_2O$ [6]
17 L-Leucinamide + $H_2O$ [6]
18 L-Phenylalanine amide + $H_2O$ [6]
19 Succinylmonohydroxamate + $H_2O$ [6]
20 More (monomethyl and monoethyl esters of alpha-ketoglutarate,
glutarate, succinate, and p-chloro, p-methyl, and unsubstituted
phenylesters of glutarate) [1, 5]

**Product spectrum**

1 Alpha-ketoglutarate + $NH_3$ [1]
2 Alpha-keto-succinate + $NH_3$
3 Hydroxamate [3, 6]
4 Succinic acid + $NH_3$
5 Glutamate + $NH_3$
6 ?
7 Succinate + $NH_3$
8 ?
9 Aspartate + $NH_3$
10 ?
11 ?
12 ?
13 Valeric acid + $NH_3$
14 Caproic acid + $NH_3$
15 L-Isoglutamate + $NH_3$
16 L-Isoaspartate + $NH_3$
17 L-Leu + $NH_3$
18 L-Phe + $NH_3$
19 Succinate + $NH_2OH$
20 ?

**Inhibitor(s)**

5,5-Dithiobis-(2-nitrobenzoic acid) [1]; N-Ethylmaleimide [1]; Iodoacetate
[1]; Iodoacetamide [1]; $NH_4^+$ [1, 4]; Glycylglycine (competitive) [1];
6-Diazo-5-oxo-L-norleucine [2]; Methylamine [5]; p-Substituted mercuribenzoate [1]

**Cofactor(s)/prostethic group(s)**

**Metal compounds/salts**

Turnover number (min$^{-1}$)

Specific activity (U/mg)
6.9 (p-methyl-phenyl-glutarate) [5]; 2545 [1]; 10.9 [3]; 66.0 [3]; 17.4 (alpha-ketoglutarate) [5]

K$_m$-value (mM)
3.0 (alpha-ketoglutaramate) [1]; 3.8 (alpha-ketosuccinamate) [1]; 1.3 (succinamate) [3]; 30.0 (succinate) [3]; 63.0 (succinate) [3]; 100.0 (hydroxylamine, Thermus) [3]; 50.0 (hydroxylamine, Bacillus subtilis) [3]

pH-optimum
6.5 (acyltransferase activity, Bacillus subtilis) [3] 7.0–9.0 [3]; 6.5–7.5 [6]

pH-range

Temperature optimum (°C)
46–56 (Bacillus subtilis) [3]; 80 (Thermus aquaticus) [3]

Temperature range (°C)

---

## 3 ENZYME STRUCTURE

Molecular weight
58000 (rat) [1]
120000 (gel filtration, Bacillus subtilis) [3]
36000–38500 (gel filtration, Thermus aquaticus) [3]

Subunits
Dimer (2 × 28000, rat) [1]

Glycoprotein/Lipoprotein
–

---

## 4 ISOLATION/PREPARATION

Source organism
Rat [1, 5, 6]; Mouse [1]; Plants [1]; Neurospora crassa [2]; Bacillus subtilis [3, 4]; Thermus aquaticus [3]

Source tissue
Liver [1]

Localisation in source
Mitochondria [1]; Cytoplasm [1]; Membrane (associated) [3]

Purification
Rat [1, 5]; Bacillus subtilis [3]; Thermus aquaticus [3]

---

Crystallization
–

Cloned
–

Renaturated
–

---

# 5 STABILITY

pH

Temperature (°C)
60 (rapid loss of activity) [1]

Oxidation

Organic solvent

General stability information

Storage
3 months, –20°C [3]; More [1]

---

# 6 CROSSREFERENCES TO STRUCTURE DATABANKS

PIR/MIPS code

Brookhaven code

---

# 7 LITERATURE REFERENCES

[1] Cooper, A.J.L., Duffy, T.E., Meister, A.: Methods Enzymol., 113, 350–358 (1985) (Review)
[2] Calderon, J., Morett, E., Mora, J.: J. Bacteriol., 161, 807–809 (1985)
[3] Fernald, N.J., Ramaley, R.F.: Arch. Biochem. Biophys., 153, 95–104 (1972)
[4] Ramaley, R.F., Fernald, N., DeVries, T.: Arch. Biochem. Biophys., 153, 88–94 (1972)
[5] Hersh, L.B.: Biochemistry, 11, 2251–2256 (1972)
[6] Meister, A., Levintow, L., Greenfield, R.E., Abendschein, P.A.: J. Biol. Chem., 251, 441–460 (1955)

# 1 NOMENCLATURE

**EC number**
3.5.1.4

**Systematic name**
Acylamide amidohydrolase

**Recommended name**
Amidase

**Synonymes**
Acylamidase
Acylase
Amidohydrolase
Deaminase
Fatty acylamidase
N-Acetylaminohydrolase

**CAS Reg. No.**
9012-56-0

# 2 REACTION AND SPECIFICITY

**Catalysed reaction**
A monocarboxylic acid amide + $H_2O$ →
→ a monocarboxylate + $NH_3$; Amide + hydroxylamine →
→ hydroxamic acid; Acid + hydroxylamine →
→ hydroxamic acid; Ester + hydroxylamine →
→ hydroxamic acid + alcohol [3, 5]

**Reaction type**
Carboxylic acid amide hydrolysis
Amide transfer
Acid transfer
Ester transfer

**Natural substrates**
Acetamide + ? [2, 3, 6, 7]
Acetanilide + ? [6, 11]
2-Chloropropionamide + ? [4]
Propionamide + ? [4]
Chloroacetamide + ? [4]

**Substrate spectrum**

1 Acetamide + $H_2O$ [2–6, 8, 9, 10, 13]
2 Acetate + hydroxylamine [3]
3 Ethylacetate + hydroxylamine [3]
4 Propionamide + ?
5 Acrylamide + ?
6 Propionate + ?
7 Acrylate + ?
8 Ethyl propionate + ?
9 Ethylacrylate + ?
10 N-Methylacetamide + ?
11 Acylamides + ?
12 Acyl anilides + ?
13 Acyl hydrazides + ?
14 Acyl hydroxamates + ?
15 Phenacetin + ?
16 Methylbutyrate + ?
17 Phenylacetate + ?

**Product spectrum**

1 Acetate + $NH_4^+$ (hydrolysis)
2 Acetohydroxamic acid + $NH_4^+$ [3]
3 Acetohydroxamic acid + ethanol [3]
4 Acetohydroxamic acid + $H_2O$ [3]
5 ?
6 ?
7 ?
8 ?
9 ?
10 ?
11 ?
12 ?
13 ?
14 ?
15 ?
16 ?
17 ?

## Inhibitor(s)

Diisopropylfluorophosphate [10]; HgCl$_2$ [10]; 4-Chloromercuribenzoate [10]; Diethyl-4-nitrophenyl phosphate [10, 11]; p-Hydroxymercuribenzoate [4]; N-Ethylmaleimide [4]; Acylnitriles [7]; Acetic acid [7, 9]; Urea [8, 9]; Hydroxyurea [8]; Cyanate [8]; Acetaldehyde [8]; Acetaldehyde ammonia [8, 9]; DL-Lactimide [9]; NH$_3$ (pH 9) [9]; Phenacetin [10]; Acetanilide [10]; 2-Nitrophenol [10]; Phenitidine [10]; Physostigmine [10]; 5,5'-Dithiobis-(2-nitrobenzoic acid) [10]; Iodoacetamide [10, 12]; Hydroxylamine [9, 12]

## Cofactor(s)/prostethic group(s)

## Metal compounds/salts

---

## Turnover number (min$^{-1}$)

2900–8160 (acylanilides) [10]; 4090–8830 (phenylacetates) [10]; 2840 (methylbutyrate) [10]; 600 (acetohydrazide hydrolysis) [9]; More [9]

## Specific activity (U/mg)

76000 (transformed E. coli cells) [1]; 0.91 [4]; 131.5 [11]

## K$_m$-value (mM)

0.0034–0.042 (acylanilides) [10]; 0.006–0.1 (phenylacetate) [10]; 41 (methyl-butyrate) [10]; 0.23–44 (acylamides) [6]; 131 (hydroxylamine) [5]; 31 (acetamide) [3]; 83 (acetate) [3]; 628 (propionate) [3]; More [3, 4, 6, 9, 10]

## pH-optimum

9.7–10.5 [10]; 7.2 [8, 9]

## pH-range

## Temperature optimum (°C)

## Temperature range (°C)

---

## 3 ENZYME STRUCTURE

## Molecular weight

230400 (amino acid sequence, Pseudomonas aeruginosa) [2]
59000 (gel filtration, Pseudomonas putida) [4]
56700–59100 (gel filtration, SDS-PAGE, amino acid composition, Pseudomonas acidovorans) [10]

## Subunits

Hexamer (6 × 38400, Pseudomonas aeruginosa) [2]
Dimer (2 × 29000, SDS-PAGE, Pseudomonas putida) [4]
Monomer [10]

Glycoprotein/Lipoprotein
–

---

## 4 ISOLATION/PREPARATION

### Source organism
Streptococcus pneumoniae [1]; Pseudomonas aeruginosa [2, 4]; Brevibac-
terium sp. [3, 5, 6, 7]; Pseudomonas putida [4]; Aspergillus nidulans [6, 15];
Pseudomonas acidovorans [10]

### Source tissue

### Localisation in source
Soluble [4, 6]

### Purification
E. coli (transformed cells) [1]; Pseudomonas putida [4]; Pseudomonas
aeruginosa [8]; Pseudomonas acidovorans [11]

### Crystallization
–

### Cloned
[1, 14]

### Renaturated
–

---

## 5 STABILITY

### pH

### Temperature (°C)
50 (above 50°C denaturation) [10]

### Oxidation

### Organic solvent
Acetone (rapid denaturation) [11]; Methanol (rapid denaturation) [11];
Ethanol (rapid denaturation) [11]

### General stability information

### Storage
–30°C, pH 7.2 [8]; 23% loss of activity after 3 months at –18°C [11]

# 6 CROSSREFERENCES TO STRUCTURE DATABANKS

**PIR/MIPS code**
 A26741 (aliphatic Pseudomonas aeruginosa)

**Brookhaven code**

---

# 7 LITERATURE REFERENCES

[1] Garcia, J.L., Garcia, E., Lopez, R.: Arch. Microbiol., 149, 52–56 (1987)

[2] Ambler, R.P., Auffret, A.D., Clarke, P.H.: FEBS Lett., 215, 285–290 (1987)

[3] Thiery, A., Maestracci, M., Arnaud, A., Galzy, P.: J. Gen. Microbiol., 132, 2205–2208 (1986)

[4] Wyndham, R.C., Slater, J.H.: J. Gen. Microbiol., 132, 2195–2204 (1986)

[5] Maestracci, M., Thiery, A., Arnaud, A., Galzy, P.: Agric. Biol. Chem., 50, 2237–2241 (1986)

[6] Maestracci, M., Thiery, A., Bui, K., Arnauld, A., Galzy, P.: Arch. Microbiol., 138, 315–320 (1984)

[7] Maestracci, M., Bui, K., Thiery, A., Arnauld, A., Galzy, P.: Biotechnol. Lett., 6, 149–154 (1984)

[8] Gregoriou, M., Brown, P.R.: Eur. J. Biochem., 96, 101–108 (1979)

[9] Woods, M.J., Findlater, J.D., Orsi, B.A.: Biochim. Biophys. Acta, 567, 225–237 (1979)

[10] Alt, J., Heyman, E., Krisch, K.: Eur. J. Biochem., 53, 357–369 (1975)

[11] Alt, J., Krisch, K.: J. Gen. Microbiol., 87, 260–272 (1975)

[12] Woods, M.J., Orsi, B.A.: Biochem. Soc. Trans., 12, 552nd Meet., Galway, 1344–1346 (1974)

[13] Brown, P.R., Smyth, M.J., Clarke, P.H., Rosemeyer, M. A.: Eur. J. Biochem., 34, 177–187 (1973)

[14] Drew, R.E., Clarke, P.H., Brammar, W.J.: Mol. Gen. Genet., 177, 311–320 (1980)

[15] Hynes, M.J.: J. Bacteriol., 103, 482–487 (1970)

# 1 NOMENCLATURE

**EC number**
3.5.1.5

**Systematic name**
Urea amidohydrolase

**Recommended name**
Urease

**Synonymes**

**CAS Reg. No.**
9002-13-5

# 2 REACTION AND SPECIFICITY

**Catalysed reaction**
Urea + $H_2O \rightarrow$
$\rightarrow CO_2 + 2 NH_3$

**Reaction type**
Carboxylic acid amide hydrolysis

**Natural substrates**
Urea + $H_2O$ [16]

**Substrate spectrum**
1 Urea + $H_2O$ [16]
2 Hydroxyurea + $H_2O$ [13]
3 $(NH_4)_2CO_3 + H_2O$ [14]
4 Mesoxalic acid + $H_2O$ (at pH 2.2, not active site catalyzed) [16]

**Product spectrum**
1 $CO_2$ + ammonia [16]
2 ?
3 Urea [14]
4 Glyoxalic acid [16]

## Inhibitor(s)

EDTA (below pH 5, $Ni^{2+}$ removed) [5, 9]; Hydroxyurea [4, 6, 13]; Selenourea [4]; Phenylurea [6]; Hydroxamates (amino acid hydroxamates) [4]; Acetohydroxamate [10, 13, 15]; p-Substituted mercuribenzoate [16]; p-Chloromercuribenzoate [6]; p-Hydroxymercuribenzoate [8]; p-Choromercuribenzenesulfonate [13]; N-Ethylmaleimide [6, 13]; Phosphoramidate [10, 11, 15]; Phosphate [15]; Beta-mercaptoethanol [10, 11]; Iodine [16]; Iodosobenzoate [16]; Suramin [16]; Phenylsulfinate [16]; Furacin [16, 20]; Arsenicals (trivalent) [16, 20]; $F^-$ [10, 11]; $Hg^{2+}$ [6, 13, 16]; $Cu^{2+}$ [6, 13, 16]; $Fe^{2+}$ [6, 13, 16]; $Co^{2+}$ [6, 13, 16]; $Zn^{2+}$ [6, 13, 16]; $Ni^{2+}$ [6, 13, 16]; $Mn^{2+}$ [6, 13, 16]; $Cd^{2+}$ [6, 13, 16]; $Ag^+$ [16]; $Mg^{2+}$ (weak) [13]; $Ba^{2+}$ [13]; $Na^+$ [16]; $K^+$ [16]; Thiourea [16]; Divalent metal ions [6, 13, 16]

## Cofactor(s)/prostethic group(s)

## Metal compounds/salts

$Ni^{2+}$ [2, 5–7, 9–12, 15]

## Turnover number (min$^{-1}$)

## Specific activity (U/mg)

664 [12]; 130 [8]; 59.9 [13]; 180 [1]; 219 [5]; More [6, 8]

## $K_m$-value (mM)

2.1–5.0 [6, 13]; 0.12 [8]; 0.45–0.9 [4, 13]

## pH-optimum

6.0–8.3 [13, 19]; 7.0–7.6 [3, 4, 5]; 8.0 [13, 16]; 8.7 [8]; 5.0–5.5 [4]; 8.8 [4]

## pH-range

7.0–8.5 [13]; 7.3–9.6 [8]

## Temperature optimum (°C)

40 [5]; 60–65 [6]

## Temperature range (°C)

60 (up to) [6]

## 3 ENZYME STRUCTURE

## Molecular weight

125000–135000 (sucrose density gradient, rumen content) [13]
544740 (amino acid sequence, jack bean) [1, 18]
280000–480000 (soybean, different interconvertible forms) [2, 1]
330000–380000 (gel filtration, Ureaplasma urealyticum) [1]
More [1, 17, 2, 8, 5, 6]

## Subunits
Hexamer (6 × 90790, amino acid sequence, jack bean) [1]
Hexamer (6 × 38000, SDS-PAGE, Spirulina maxima) [8]
Multimer (SDS-PAGE, bacteria) [1, 2, 3, 6]
Pentamer (5 × 73000, subunit cloned, Providencia stuartii) [1, 17]

## Glycoprotein/Lipoprotein
–

## 4 ISOLATION/PREPARATION

### Source organism
Blue green algae; Fungi; Yeast [8, 15, 16]; Jack bean [9]; Soy bean [12];
Ureaplasma urealyticum [1, 3]; Bacillus pasteurii [2]; Arthrobacter oxydans
[5]; Brevibacterium ammoniagenes [6]; Spirulina maxima [8]

### Source tissue
Seeds [9, 12]; Cells (bacteria) [1, 2, 5 , 6, 13]

### Localisation in source

### Purification
Jack bean [9]; Soy bean [12]; Ureaplasma urealyticum [1, 3]; Bacillus
pasteurii [2]; Arthrobacter oxydans [5]; Brevibacterium ammoniagenes [6];
Spirulina maxima [8]

### Crystallization
[16]

### Cloned
(subunit, Providencia stuartii) [1, 17]

### Renaturated
–

## 5 STABILITY

### pH
7–10 (Brevibacterium ammoniagenes) [6]

### Temperature (°C)

### Oxidation

### Organic solvent

### General stability information

## Storage
70% loss of activity after 2 months at 20°C, pH 7.5 [6]

## 6 CROSSREFERENCES TO STRUCTURE DATABANKS

### PIR/MIPS code
URJB (Jack bean); A35306 (large chain, Helicobacter pylori, fragment); A35389 (63K chain, Morganella morganii, fragment); B35306 (small chain, Helicobacter pylori, fragment); B35389 (15K chain, Morganella morganii, fragment); C35389 (6K chain, Morganella morganii, fragment); S10030 (gamma chain, Ureaplasma urealyticum, SGC3); S10031 (beta chain, Ureaplasma urealyticum, SGC3); S10032 (alpha chain, Ureaplasma urealyticum SGC3); S01021 (Jack bean)

### Brookhaven code

## 7 LITERATURE REFERENCES

[1] Stemke, G.W., Robertson, J.A., Nhan, M.: Can. J. Microbiol., 33, 857–862 (1987)
[2] Christians, S., Kaltwasser, H.: Arch. Microbiol., 145, 51–55 (1986)
[3] Eng, H., Robertson, J.A., Stemke, G.W.: Can. J. Microbiol., 32, 487–493 (1986)
[4] Davis, H.M., Shih, L.-M.: Phytochemistry, 23, 2741–2745 (1984)
[5] Schneider, J., Kaltwasser, H.: Arch. Microbiol., 139, 355–360 (1984)
[6] Nakano, H., Takenishi, S., Watanabe, Y.: Agric. Biol. Chem., 48, 1495–1502 (1984)
[7] Alagna, L., Hasnain, S.S., Piggott, B., Williams, D.J. : Biochem. J., 220, 591–595 (1984)
[8] Carvajal, N., Fernández, M., Rodriguez, J.P., Donoso, M.: Phytochemistry, 21, 2821–2823 (1982)
[9] Dixon, N.E., Blakeley, R.L., Zerner, B.: Can. J. Biochem., 58, 469–473 (1979)
[10] Dixon, N.E., Gazzola, C., Asher, C.J., Lee, D.S.W., Blakeley, R.L., Zerner, B.: Can. J. Biochem., 58, 474–480 (1979)
[11] Dixon, N.E., Blakeley, R.L., Zerner, B.: Can. J. Biochem., 58, 481–488 (1979)
[12] Polacco, J.C., Havir, E.A.: J. Biol. Chem., 254, 1707–1715 (1979)
[13] Mahadevan, S., Sauer, F.D., Erfle, J.D.: Biochem. J., 163, 495–501 (1977)
[14] Butler, L.G., Reithel, F.J.: Arch. Biochem. Biophys., 178, 43–50 (1977)
[15] Dixon, N.E., Gazzola, C., Blakeley, R.L., Zerner, B.: Science, 191, 1144–1150 (1976)
[16] Varner, J.E. in "The Enzymes", 2nd Ed. (Boyer, P.D., Ed.), 4, 247–256 (1960) (Review)
[17] Mobley, H.L., Jones, D.J., Jerse, A.E.: Infect. Immun., 54, 161–169 (1986)
[18] Mamiya, G., Takishima, K., Masakuni, M., Kayumi, T., Ogawa, K., Sekita, T.: Proc. Jpn. Acad. Ser. B Phys. Biol. Sci., 61, 395–398 (1985)
[19] Andersen, J.A., Kopko, F., Siedler, A.J., Nohle, E.G.: Fed. Proc. (Fed. Am. Soc. Exp. Biol.), 28, 764 (1969)
[20] Yall, I., Green, M.N.: Proc. Soc. Exp. Biol. Med., 79, 306 (1952)

## 1 NOMENCLATURE

**EC number**
3.5.1.6

**Systematic name**
N-Carbamoyl-beta-alanine amidohydrolase

**Recommended name**
Beta-ureidopropionase

**Synonymes**

**CAS Reg. No.**
9027-27-4

## 2 REACTION AND SPECIFICITY

**Catalysed reaction**
N-Carbamoyl-beta-alanine + $H_2O$ →
→ beta-alanine + $CO_2$ + $NH_3$

**Reaction type**
Carboxylic acid amide hydrolysis

**Natural substrates**
N-Carbamoyl-beta-alanine

**Substrate spectrum**
1 N-Carbamoyl-beta-alanine + $H_2O$
2 N-Carbamoyl-DL-beta-aminoisobutyrate + $H_2O$ [1, 8]
3 N-Carbamoyl-DL-alanine + $H_2O$ [1]
4 N-Carbamoyl-glycine + $H_2O$ [1]
5 Beta-ureidoisobutyrate + $H_2O$ (animal enzyme)

**Product spectrum**
1 Beta-alanine + $CO_2$ + $NH_3$
2 DL-Beta-aminobutyrate + $CO_2$ + $NH_3$
3 Alanine + $CO_2$ + $NH_3$
4 Glycine + $CO_2$ + $NH_3$
5 ?

**Inhibitor(s)**

$Zn^{2+}$ [1]; Beta-alanine [2]; Gamma-aminobutyrate [2];
N-Amidino-beta-alanine [2]; N-Carbamoyl-glycine [2]; $Cu^{2+}$ [6];
Beta-ureidoisobutyric acid [6]; Isobutyrate [6]; Propionate [6]; Alpha-fluoroacetate [6]; Acetate [6]

**Cofactor(s)/prostethic group(s)**

**Metal compounds/salts**

$Mg^{2+}$ (Euglena gracilis: stimulation, other organisms: not) [6]

---

**Turnover number** (min$^{-1}$)

**Specific activity** (U/mg)

1.59 (rat liver) [1]; 0.0147 (Euglena gracilis) [6]

**$K_m$-value** (mM)

0.0065–0.032 (N-carbamoyl-beta-alanine, rat liver) [2, 3]; 0.17
(beta-ureidopropionate, $K_{1/2}$, no Michaelis-Menten kinetics, Hill coefficient
n: 20, rat liver) [1]; 0.23 ($K_{1/2}$, DL-beta-aminoisobutyrate, rat liver) [1]; More
[1–3, 5–7]

**pH-optimum**

8.2–8.6 (Rhodopseudomonas capsulata) [5]; 7.0–7.5 (rat liver) [1, 8]; 6.25
(rat liver [4], Euglena gracilis [6]) [4, 6]; 7.4–7.8 (Clostridium uracilium) [7];
6.75 (DL-beta-amino isobutyrate, rat liver) [8]

**pH-range**

4.5–8.3 (rat liver) [1]; 7.0–8.3 (Clostridium uracilium) [7]

**Temperature optimum** (°C)

30–35 (Clostridium uracilium) [7]; 45–50 (rat liver); 60 (Euglena gracilis) [6]

**Temperature range** (°C)

---

## 3 ENZYME STRUCTURE

**Molecular weight**

1500000–2000000 (molecular sieve chromatography, Euglena gracilis) [6]
323000–340000 (gel filtration, sucrose density gradient, rat) [1, 3]

**Subunits**

Hexamer (3 dimers, 6 × 54000, SDS-PAGE, rat) [1]

**Glycoprotein/Lipoprotein**

–

## 4 ISOLATION/PREPARATION

**Source organism**
Rat [1, 2]; Man [2]; Mouse [2]; Clostridium uracilium [7]; Euglena gracilis [6]; Rhodopseudomonas capsulata [5]

**Source tissue**
Liver [1, 2]

**Localisation in source**
Cytoplasm [1, 3]

**Purification**
Rat [1, 2]; Euglena gracilis [6]; Clostridium uracilium [7]

**Crystallization**
–

**Cloned**
–

**Renaturated**
–

## 5 STABILITY

**pH**

**Temperature (°C)**
45 (inactivated at, Clostridium uracilium) [7]

**Oxidation**

**Organic solvent**

**General stability information**

**Storage**
At least 3 months at –20°C [7]; More [1, 3]

## 6 CROSSREFERENCES TO STRUCTURE DATABANKS

**PIR/MIPS code**

**Brookhaven code**

# 7 LITERATURE REFERENCES

[1] Tamaki, N., Mizutani, N., Kikugawa, M., Fujimoto, S., Mizota, C.: Eur. J. Biochem., 169, 21–26 (1987)

[2] Matthews, M.M., Traut, T.W.: J. Biol. Chem., 262, 7232–7237 (1987)

[3] Traut, T.W., Loeckel, S.: Biochemistry, 23, 2593–2539 (1984)

[4] Hardiman, M.K., Alfant, M., Wakelin, V.P., Tremblay, G.C.: Arch. Biochem. Biophys., 224, 326–331 (1983)

[5] Kaspari, H.: J. Gen. Microbiol., 122, 95–100 (1981)

[6] Wasternack, C., Lippmann, G., Reinbotte, H.: Biochim. Biophys. Acta, 570, 341–351 (1979)

[7] Campbell, L.L.: J. Biol. Chem., 235, 2375–2378 (1960)

[8] Caravaca, J., Grisolia, S.: J. Biol. Chem., 231, 357–365 (1958)

## 1 NOMENCLATURE

**EC number**
3.5.1.7

**Systematic name**
N-Carbamoyl-L-aspartate amidohydrolase

**Recommended name**
Ureidosuccinase

**Synonymes**

**CAS Reg. No.**
9024-81-1

## 2 REACTION AND SPECIFICITY

**Catalysed reaction**
N-Carbamoyl-L-aspartate + $H_2O$ →
→ L-aspartate + $CO_2$ + $NH_3$

**Reaction type**
Carboxylic acid amide hydrolysis

**Natural substrates**
N-Carbamoyl-L-aspartate + $H_2O$ [1]

**Substrate spectrum**
1 N-Carbamoyl-L-aspartate + $H_2O$ [1]
2 N-Carbamoyl-L-glutamate + $H_2O$ (weak) [1]

**Product spectrum**
1 L-Aspartate + $CO_2$ + $NH_3$ [1]
2 Glutamate + $CO_2$ + $NH_3$

**Inhibitor(s)**
Sodium thioglycolate [1]

**Cofactor(s)/prostethic group(s)**

**Metal compounds/salts**
$Mn^{2+}$ [1]; $Fe^{2+}$ [1]

Turnover number (min$^{-1}$)

Specific activity (U/mg)
7.4 [1]

K$_m$-value (mM)
2.8–13 (N-carbamoyl-L-aspartate) [1]

pH-optimum
7.8–8.5 [1]

pH-range
5.1–9.6 [1]

Temperature optimum (°C)

Temperature range (°C)

---

3 ENZYME STRUCTURE

Molecular weight

Subunits

Glycoprotein/Lipoprotein
–

---

4 ISOLATION/PREPARATION

Source organism
Zymobacterium oroticum [1]

Source tissue

Localisation in source

Purification
Zymobacterium oroticum [1]

Crystallization
–

Cloned
–

Renaturated
–

## 5 STABILITY

pH

Temperature (°C)

Oxidation

Organic solvent

**General stability information**
Cysteine retards inactivation [1]

**Storage**
0°C, in vacuum, at least 10 days [1]

---

## 6 CROSSREFERENCES TO STRUCTURE DATABANKS

PIR/MIPS code

Brookhaven code

---

## 7 LITERATURE REFERENCES

[1] Lieberman, I., Kornberg, A.: J. Biol.Chem., 212, 909–920 (1955)

## 1 NOMENCLATURE

**EC number**
3.5.1.8

**Systematic name**
N-Formyl-L-aspartate amidohydrolase

**Recommended name**
Formylaspartate deformylase

**Synonymes**
Deformylase, formylaspartate
Formylaspartic formylase (Formylase I, Formylase II) [1]

**CAS Reg. No.**
9025-09-6

## 2 REACTION AND SPECIFICITY

**Catalysed reaction**
N-Formyl-L-aspartate + $H_2O$ →
→ formate + L-aspartate

**Reaction type**
Carboxylic acid amide hydrolysis

**Natural substrates**

**Substrate spectrum**
1 N-Formyl-L-aspartate + $H_2O$ (formylase I/II) [1]
2 Formylglutamic acid + $H_2O$ (formylase II) [1]
3 Acetyl-L-glutamic acid + $H_2O$ (formylase II) [1]
4 Chloroacetyl-L-aspartic acid + $H_2O$ (formylase II) [1]

**Product spectrum**
1 Formate + L-aspartate
2 Formate + glutamate
3 Acetate + L-glutamate
4 Chloroacetate + L-glutamate

**Inhibitor(s)**
$Fe^{2+}$ (formylase II) [1]

**Cofactor(s)/prostethic group(s)**

**Metal compounds/salts**
$Fe^{2+}$ (formylase I) [1]; $Co^{2+}$ (formylase I/II) [1]; $Mn^{2+}$ (formylase II) [1]

**Turnover number** (min$^{-1}$)

**Specific activity** (U/mg)
More [1]

**$K_m$-value** (mM)
1.32 (N-formyl-L-aspartate, formylase I) [1]; 1.25 (N-formyl-L-aspartate, formylase II) [1]

**pH-optimum**
7.0 (formylase I) [1]; 8.0 (formylase II) [1]

**pH-range**
6.5–8.5 (formylase I/II) [1]

**Temperature optimum** (°C)

**Temperature range** (°C)

## 3 ENZYME STRUCTURE

**Molecular weight**

**Subunits**

**Glycoprotein/Lipoprotein**
–

## 4 ISOLATION/PREPARATION

**Source organism**
Pseudomonas sp. (cells adapted to beta-imidazolyl-4/5/acetic acid produce formylase I, cells adapted to histidine produce formylase II which is not identical with formylase I) [1]

**Source tissue**

**Localisation in source**

**Purification**
Pseudomonas sp. [1]

**Crystallization**
–

## Cloned

–

## Renaturated

–

## 5 STABILITY

pH

Temperature (°C)

Oxidation

Organic solvent

General stability information

Storage

## 6 CROSSREFERENCES TO STRUCTURE DATABANKS

PIR/MIPS code

Brookhaven code

## 7 LITERATURE REFERENCES

[1] Einosuke, O., Hayaishi, O.: J. Biol. Chem., 227, 181–190 (1957)

# 1 NOMENCLATURE

**EC number**
3.5.1.9

**Systematic name**
Aryl-formylamine amidohydrolase

**Recommended name**
Arylformamidase

**Synonymes**
Kynurenine formamidase
Formylase
Formylkynureninase
Formylkynurenine formamidase
Formamidase I
Formamidase II

**CAS Reg. No.**
9013-59-6

# 2 REACTION AND SPECIFICITY

**Catalysed reaction**
N-Formyl-L-kynurenine + $H_2O$ →
→ formate + L-kynurenine

**Reaction type**
Carboxylic acid amide hydrolysis

**Natural substrates**
N-Formyl-L-kynurenine + $H_2O$ [16]

**Substrate spectrum**
1 N-Formyl-L-kynurenine + $H_2O$
2 Formylanthranilic acid + $H_2O$
3 N'N$^{alpha}$-Diformyl-L-kynurenine + $H_2O$ [10]
4 Formyl amines (aromatic) + $H_2O$
5 Acetylamines (aromatic) + $H_2O$ [6]
6 Propionylamines (aromatic) + $H_2O$ [6]
7 More [8, 15]

## Product spectrum
1 Formate + L-kynurenine
2 Formate + anthranilic acid
3 Formate + $N^{alpha}$-formyl-L-kynurenine
4 Formate + amine (aromatic)
5 Acetate + amine (aromatic)
6 Propionate + amine (aromatic)
7 ?

## Inhibitor(s)
Cyanide [15]; Bisulfite [15]; $Zn^{2+}$ [6]; $Hg^{2+}$ [12, 13]; $Ag^+$ [12]; $Cu^{2+}$ [12]; NaF [6]; p-Chloromercuribenzoate [6]; Iodoacetamide [6]; O,O-Diethyl-O-(4-nitrophenyl)phosphate [13]; Phenylmethylsulfonylfluoride [12]; Urea [12]; Na-Metaarsenite [8]; Aniline [8]; 1-Naphthylamine [8]; 2-Naphthylamine [8]; Anthranilic acid [8]; Formylanthranilic acid [8]; Organophosphates [5]; Bromophenol Blue [5]; More [2, 5]

## Cofactor(s)/prostethic group(s)

## Metal compounds/salts

## Turnover number (min$^{-1}$)

## Specific activity (U/mg)
303 [13]; 80.4 [12]; 1.13 (formamidase I) [11]; 2.1 (formamidase II) [11]; 0.589 (formamidase I) [6]; 0.576 (formamidase II) [6]; 8333 (formamidase I) [1]; 26818 (formamidase II) [1]

## $K_m$-value (mM)
14 (formylkynurenine) [12]; 5 (formylkynurenine, formamidase I) [11]; 0.83 (formylkynurenine, formamidase II) [11]; 0.57 (formylanthranilic acid) [8]; 39 (N',$N^{alpha}$-diformyl-L-kynurenine) [5]; 1 (2-chloro-N-(1-naphthyl)acetamide) [5]; 8.1 (N-formyl-aminoacetophenone) [5]

## pH-optimum
7.3–7.8 [15]; 6.0–8.5 [12]; 6.0–9.0 [8]; 6.5–8.0 (formamidase II) [9, 11]; 5.5–7.0 (formamidase II) [6]; 7.0 (formamidase II) [2]; 6.7–7.8 (formamidase I) [11]; 6.7–7.6 (formamidase I) [9]; 7.0–8.0 (formamidase I) [6]

## pH-range

## Temperature optimum (°C)
60 [13]; 30 (formamidase II) [2]

## Temperature range (°C)

## 3 ENZYME STRUCTURE

**Molecular weight**
56000–60000 (formylase I, gel filtration, overview) [11, 12]
46000 (guinea pig) [17]
42000 (Streptomyces parvulus, gel filtration, formamidase I) [2, 4]
28000–31000 (formylase II, gel filtration, overview) [11, 12]
24000 (Streptomyces parvulus, gel filtration, formamidase II) [2, 4]
34700 (microheterogenous) [8]
35000 (rat) [13]
More (overview) [11, 12]

**Subunits**
Monomer (1 × 28000–31000, formylase II) [9, 12, 13]
Monomer (1 × 24000, Streptomyces parvulus, formamidase II) [2, 4]
Dimer (2 × 28000–34000, formylase I) [9, 11, 12, 13]
Dimer (2 × 24000, Streptomyces parvulus, formamidase I) [2, 4]

**Glycoprotein/Lipoprotein**
–

## 4 ISOLATION/PREPARATION

**Source organism**
Dog [12]; Rat [8, 12, 13]; Rabbit [12]; Guinea pig [12, 17]; Pigeon [12];
Chicken (formylase I) [12]; Horse [11]; Frog (formylase II) [11]; Mouse
(formylase II) [11]; Cow (formylase I); Drosophila melanogaster (formylase
I/II) [9, 10, 11]; Drosophila viridis (formylase I/II) [9]; Yeast (formylase I/II)
[11]; Neurospora sp. [15]; Anagasta kuhnilla [11]; Hansenula henricii
(formylase I/II) [6, 7]; Streptomyces parvulus (formylase I: constitutive,
formylase II: inducible) [1, 2, 4]; Alcaligenes eutrophus [3]; More (overview,
formamidase I (constitutive enzyme), formamidase II (inducible enzyme))
[11, 12]

**Source tissue**
Liver [12, 13, 14, 16]; Mycelium [15]; Cell [4, 6]; Kidney [16]; Speen [16]; Intestine [16]

**Localisation in source**
Cytoplasm [8, 12, 13, 14, 16]

**Purification**
Rat [8, 13]; Guinea pig [17]; Chicken [12]; Drosophila melanogaster [9];
Hansenula henricii [6]; Streptomyces parvulus [2]

**Crystallization**
–

Cloned
−

Renaturated
−

## 5 STABILITY

pH

Temperature (°C)
56 (2 minutes, 30% loss of activity) [15]; 45 (inactivated after 40 minutes, formylase II) [16]; More (formamidase I is more heat stable than formamidase II) [6, 11]

Oxidation

Organic solvent

General stability information

Storage
Lyophilized, 0°C, several months [16]; 5°C [12]

## 6 CROSSREFERENCES TO STRUCTURE DATABANKS

PIR/MIPS code

Brookhaven code

## 7 LITERATURE REFERENCES

[1] Katz, E., Brown, D., Hitchcock, M.J.M.: Methods Enzymol., 142, 225–234 (1987)
[2] Brown, D., Hitchcock, M.J.M., Katz, E.: Can. J. Microbiol., 32, 465–472 (1986)
[3] Friedrich, C.G., Mitrenga, G.: J. Gen. Microbiol., 125, 367–374 (1981)
[4] Brown, D.D., Hitchcock, M.J.M., Katz, E.: Arch. Biochem. Biophys., 202 (1), 18–22 (1980)
[5] Seifert, J., Casida, J.E.: Pestic. Biochem. Physiol., 12, 273–279 (1979)
[6] Bode, R., Birnbaum, D.: Biochem. Physiol. Pflanz., 174, 26–38 (1979)
[7] Bode, R., Birnbaum, D.: Z. Allg. Mikrobiol., 19 (3), 221–222 (1979)
[8] Menge, U.: Hoppe-Seyler's Z. Physiol. Chem., 360, 185–196 (1979)
[9] Moore, G.P., Sullivan, D.T.: Biochem. Genet., 16, 619–634 (1978)
[10] Jacobson, K.B.: Arch. Biochem. Biophys., 186 (1), 84–88 (1978)
[11] Moore, G.P., Sullivan, D.T.: Biochim. Biophys. Acta, 397, 468–477 (1975)

[12] Bailey, C.B., Wagner, C.: J. Biol. Chem., 249 (14) , 4439–4444 (1974)
[13] Arndt, R., Junge, W., Michelssen, K., Krisch, K.: Hoppe-Seyler's Z. Physiol. Chem.,
     354, 1583–1590 (1973)
[14] Santti, R., Soini, J.: Acta Chem. Scand., 22, 3321–3323 (1968)
[15] Jakoby, W.B.: J. Biol. Chem., 207, 657–663 (1954)
[16] Mehler, A.H., Knox, W.E.: J. Biol. Chem., 187, 431–438 (1950)
[17] Santti, R.S., Hopsu-Havu, V.K.: Hoppe-Seyler's Z. Physiol. Chem., 349, 753–766
     (1968)

Enzyme Handbook © Springer-Verlag Berlin Heidelberg 1991
Duplication, reproduction and storage in data banks are only
allowed with the prior permission of the publishers

## 1 NOMENCLATURE

**EC number**
3.5.1.10

**Systematic name**
10-Formyltetrahydrofolate amidohydrolase

**Recommended name**
Formyltetrahydrofolate deformylase

**Synonymes**

**CAS Reg. No.**
9025-08-5

## 2 REACTION AND SPECIFICITY

**Catalysed reaction**
10-Formyltetrahydrofolate + $H_2O$ →
→ formate + tetrahydrofolate

**Reaction type**
Carboxylic acid amide hydrolysis

**Natural substrates**
10-Formyltetrahydrofolate + $H_2O$

**Substrate spectrum**
1 10-Formyltetrahydrofolate + $H_2O$

**Product spectrum**
1 Formate + tetrahydrofolate

**Inhibitor(s)**

**Cofactor(s)/prostethic group(s)**

**Metal compounds/salts**

**Turnover number** $(min^{-1})$

**Specific activity** (U/mg)

**$K_m$-value** (mM)

Enzyme Handbook © Springer-Verlag Berlin Heidelberg 1991
Duplication, reproduction and storage in data banks are only
allowed with the prior permission of the publishers

pH-optimum

pH-range

Temperature optimum (°C)

Temperature range (°C)

---

## 3 ENZYME STRUCTURE

Molecular weight

Subunits

Glycoprotein/Lipoprotein
  –

---

## 4 ISOLATION/PREPARATION

Source organism
  Bovine [1]

Source tissue
  Liver [1]

Localisation in source

Purification

Crystallization
  –

Cloned
  –

Renaturated
  –

---

## 5 STABILITY

pH

Temperature (°C)

Oxidation

Organic solvent

General stability information

Storage

---

## 6 CROSSREFERENCES TO STRUCTURE DATABANKS

PIR/MIPS code

Brookhaven code

---

## 7 LITERATURE REFERENCES

[1] Huennekens, F.M.: Fed. Proc., 16, 199 (1957)

## 1 NOMENCLATURE

**EC number**
3.5.1.11

**Systematic name**
Penicillin amidohydrolase

**Recommended name**
Penicillin amidase

**Synonymes**
Penicillin acylase [6]
Benzylpenicillin acylase
Acylase, penicillin
Novozym 217
Semacylase
Alpha-acylamino-beta-lactam acylhydrolase
Ampicillin acylase

**CAS Reg. No.**
9014-06-6

---

## 2 REACTION AND SPECIFICITY

**Catalysed reaction**
Penicillin + $H_2O$ →
→ fatty acid anion + 6-aminopenicillanate (6-APA); Penicillin G + $H_2O$ →
→ phenylacetic acid + 6-aminopenicillanate [1]

**Reaction type**
Carboxylic acid amide hydrolysis

**Natural substrates**
Penicillin G + $H_2O$ [1–3]
More (other natural penicillins) [23, 24]

**Substrate spectrum**
1 Penicillin G + $H_2O$ (r) [1–3, 12]
2 Penicillin N + $H_2O$ [6, 7]
3 Penicillin V + $H_2O$ [6, 7]
4 Ampicillin + $H_2O$ [6, 7, 15]
5 Cloxacillin + $H_2O$ [8]
6 Methicillin + $H_2O$ [8]

7  Phenylacetyl-L-asparagin + $H_2O$ [1]
8  Phenylacetamide + $H_2O$ [9]
9  6-Aminopenicillanate + phenylacetylglycine [12]
10  More [7, 20, 21, 10]

## Product spectrum

1  6-Aminopenicillanate + phenylacetic acid (r) [1, 2, 3, 12]
2  ?
3  ?
4  ?
5  ?
6  ?
7  L-Asparagin + phenylacetic acid [1]
8  ?
9  ?
10  More [7, 8, 10, 20]

## Inhibitor(s)

Benzylpenicillin [2, 3, 7]; Phenylacetic acid [2, 3, 7]; 6-Aminopenicillanate
(6-APA) [2, 3, 7]; Phenylmethylsulfonyl fluoride [9, 14]; Phenylmethylsulfonyl
chloride [14]; Phenylmethylsulfonyl azide [14]; N-Hydroxysuccinimide ester
[14]; Alcohols [16]; Benzylisocyanate [14]; More (enzyme of E. coli only inhibited by substrate) [7]; 8-Hydroxyquinoline (complexes zinc, inhibition is
restored by addition of $ZnSO_4$, $MnSO_4$, $MgSO_4$, $CoSO_4$ or $MgSO_4$) [8]

## Cofactor(s)/prostethic group(s)

## Metal compounds/salts

$ZnSO_4$ (8-hydroxyquinoline inhibition is restored by addition of) [8]; $MnSO_4$
(8-hydroxyquinoline inhibition is restored by addition of) [8]; $MgSO_4$
(8-hydroxyquinoline inhibition is restored by addition of) [8]; $CoSO_4$
(8-hydroxyquinoline inhibition is restored by addition of) [8]; $FeSO_4$
(8-hydroxyquinoline inhibition is restored by addition of) [8]; Zinc (enzyme
contains 2 zinc atoms per molecule) [8]

## Turnover number ($min^{-1}$)

984 [5]; 32.4 [11]; 660 [11]; 1500 [1]; 1980 [11]; More [11, 12, 22]

## Specific activity (U/mg)

6.35 [2, 7]; 12 [2]; 12.45 [4]; 48 [7]; 31.5 [7]; 171.2 (immobilized enzyme) [7];
3.02 [8]; 60 [14]

## $K_m$-value (mM)

30 [1] (penicillin G) [2]; 0.67 (penicillin G) [2]; 0.63 (penicillin G) [5]; More
[18, 19, 3, 7, 8, 9, 10, 22]; 2.5 (phenoxymethylpenicillin) [8]; 0.18
(phenylacetyl-4-aminobenzoic acid) [10]; 0.4 (phenylacetyl-3-aminobenzoic
acid) [10]; 0.95 (phenylacetylanthranilic acid) [10]; 3.2

(D-(-)alpha-aminophenylacetic acid p-nitroanilide) [11]; 2.1 (cephalexin) [11]; 0.042 (cephalothin) [11]; 0.031 (p-nitrophenylphenylacetate) [11]; 0.045 (ethylphenylacetate) [11]; 0.08 (phenylacetylglycine) [11]; 0.097 (phenylacetate p-nitroanilide) [11]; 0.0046 [11] (benzylpenicillin) [22]; 0.009 (benzylpenicillin) [22]; 0.008 (benzylpenicillin) [22]; 0.01 (7-phenylacetamidodeacetoxycephalosporanic acid) [11]

**pH-optimum**
7.2 [1]; 7.0 [18]; 8.0–8.5 [2]; 7.65 [3]; 7.4 [3]; 8.15 [5]; 8.2 [7]; 8.5 [7]; 5.0–7.0 (reverse reaction) [7]; 7.0–8.0 [8]; 10.0 [8, 22]; 4.5–5.5 [8]; 8.0 [8]; 6.5–7.6 [9]; 5.0–6.0 [12]; 7.5–8.0 [13]

**pH-range**
7.0–8.5 [2]; 6–9 [3]; 5–8.5 [6]; 7–11 [22]

**Temperature optimum (°C)**
55 (benzylpenicillin, E.coli) [7]; 45 (benzylpenicillin, Bacillus megaterium) [7]; 37 (generally chosen, reason: enzyme unstable at 45 and 55°C) [7]; 50 [8]

**Temperature range (°C)**
20–50 [2]; 25–50 [8]; 10–23 [13]

---

## 3 ENZYME STRUCTURE

**Molecular weight**
71000 (E. coli, sedimentation equilibrium) [7]
120000 (Bacillus megaterium, sedimentation equilibrium) [7]
70000 (E. coli, gel filtration) [7]

**Subunits**

**Glycoprotein/Lipoprotein**
–

---

## 4 ISOLATION/PREPARATION

**Source organism**
E. coli [1–3, 15, 17]; Bacillus megaterium [7]; Proteus rettgeri [7]; Alcaligenes faecalis [7]; Erwinia aroideae [7, 8]; Achromobacter [7]; Kluyvera citrophila [7]; Pseudomonas melanogenum [7]; Micrococcus roseus [7]; Streptomyces lavendulae [7, 8]; Nocardia [7]; Penicillium chrysogenum [8]; Fusarium [8]; Bovista plumbea [7]; Aspergillus ochraceus [8]; Cephalosporium CMI 49137 [8]; Emericellopsis minima [8]; Epidermophyton floccosum [8]; Trychophyton mentagrophyta [8]; Calonectria [8]; Nectria [8]; Pleurotus ostreatus [8]; Streptomyces noursei [8]; Streptomyces erythreus [8]; Streptomyces netropsis [8]

#### Source tissue
Cell [2]; Mycelium [8]; Spores [8]

#### Localisation in source
Intracellular (E. coli) [7]; Extracellular (Bacillus megaterium) [7]

#### Purification
E. coli [2]; Bacillus megaterium [7]; Penicillium chrysogenum [8]; Fusarium [8]

#### Crystallization
[7]

#### Cloned
–

#### Renaturated
–

---

## 5 STABILITY

#### pH
10 (highest stability, Streptococcus lavendulae, 20°C, 5 hours) [7]; 6–10 (above 24 hours, 32°C) [8]; 4.2 (loss of activity) [8]

#### Temperature (°C)
40 (unstable above 40°C) [7]; 32 [8]; More (higher thermal stability of immobilized enzyme) [9]

#### Oxidation

#### Organic solvent
Methanol (in absence of methanol very stable, in 40% methanol, inactivation) [16]

#### General stability information
Immobilized enzyme (higher stability than soluble enzyme) [7]

#### Storage
Purified enzyme, –20°C [2, 7]; Immobilized enzyme, 2°C [3]; Immobilized enzyme, 4°C [7]

---

## 6 CROSSREFERENCES TO STRUCTURE DATABANKS

#### PIR/MIPS code
PNECA (precursor, Escherichia coli, fragment); B28392 (I, precursor, Pseudomonas sp.); A28392 (II, precursor, Pseudomonas sp.); A23593 (precursor, Escherichia coli); A26528 (Kluyvera cryocrescens); A25559 (V, Bacillus sphaericus)

**Brookhaven code**

# 7 LITERATURE REFERENCES

[1] Bauer, K., Kaufmann, W., Ludwig, S.A.: Hoppe-Seyler's Z. Physiol. Chem., 352, 1723–1724 (1971)

[2] Balasingham, K., Warburton, O., Dunnil, P., Lilly, M. D.: Biochim. Biophys. Acta, 276, 250–256 (1972)

[3] Warburton, D., Balasingham, K., Dunnil, P., Lilly, M. D.: Biochim. Biophys. Acta, 284, 278–284 (1972)

[4] Kutzbach, C.: (Bayer A.-G.) , Appl. P2217745.3 (1972)

[5] Berezin, I.V., Klibanov, A.M., Klyosov, A.A., Martinek, K., Svedas, V.K.: FEBS Lett., 49, 325–328 (1975)

[6] Cole, M., Savidge, T., Vander Haeghe, H.: Methods Enzymol., 43, 698–705 (1975)

[7] Savidge, T.A., Cole, M.: Methods Enzymol., 43, 705–721 (1975)

[8] Vander Haeghe, H.: Methods Enzymol., 43, 721–728 (1975)

[9] Szewczuk, A., Ziomek, E., Mordarski, M., Siewinski, M. , Wieczorek, J.: Biotechnol. Bioeng., 21, 1543–1552 (1979)

[10] Szewczuk, A., Siewinski, M., Slowinska, R.: Anal. Biochem., 103, 166–169 (1980)

[11] Morgolin, A.L., Svedas, V.K., Berezin, I.V.: Biochim. Biophys. Acta, 616, 283–289 (1980)

[12] Svedas, V.K., Margolin, A.L., Borisov, I.L., Berezin, I.V.: Enzyme Microb. Technol., 2, 313–317 (1980)

[13] McDougall, B., Dunnill, P., Lilly, M.D.: Enzyme Microb. Technol., 4, 114–115 (1982)

[14] Siewinski, M., Kuropatwa, M., Szewczuk, A.: Hoppe-Seyler's Z. Physiol. Chem., 365, 829–837 (1984)

[15] Kasche, V., Haufler, U., Zöllner, R.: Hoppe-Seyler's Z. Physiol. Chem., 365, 1435–1443 (1984)

[16] Kasche, V.: Biotechnol. Lett., 7, 877–882 (1985)

[17] Sudhakaran, V.K., Shewale, J.G.: Biotechnol. Lett., 9, 539–542 (1987)

[18] Brandl, E.: Hoppe-Seyler's Z. Physiol. Chem., 342 , 86 (1965)

[19] Self, D.A., Kay, G., Lilly, M.D., Dunnill, P.: Biotechnol. Bioeng., 11, 337 (1969)

[20] Kutzbach, C., Rauenbusch, E.: Hoppe-Seyler's Z. Physiol. Chem., 355, 45–53 (1974)

[21] Nys, P.S., Kolygina, T.S., Garaev, M.M.: Antibiotiki, 22, 211–216 (1977)

[22] Margolin, A.L., Izumrov, V.A., Svedas, V.K., Zezin, A.B., Kabanov, V.A., Berezin, I.V.: Biochim. Biophys. Acta, 660, 359–365 (1981)

[23] Kaufmann, W., Bauer, K.: Naturwissenschaften, 40, 474–475 (1960)

[24] Rolinson, G.N., Batchelor, F.R., Butterworth, K., Cameron-Wood, J., Cole, M., Eustace, G.C., Hart, M.V., Richards, D., Chain, E.B.: Nature (London) , 187, 236 (1960)

# 1 NOMENCLATURE

**EC number**
3.5.1.12

**Systematic name**
Biotin amide amidohydrolase

**Recommended name**
Biotinidase

**Synonymes**
Amidohydrolase biotinidase

**CAS Reg. No.**
9025-15-4

# 2 REACTION AND SPECIFICITY

**Catalysed reaction**
Biotin amide + $H_2O$ →
→ biotin + $NH_3$; More (biotinyl peptide + acceptor →
→ biotin + apopeptide [5], acceptors: $H_2O$ [2], hydroxylamine [2]) [2, 5]

**Reaction type**
Carboxylic acid amide hydrolysis

**Natural substrates**
Biocytin (i.e. epsilon-N-(d-biotinyl)-L-lysine) + $H_2O$ [6, 7]

**Substrate spectrum**
1  Triacetin + $H_2O$ [1]
2  Ethylacetate + $H_2O$ [1]
3  Glycyl-DL-leucylglycine + $H_2O$ [1]
4  Biotin methylester + $H_2O$ [1, 2]
5  Biotin amide + $H_2O$ [2]
6  N-Biotinyl-p-aminobenzoic acid + $H_2O$ [1, 2]
7  N-(1-N-Methoxycarbonyl-biotinyl)p-aminobenzoic acid + $H_2O$ [2]
8  1-(N-Methoxycarbonyl)-biocytin + $H_2O$ [2]
9  Biocytin + $H_2O$ [2, 5, 8]
10  N-(+)-Biotinyl-beta-alanine + $H_2O$ [3]
11  (+)-Biotinyl-L-aspartate + $H_2O$ [3]
12  Biotinyl-6-aminoquinoline + $H_2O$ [4, 6]

## Product spectrum

    1  ?
    2  ?
    3  ?
    4  Biotin + methanol [1, 2]
    5  Biotin + $NH_3$ [2]
    6  Biotin + p-aminobenzoic acid [1, 2]
    7  ?
    8  ?
    9  Biotin + lysine [2, 5, 8]
    10 Biotin + Ala [3]
    11 Biotin + L-Asp [3]
    12 6-Aminoquinoline + biotin [4, 6]

## Inhibitor(s)

p-Hydroxymercuribenzoate (reactivation with mercaptoethanol) [2];
Iodoacetamide [2]; Diisopropylfluorophosphate [2, 3]; Phenylmethanesul-
fonyl fluoride [5]; Phenylmethanesulfonamide [5]; Biotin [4]; Guanidine
chloride [7]; $ZnCl_2$ (1 mM, 30% inhibition) [7]; $CuCl_2$ (0.5 mM, 40% inhibition) [7]

## Cofactor(s)/prostethic group(s)

Biotin [3, 5]

## Metal compounds/salts

More (enzyme activity not affected by salt concentrations: $MnCl_2$, $MgCl_2$,
$CaCl_2$ /below 1 mM, EDTA-Na/below 25 mM, NaCl, NaF, KCl/below 50 mM)
[7]

## Turnover number ($min^{-1}$)

61.2 (N-biotinyl-p-aminobenzoate, pH 5) [7]; 73. 88
(N-biotinyl-p-aminobenzoate, pH 7) [7]; 97.2 (biocytin, pH 5) [7]; 81.6
(biocytin, pH 7) [7]

## Specific activity (U/mg)

0.187 [2]; 0.292 [2]; 0.1 [1]; 0.15 [1]; 0.36 [5]; 1927 [7]

## $K_m$-value (mM)

5 (1-N-methoxycarbonyl-biocytin) [2]; 0.57 (N-(-)-Biotinyl-p-aminobenzoic
acid) [2]; 0.85 (N-1-N-methoxycarbonyl-(biotinyl)-p-aminobenzoic acid) [2];
0.016 (biocytin) [2]; More [4, 5, 6]

## pH-optimum

6 (kidney) [2]; 7 (bacterial) [2]; 4.5–6.6 (natural substrate) [6, 7]; 6.0–7.5
(synthetic substrate) [6, 7]

**pH-range**
4.5–7.5 (kidney) [2]; 5.3–9.5 (bacterial) [2]

**Temperature optimum (°C)**
37 [1, 2]

**Temperature range (°C)**

## 3 ENZYME STRUCTURE

**Molecular weight**
68000 (sedimentation analysis, human) [7]
76000 (gel electrophoresis, human) [5]
78000 (chromatography, human) [5]
115000 [9]

**Subunits**
Monomer (1 × 80000, human, SDS-PAGE) [5, 7]

**Glycoprotein/Lipoprotein**
Glycoprotein [5, 7]

## 4 ISOLATION/PREPARATION

**Source organism**
Pig [2]; Rat [2]; Lactobacillus casei [2]; Chicken [1]; Streptococcus faecalis [3]; Human [5, 7]

**Source tissue**
Kidney [2]; Serum [7, 9]; Intestine [9]; Adrenal gland [9]; Liver [2]; Heart [2]; Pancreas [1]; Plasma [5]

**Localisation in source**
Microsomes [9]

**Purification**
Streptococcus faecalis [3]; Pig [2]; Lactobacillus casei [2]; Human [5]

**Crystallization**
–

**Cloned**
–

**Renaturated**
–

## 5 STABILITY

**pH**
5–7 (highest stability) [7]

**Temperature (°C)**
60 (60% loss of activity after 15 minutes, pH 5.5 and pH 7.0) [7]; 70 (complete loss of activity after 15 minutes, pH 5.5 and pH 7.0) [7]; More (heat stability) [2]

**Oxidation**

**Organic solvent**

**General stability information**
Instable in buffers of low ionic strength [7]; EDTA stabilizes [7]

**Storage**
0.1 M phosphate buffer, 4°C [7]; 1 week, 15°C, aqueous solution [1]; Some months, –20°C, 0.01 M phosphate buffer, pH 6 [2]

## 6 CROSSREFERENCES TO STRUCTURE DATABANKS

**PIR/MIPS code**

**Brookhaven code**

## 7 LITERATURE REFERENCES

[1] Thoma, R.W., Peterson, W.H.: J. Biol. Chem., 210, 569–579 (1954)
[2] Knappe, J., Brümmer, W., Biederbick, K.: Biochem. Z., 338, 599–613 (1963)
[3] Moss, J., Lane, M.D.: Adv. Enzymol. Relat. Areas Mol. Biol., 35, 321–442 (1971)
[4] Wastell, H., Dale, G., Bartlett, K.: Anal. Biochem., 140, 69–73 (1984)
[5] Craft, D.V., Goss, N.H., Chandramouli, N., Wood, H.G.: Biochemistry, 24, 2471–2476 (1985)
[6] Ebrahim, H., Dakshinamurti, K.: Anal. Biochem., 154, 282–286 (1986)
[7] Chauhan, J., Dakshinamurti, K.: J. Biol. Chem., 261, 4268–4275 (1986)
[8] Hayakawa, K., Oizumi, J.: J. Chromatogr., 383, 148–152 (1986)
[9] Pipsa, J.: Ann. Med. Exp. Biol. Fenn., 43, 5–39 (1965)

## 1 NOMENCLATURE

**EC number**
   3.5.1.13

**Systematic name**
   Aryl-acylamide amidohydrolase

**Recommended name**
   Arylacylamidase

**Synonymes**
   E.C. 3.5.1.13, AAA-1 [6, 7, 14]
   E.C. 3.5.1.13 AAA-2 [6, 7, 14]
   Brain acetylcholinesterase (is associated with AAA-2) [2, 6, 7, 10, 11, 15]
   Pseudocholinesterase (associated with aryacylamidase) [1, 9]

**CAS Reg. No.**
   9025-18-7

## 2 REACTION AND SPECIFICITY

**Catalysed reaction**
   A fatty acid anilide →
   → a fatty acid anion + aniline

**Reaction type**
   Carboxylic acid amide hydrolysis

**Natural substrates**
   Anilides + $H_2O$

**Substrate spectrum**
   1 Fatty acid anilides + $H_2O$
   2 More (broad specificity) [17, 18, 20]
   3 Peptid-p-nitroanilides + $H_2O$
   4 Dipeptides + $H_2O$ (preferentially leucyldipeptide) [18, 12]

**Product spectrum**
   1 A fatty acid anion + aniline
   2 ?
   3 Amino acid + p-nitroaniline
   4 Amino acids

**Inhibitor(s)**

Sulfhydryl reagents [18, 20]; Puromycin [20]; Diisopropyl phosphorofluoridate [13]; $Cu^{2+}$ [20]; $Zn^{2+}$ [20]; $Ag^+$ [20]; $Hg^{2+}$ [20, 18, 17]; $Ni^{2+}$ [18]; Tryptamine (pharmacologically active tryptamine derivatives, serotonin, LSD, inhibitors for brain, erythrocyte and serum aryl acylamidase, not for liver aryl acylamidase) [16, 19, 9, 11]; Chelating agents [12]; Anticholinesterase reagents (eserine, neostygmine, inhibitors for brain, erythrocyte, serum aryl acylamidase) [10, 11, 15]; Deoxycholate [8]

**Cofactor(s)/prostethic group(s)**

**Metal compounds/salts**

---

**Turnover number** (min$^{-1}$)

**Specific activity** (U/mg)

0.144–0.249 [20]; 0.8 [18]; 7.9 [15]; 0.8–1.3 [10]; 6.6 [10]; 15.4 [10]; 0.039 [3]

**$K_m$-value** (mM)

0.09 (L-alanine-4 nitroanilide) [20]; 0.2 (L-leucine-4-nitroanilide) [20]; 0.002 (linuron) [20]; 0.03 (2, 5-dimethylfuran-3-carboxanilide) [20]; 0.0578 (N-benzoyl-DL-arginine p-nitroanilide) [18]; 0.17 (propanil) [17]; 0.069 (nitroacetanilide) [5]; 0.006 (hydroxyacetanilide) [5]

**pH-optimum**

7.0–8.5 [20]; 8.0 [18]; 7.4–7.8 [17]; 5.3 [14]; 7.5 [14]; 8.5 [12]; 8.6 [5]

**pH-range**

6.5–8.2 [18]; 8.0–11.0 [5]

**Temperature optimum** (°C)

50 [18]; 35 [17]; 45 [5]

**Temperature range** (°C)

---

## 3 ENZYME STRUCTURE

**Molecular weight**

52500 (Pseudomonas fluorescens, gel filtration) [5]
63000 (soybean, gel filtration) [18]
68000 (human, gel electrophoresis) [9]
75000 (Bacillus sphaericus, gel filtration) [20]
130000 (gel filtration) [12]
50000–70000 (E. C. 3.5.1.13, AAA-2) [6]
100000–120000 (E.C. 3.5.1.13, AAA-1) [6]

**Subunits**
   Monomer (1 × 52500, Pseudomonas fluoresecens, gel filtration) [5]

**Glycoprotein/Lipoprotein**
   –

## 4 ISOLATION/PREPARATION

**Source organism**
   Human [1, 9, 10]; Rat [2, 3, 7, 14, 16]; Sheep [8, 10]; Monkey [11]; Pig [15];
   Eel [2, 10]; Electrophorus electricus [15]; Soybean [18]; Rice [4]; Bacillus
   sphàericus (inducible) [20]; Taraxacum officinale [17]; Pseudomonas
   acidovorans (inducible) [13]; Aspergillus oryzae [12]; Pseudomonas
   fluorescens (inducible) [5]

**Source tissue**
   Cell [5, 12, 13, 17, 20]; Brain [14, 16, 19]; Liver [9, 19]; Basal ganglia [10];
   Erythrocyte [9, 10]; Root [17, 18]; Electric organ [15]; Leaf [4]

**Localisation in source**
   Mitochondrial (membrane) [4]; Membrane [9]

**Purification**
   Bacillus sphaericus [20]; Pseudomonas fluorescens [5]; Soybean [18];
   Human [9] [10]; Sheep [10]; Eel [10]

**Crystallization**
   –

**Cloned**
   –

**Renaturated**
   –

## 5 STABILITY

**pH**
   5.0–7.5 [12]

**Temperature (°C)**
   45 (5 minutes, stable up to) [18]; 50 (stable up to) [12]; 46–48 (5 minutes,
   50% loss of activity) [17]; 60 (inactivation in less than 20 minutes) [5]

**Oxidation**

**Organic solvent**

## General stability information

### Storage

Purified enzyme, −15°C , several months [18]; Purified enzyme, glycerol, −20°C, short term storage [5]; Lyophilized, in the presence of lactose, 350 days [5]

## 6 CROSSREFERENCES TO STRUCTURE DATABANKS

PIR/MIPS code

Brookhaven code

## 7 LITERATURE REFERENCES

[1] Boopathy, R., Balasubramanian, A.S.: Eur. J. Biochem., 151, 351–360 (1985)
[2] Majumdar, R., Balasubramanian, A.S.: Biochemistry, 23, 4088–4093 (1984)
[3] Tsujita, T. Okuda, H.: Eur. J. Biochem., 133, 215–220 (1983)
[4] Gaynor, J.J., Still, C.C.: Plant Physiol., 72, 80–85 (1983)
[5] Hammond, P.M., Price, C.P., Scawen, M.C.: Eur. J. Biochem., 132, 651–655 (1983)
[6] Hsu, L.L.: Int. J. Biochem., 14, 1037–1042 (1982)
[7] Hsu, L.L., Halaris, A.E., Freedman, D.X.: Int. J. Biochem., 14, 581–584 (1982)
[8] Majumdar, R., Balasubramanian, A.S.: FEBS Lett., 146 (2) , 335–338 (1982)
[9] George, S.T., Balasubramanian, A.S.: Eur. J. Biochem., 121, 177–186 (1981)
[10] George, S.T., Balasubramanian, A.S.: Eur. J. Biochem. , 111, 511–524 (1980)
[11] Oommen, A., Balasubramanian, A.S.: Eur. J. Biochem., 94, 135–143 (1979)
[12] Nakadai, T., Nasuno, S.: Agric. Biol. Chem., 42 (6) , 1291–1292 (1978)
[13] Heymann, E., Rix, H.: Int. J. Pept. Protein Res., 11, 59–64 (1978)
[14] Hsu, L.L., Paul, S.M., Halaris, A.E, Freedman, D.X.: Life Sci., 20, 857–866 (1977)
[15] Fujimoto, D.: FEBS Lett., 71 (1) , 121–123 (1976)
[16] Paul, S.M., Halaris, A. E.: Biochem. Biophys. Res. Commun. , 70 (1) , 207–211 (1976)
[17] Hoagland, R.E.: Phytochemistry, 14, 383–386 (1975)
[18] Hoagland, R.E., Graf, G.: Can. J. Biochem., 52, 903–910 (1974)
[19] Fujimoto, D.: Biochem. Biophys. Res. Commun., 61 (1) , 72–74 (1974)
[20] Engelhardt, G., Wallnöfer, P.R.: Appl. Microbiol., 26 (5) , 709–718 (1973)

# 1 NOMENCLATURE

**EC number**
3.5.1.14

**Systematic name**
N-Acyl-L-amino-acid amidohydrolase

**Recommended name**
Aminoacylase

**Synonymes**
Dehydropeptidase II
Histozyme
Hippurase
Hippuricase
Benzamidase
Acylase I
Amido acid deacylase
L-Aminoacylase
Acylase
Aminoacylase I
L-Amino-acid acylase
Alpha-N-acylaminoacid hydrolase
Short acyl amidoacylase (hydrolytic activity toward N-short chain fatty acyl amino acids) [10]
Long acyl amidoacylase (hydrolytic activity toward N-long chain fatty acyl amino acids) [10]

**CAS Reg. No.**
9012-37-7

---

# 2 REACTION AND SPECIFICITY

**Catalysed reaction**
An N-acyl-L-amino acid + $H_2O$ →
→ a fatty acid anion + L-amino acid

**Reaction type**
Carboxylic acid amide hydrolysis

**Natural substrates**

## Substrate spectrum
   1 N-Acyl-L-amino acids + $H_2O$ (wide specificity) [1, 4, 12]
   2 Dehydropeptides + $H_2O$ [9]

## Product spectrum
   1 Corresponding fatty acid anions + L-amino acids
   2 Hydrolyzed dehydropeptides

---

## Inhibitor(s)
Chelating agents [3, 5, 7]; SH-blocking agents [9, 10]; Disulfide reducing agents [9]; N-Tosyl-L-alanine [9]; D, L-Norleucine [9]; D-Phenylalanine [11]; Isocaproate [11]; $Hg^{2+}$ [10]; $Cu^{2+}$ [10]; N-Tosyl-L-lysine chloromethylketone (not Aspergillus acylase) [13, 7, 9]

## Cofactor(s)/prostethic group(s)

## Metal compounds/salts
$Zn^{2+}$ (activates) [1, 5, 7]; $Co^{2+}$ (activates) [1, 7]

---

## Turnover number (min$^{-1}$)
30000 [4]; 12000 [4]

## Specific activity (U/mg)
250 [9]; 131 [7]; 150–200 (immobilized enzyme) [6]; 3410 [1]; More [1]

## $K_m$-value (mM)
0.15 (palmitoyl-aspartic acid) [10]; 2.5 (acetyl-valine) [10]; 6.3 (chloroacetylalanine) [7]; 6.6 (chloroacetylalanine) [7]; 6.0 (N-acetyl-L-methionine) [6]; 1.4 (N-acetyl-L-methionine, immobilized enzyme) [6]; 7.9 (N-acetyl-L-methionine) [1]; More [1, 3]

## pH-optimum
7.2 [13]; 5.6–7.2 [11]; 8.5 [4, 7]; 8.3 [1]; 7.3–7.6 (immobilized enzyme) [6]; 5.7 (in absence of $K^+$) [10]; 6.8 (in presence of $K^+$) [10]

## pH-range
6.0–11.5 [1]

## Temperature optimum (°C)
60 [6]; 65 (immobilized enzyme) [6]; 70 [1]

## Temperature range (°C)

---

## 3 ENZYME STRUCTURE

## Molecular weight
40000–48000 (Mycobacterium smegmatis) [10]
85500 (chemical analysis, pig) [9]
175000 (gel filtration, Bacillus thermoglucosidius) [1]

## Subunits
Dimer (2 × 36000–43000, bovine [4], Aspergillus oryzae [7], pig [9]) [4, 7, 9]
Tetramer (4 × 42500, Bacillus thermoglucosidius) [1]
Monomer (1 × 40000–48000, Mycobacterium smegmatis) [10]

## Glycoprotein/Lipoprotein
–

## 4 ISOLATION/PREPARATION

## Source organism
Hog [2, 13]; Pig [9]; Bull [4]; Lactobacillus arabinosus [11]; Mycobacterium
smegmatis [10]; Streptomyces sp. [8]; Aspergillus oryzae [7]; Bacillus
thermoglucosidius (inducible enzyme) [1]

## Source tissue
Kidney [13]; Liver [4]

## Localisation in source

## Purification
Pig [9]; Bovine [4]; Mycobacterium smegmatis [10]; Aspergillus oryzae [7];
Bacillus thermoglucosidius [1]

## Crystallization
–

## Cloned
–

## Renaturated
–

## 5 STABILITY

## pH
6.5–6.6 [6]; 7.2 (immobilized enzyme) [6]

## Temperature (°C)
37–55 [1]; 60 (1 hour) [13]; 70 (10 minutes) [1]

## Oxidation

## Organic solvent
Ethanol (stable up to 50 %) [1]; Iso-propanol (stable up to 30 %) [1]

## General stability information
Purified enzyme resistant to urea [1]

## Storage

1 mg/ml enzyme concentration, $-20°C$, 6 months [10]; Purified enzyme
stable in presence of $Zn^{2+}$, $4°C$ [4]

## 6 CROSSREFERENCES TO STRUCTURE DATABANKS

### PIR/MIPS code

### Brookhaven code

## 7 LITERATURE REFERENCES

[1] Cho, H.-Y., Tanizawa, K., Tanaka, H., Soda, K.: Agric. Biol. Chem., 51 (10),
     2793–2800 (1987)
[2] Henseling, J., Roehm, K.-H.: FEBS Lett., 219 (1), 27–30 (1987)
[3] Gilles, J., Loeffler, H.-G., Schneider, F.: Z. Naturforsch., 36c, 751–754 (1981)
[4] Gade, W., Brown, J.L.: Biochim. Biophys. Acta, 662, 86–93 (1981)
[5] Kumpe, E., Loeffler, H.-G., Schneider, F.: Z. Naturforsch., 36c, 951–955 (1981)
[6] Szajáni, B., Ivony, K., Boross, L.: J. Appl. Biochem., 2, 72–80 (1980)
[7] Gentzen, J., Loeffler, H.-G., Schneider, F.: Z. Naturforsch., 35c, 544–550 (1980)
[8] Sugie, M., Suzuki, H.: Agric. Biol. Chem., 44 (5), 1089–1095 (1980)
[9] Koerdel, W., Schneider, F.: Biochim. Biophys. Acta, 445, 446–457 (1976)
[10] Matsumoto, J., Nagai, S.: J. Biochem., 72, 269–279 (1972)
[11] Park, R.W., Fox, S.W.: J. Biol. Chem., 235 (1), 3193–3197 (1960)
[12] Fones, W.S., Lee, M.: J. Biol. Chem., 201, 847–856 (1953)
[13] Birnbaum, S.M., Levintow, L., Kingsley, R.B., Greenstein, J.P.: J. Biol. Chem., 194,
     455–470 (1952)

# 1 NOMENCLATURE

**EC number**
3.5.1.15

**Systematic name**
N-Acyl-L-aspartate amidohydrolase

**Recommended name**
Aspartoacylase

**Synonymes**
Aminoacylase II
N-Acetylaspartate amidohydrolase
Acetyl-aspartic deaminase
Acylase II

**CAS Reg. No.**
9031-86-1

# 2 REACTION AND SPECIFICITY

**Catalysed reaction**
N-Acetyl-L-aspartate + $H_2O$ →
→ a fatty acid anion + L-aspartate

**Reaction type**
Carboxylic acid amide hydrolysis

**Natural substrates**
N-Acetyl-L-aspartate + $H_2O$ [3]

**Substrate spectrum**
1 N-Acetyl-L-aspartate + $H_2O$
2 N-Chloroacetyl-L-aspartate + $H_2O$
3 N-Formyl-L-aspartate + $H_2O$ [2, 3]
4 2-Methyl-L-aspartate + $H_2O$ [3]
5 Glycyl-L-aspartate + $H_2O$ [6]
6 More (not: N-carbamyl-L-aspartate) [3]

## Product spectrum
1 Acetate + L-aspartate
2 Chloroacetate + L-aspartate
3 Formate + L-aspartate
4 ?
5 Glycine + L-aspartate
6 ?

## Inhibitor(s)
p-Hydroxymercuribenzoate [3]; N-Ethylmaleimide [3]; Aspartate (2 mM) [3]; Glutamate (2 mM) [3]

## Cofactor(s)/prostethic group(s)

## Metal compounds/salts

## Turnover number (min$^{-1}$)

## Specific activity (U/mg)
0.47 [3]

## $K_m$-value (mM)
5.1 (N-acetyl-L-aspartate) [4]; 0.2 (N-acetyl-L-aspartate) [8]; 0.1 (N-acetyl-L-aspartate) [1]

## pH-optimum
8.5 [4]; 7.5–8.5 [3]

## pH-range

## Temperature optimum (°C)

## Temperature range (°C)

## 3 ENZYME STRUCTURE

## Molecular weight

## Subunits

## Glycoprotein/Lipoprotein
–

## 4 ISOLATION/PREPARATION

**Source organism**
  Hog [4, 6, 7]; Rat [2–5]; Pigeon [4]; Hamster [4]; Cat [4]; Guinea pig [2]; Rabbit [4]; Mouse [4]; Bovine [4]; Monkey [4]; Chicken [1]

**Source tissue**
  Kidney [6, 7]; Brain [4, 5]; Liver [5]; Spinal cord [5]; Mammary gland [5]

**Localisation in source**
  Cytoplasm [1–7]; Membrane (bound) [1–7]

**Purification**
  Rat [3]

**Crystallization**
  –

**Cloned**
  –

**Renaturated**
  –

## 5 STABILITY

**pH**

**Temperature (°C)**
  57 (stable up to) [3]; 70 (stable up to) [9]

**Oxidation**

**Organic solvent**

**General stability information**

**Storage**
  Dialysed, at –20°C, several months [3]

## 6 CROSSREFERENCES TO STRUCTURE DATABANKS

**PIR/MIPS code**

**Brookhaven code**

# 7 LITERATURE REFERENCES

[1] D'Adamo, A.F., Wertman, E., Foster, F., Schneider, H.: Life Sci., 23, 791–796 (1978)
[2] Endo, Y.: FEBS Lett., 95 (2), 281–283 (1978)
[3] D'Adamo, A.F., Peisach, J., Manner, G., Weiler, C.T.: J. Neurochem., 28, 739–744 (1977)
[4] Goldstein, F.B.: J. Neurochem., 26, 45–49 (1976)
[5] D'Adamo, A.F., Smith, J.C., Weiler, C.: J. Neurochem., 20, 1275–1278 (1973)
[6] Birnbaum, S.M.: Methods Enzymol., 2, 115–119 (1955)
[7] Birnbaum, S.M., Levintow, L., Kinsley, R.B., Greenstein, J.P.: J. Biol. Chem., 194, 455–470 (1952)
[8] Woiler, C.T., D'Adamo, A.F.: Pharmacologist, 13, 283 (1971)
[9] Fleming, M., Lowry, O.: J. Neurochem., 13, 779–783 (1966)

# 1 NOMENCLATURE

**EC number**
3.5.1.16

**Systematic name**
$N^2$-Acetyl-L-ornithine amidohydrolase

**Recommended name**
Acetylornithine deacetylase

**Synonymes**
N-Acetylornithinase

**CAS Reg. No.**
9025-12-1

---

# 2 REACTION AND SPECIFICITY

**Catalysed reaction**
$N^2$-Acetyl-L-ornithine + $H_2O$ →
→ acetate + L-ornithine

**Reaction type**
Carboxylic acid amide hydrolysis

**Natural substrates**
$N^2$-Acetyl-L-ornithine + $H_2O$ [1]
Acetyl-L-methionine + $H_2O$ [1]

**Substrate spectrum**
1 $N^2$-Acetyl-L-ornithine + $H_2O$ [1]
2 Acetyl-L-methionine + $H_2O$ [1]
3 $N^2$-Acetyl-D-ornithine + $H_2O$ [1]
4 Acetyl-D-methionine + $H_2O$ [1]

**Product spectrum**
1 L-Ornithine + acetate [1]
2 L-Methionine + acetate [1]
3 D-Ornithine + acetate [1]
4 D-Methionine + acetate [1]

---

**Inhibitor(s)**
$Cu^{2+}$ [1]; $Zn^{2+}$ [1]; $Ni^{2+}$ [1]; p-Chloromercuribenzoate [1]; EDTA [1]

---

**Cofactor(s)/prostethic group(s)**

**Metal compounds/salts**
$Co^{2+}$ [1]

---

**Turnover number** (min$^{-1}$)

**Specific activity** (U/mg)
800 [1]

**$K_m$-value** (mM)
2.8 ($N^2$-acetylornithine) [1]

**pH-optimum**
7.0 ($N^2$-acetylornithine) [1]

**pH-range**

**Temperature optimum** (°C)

**Temperature range** (°C)

---

## 3 ENZYME STRUCTURE

**Molecular weight**

**Subunits**

**Glycoprotein/Lipoprotein**
–

---

## 4 ISOLATION/PREPARATION

**Source organism**
Enterobacteriaceae [1]

**Source tissue**

**Localisation in source**

**Purification**
Escherichia coli [1]

**Crystallization**
–

**Cloned**
–

Renaturated

—

## 5 STABILITY

pH

Temperature (°C)

Oxidation

Organic solvent

General stability information

Storage

## 6 CROSSREFERENCES TO STRUCTURE DATABANKS

PIR/MIPS code

Brookhaven code

## 7 LITERATURE REFERENCES

[1] Vogel, H.J., Bonner, D.M.: J. Biol. Chem., 218, 97–106 (1956)

# 1 NOMENCLATURE

**EC number**
  3.5.1.17

**Systematic name**
  $N^6$-Acyl-L-lysine amidohydrolase

**Recommended name**
  Acyl-lysine deacylase

**Synonymes**
  Epsilon-lysine acylase

**CAS Reg. No.**
  9025-11-0

# 2 REACTION AND SPECIFICITY

**Catalysed reaction**
  $N^6$-Acyl-L-lysine + $H_2O$ →
  → a fatty acid anion + L-lysine

**Reaction type**
  Carboxylic acid amide hydrolysis

**Natural substrates**
  Epsilon-N-acyl-L-lysine + $H_2O$ [1]

**Substrate spectrum**
  1 Epsilon-N-acyl-L-lysine + $H_2O$ [1]

**Product spectrum**
  1 L-Lysine + a fatty acid anion [1]

**Inhibitor(s)**
  $Ag^+$ [1]; $Hg^+$ [1]; $Hg^{2+}$ [1]; Monoiodoacetic acid [1];
  p-Chloromercuribenzoate [1]; Omega-chloroacetophenon [1]; Oxalate [1];
  Pyrophosphate [1]

**Cofactor(s)/prostethic group(s)**

**Metal compounds/salts**

**Turnover number ($min^{-1}$)**

**Specific activity** (U/mg)
  240 [1]; 0.09 [2]

**K$_m$-value** (mM)
  0.29 (epsilon-N-chloroacetyl-L-lysine) [1]; 0.5 (epsilon-N-acetyl-L-lysine) [1]

**pH-optimum**
  4.8–5.2 (epsilon-N-benzoyl-L-lysine, Achromobacter) [1]; 8.0 (epsilon-N-acetyl-L-lysine, hog or rat kidney) [2]; 9.0 (epsilon-N-acetyl-L-lysine,
  chicken kidney) [2]; 6.0 (Pseudomonas) [1]; 8.2–8.4 (Aspergillus) [1]; 7.0 (rat
  kidney) [1]

**pH-range**
  5.0 (not active below, epsilon-N-acetyl-L-lysine, enzyme from rat kidney) [2]

**Temperature optimum** (°C)

**Temperature range** (°C)

---

## 3 ENZYME STRUCTURE

**Molecular weight**

**Subunits**

**Glycoprotein/Lipoprotein**
  –

---

## 4 ISOLATION/PREPARATION

**Source organism**
  Achromobacter pestifer [1]; Animals (overview) [1]; Bacteria (overview) [1];
  Fungi (overview) [1]

**Source tissue**
  Kidney [1]

**Localisation in source**

**Purification**
  Achromobacter pestifer [1]; Hog kidney [2]

**Crystallization**
  –

**Cloned**
  –

**Renaturated**
  –

## 5 STABILITY

pH

Temperature (°C)

Oxidation

Organic solvent

General stability information

**Storage**
0°C, 60% ammonium sulfate (several months)

## 6 CROSSREFERENCES TO STRUCTURE DATABANKS

PIR/MIPS code

Brookhaven code

## 7 LITERATURE REFERENCES

[1] Chibata, I., Ishikawa, T., Tosa, T.: Methods Enzymol., 19, 756–762 (1970)
[2] Leclerc, J., Benoiton, L.: Can. J. Biochem., 46, 471–475 (1968)

# 1 NOMENCLATURE

**EC number**
3.5.1.18

**Systematic name**
N-Succinyl-LL-2,6-diaminoheptanedioate amidohydrolase

**Recommended name**
Succinyldiaminopimelate desuccinylase

**Synonymes**
N-Succinyl-L-alpha, epsilon-diaminopimelic acid deacylase [1]

**CAS Reg. No.**
9024-94-6

# 2 REACTION AND SPECIFICITY

**Catalysed reaction**
N-Succinyl-LL-2,6-diaminoheptanedioate + $H_2O$ →
→ succinate + LL-2,6-diaminoheptanedioate

**Reaction type**
Carboxylic acid amide hydrolysis

**Natural substrates**
N-Succinyl-L-diaminopimelic acid + $H_2O$ [1]

**Substrate spectrum**
1 N-Succinyl-L-diaminopimelic acid + $H_2O$ [1]

**Product spectrum**
1 L-Diaminopimelic acid + succinate [1]

**Inhibitor(s)**
EDTA [1]

**Cofactor(s)/prostethic group(s)**

**Metal compounds/salts**
$Co^{2+}$ (activates) [1]; $Mn^{2+}$ (activates); $Zn^{2+}$ (activates); $Fe^{3+}$ (activates, less active than $Co^{2+}$) [1]

**Turnover number (min⁻¹)**

**Specific activity (U/mg)**

**K$_m$-value (mM)**
1.3 (N-succinyl-L-diaminopimelate) [1]; 0.015 (Co$^{2+}$) [1]

**pH-optimum**
8 [1]

**pH-range**
6–9 [1]

**Temperature optimum (°C)**

**Temperature range (°C)**

---

## 3 ENZYME STRUCTURE

**Molecular weight**

**Subunits**

**Glycoprotein/Lipoprotein**
–

---

## 4 ISOLATION/PREPARATION

**Source organism**
Escherichia coli [1]; Corynebacterium diphteriae [1]; Bacillus cereus [1]; Micrococcus lysodeikticus [1]

**Source tissue**
Cell [1]

**Localisation in source**

**Purification**
Escherichia coli [1]

**Crystallization**
–

**Cloned**
–

**Renaturated**
–

## 5 STABILITY

pH

Temperature (°C)

Oxidation

Organic solvent

General stability information

**Storage**
Dialyzed enzyme, −15°C, 6 months [1]

---

## 6 CROSSREFERENCES TO STRUCTURE DATABANKS

PIR/MIPS code

Brookhaven code

---

## 7 LITERATURE REFERENCES

[1] Kindler, S.H., Gilvarg, C.: J. Biol. Chem., 235 (12) , 3532–3535 (1960)

# 1 NOMENCLATURE

**EC number**
3.5.1.19

**Systematic name**
Nicotinamide amidohydrolase

**Recommended name**
Nicotinamidase

**Synonymes**
Nicotinamide deaminase
Nicotinamide amidase [2]
YNDase [15]

**CAS Reg. No.**
9033-32-3

# 2 REACTION AND SPECIFICITY

**Catalysed reaction**
Nicotinamide + $H_2O \rightarrow$
$\rightarrow$ nicotinate + $NH_3$

**Reaction type**
Carboxylic acid amide hydrolysis

**Natural substrates**
Nicotinamide + $H_2O$

**Substrate spectrum**
1 Nicotinamide + $H_2O$
2 Nicotinamide (structural analogs) + $H_2O$ [8]
3 p-Nitrophenylacetate + $H_2O$ [4, 6]
4 p-Nitrophenylnicotinate + $H_2O$ [4, 6]

**Product spectrum**
1 Nicotinate + $NH_3$
2 Acid (corresponding) + $NH_3$ [8]
3 p-Nitrophenol + acetate [4, 6]
4 p-Nitrophenol + nicotinic acid [4, 6]

## Inhibitor(s)

Chelating agents [1]; Nicotinamide (analogs with a trivalent nitrogen) [5];
Carbobenzoxyamido-2-phenyl-ethyl-chloromethyl-ketone [4, 6];
Diisopropylfluorophosphate [4, 6]; p-Chloromercuribenzoate [7]; Sulfhydryl
agents [10]; $Cu^{2+}$ [10]; $Zn^{2+}$ [10]; $Fe^{2+}$ [10]; N-Ethylmaleimide [15]

## Cofactor(s)/prostethic group(s)

## Metal compounds/salts

$Mg^{2+}$ (necessary) [5, 10]; $Mn^{2+}$ (increases activity) [5, 10]; $Hg^{2+}$ (activates)
[5]; $Ca^{2+}$ (activates) [10]

---

## Turnover number (min⁻¹)

More [3, 4, 15]

## Specific activity (U/mg)

More [1, 2, 3, 5, 15]

## $K_m$-value (mM)

0.65 (nicotinamide) [1]; More [2, 3, 5, 7, 10, 15]

## pH-optimum

9.0–10.0 [2]; 7.2 [3]; 6.5 [5]; 6.6 [12]; 8–10 (esterase) [4]; 7.5 [1, 10]

## pH-range

6–8 [3]; 5–10 [4]; 5–10.5 (esterase) [4]; 6–8.5 [15]

## Temperature optimum (°C)

40 [1]; 37 [10]

## Temperature range (°C)

50 (not active above) [3, 7]; 45 (not active above) [10]

---

## 3 ENZYME STRUCTURE

## Molecular weight

30000 (E. coli, gel filtration) [3]
48000 (Flavobacterium peregrinum, gel filtration) [5, 12]
211000 (rabbit, gel electrophoresis) [6]
230000 (gel electrophoresis) [10]
34000 (HPLC) [15]
More (in bacteria : about 40000, in higher organisms: about 220000)

## Subunits

Monomer (bacteria) [1, 5]
Tetramer (2 × 65000, 2 × 50000, SDS-PAGE [10], 4 × 60000, rabbit, SDS-
PAGE [6]) [6, 10]

## Glycoprotein/Lipoprotein
Glycoproteine (mannose/N-acetylglucosamine) [10]

---

## 4 ISOLATION/PREPARATION

### Source organism
Aspergillus niger [1]; Rat [2, 14]; Rabbit [2, 6]; E. coli [3]; Flavobacterium peregrinum [5]; Micrococcus lysodeiktikus [7]; Salmonella typhimurium [9]; Yeast [11, 15]; Vibrio cholera [13]

### Source tissue
Cell [1, 5]; Liver [2, 6, 14]; Small intestine [14]

### Localisation in source
Microsomes [2]; Mitochondria [2]

### Purification
Aspergillus niger [1]; Rat [2]; Rabbit [2]; E. coli [3]; Flavobacterium peregrinum [5]; Yeast [15]

### Crystallization
–

### Cloned
–

### Renaturated
–

---

## 5 STABILITY

### pH
5.0–10.0 [4]; 7.0–9.5 [7]

### Temperature (°C)
50 (up to) [3, 7]; 37 (half-life: 2 hours) [3]

### Oxidation

### Organic solvent

### General stability information
Instable without $Mn^{2+}$ [5]; Stable with 0.005 mM nicotinamide [7]; Low stability at low salt concentrations [3]; $HgCl_2$ stabilizes [5]

### Storage
–20°C, 50% glycerol [2]

---

# 6 CROSSREFERENCES TO STRUCTURE DATABANKS

PIR/MIPS code

Brookhaven code

# 7 LITERATURE REFERENCES

[1] Sarma, D.S., Rajalakshmi, S., Sarma, P.: Biochim. Biophys. Acta, 81, 311–322 (1964)
[2] Petrack, B., Greengard, P., Craston, N., Sheppy, F.: J. Biol. Chem., 240, 1725–1730 (1965)
[3] Pardee, B.P.: J. Biol. Chem., 246 (22) , 6792–6796 (1971)
[4] Albizati, L.D., Hedrick, J.L.: Biochemistry, 11 (8) , 1508–1516 (1972)
[5] Tanigawa, J., Shimiyama, M., Dohi, K., Ueda, I.: J. Biol. Chem., 247 (24) , 8036–8042 (1972)
[6] Gillam, S.S., Watson, J.G., Chaykin, S.: Arch. Biochem. Biophys., 157, 268–284 (1973)
[7] Gadd, R.E.A., Johson, W.J.: Int. J. Biochem., 5, 397–407 (1974)
[8] Johnson, W.J., Gadd, R.E.A.: Int. J. Biochem., 5, 633–641 (1974)
[9] Foster, W.J., Kinney, D.M., Moat, A.G.: J. Bacteriol., 138 (3) , 957–961 (1979)
[10] Wintzerith, M., Dierich, A., Mandel, P.: Biochim. Biophys. Acta, 613, 191–202 (1980)
[11] Fuller, L.: Methods Enzymol., 66, 3–4 (1980) (Review)
[12] Tanigawa, Y., Shimiyama, M., Ueda, I.: Methods Enzymol., 66, 132–136 (1980) (Review)
[13] Foster, J.W., Brestel, C.: J. Bacteriol., 149 (1) , 368–371 (1982)
[14] Shibata, K., Hayakawa, T., Iwai, K.: Agric. Biol. Chem. , 50 (12) , 3037–3041 (1986)
[15] Yan, C., Sloan, D.L.: J. Biol. Chem., 262 (19) , 9082–9087 (1987)

# 1 NOMENCLATURE

**EC number**
3.5.1.20

**Systematic name**
L-Citrulline $N^5$-carbamoyldihydrolase

**Recommended name**
Citrullinase

**Synonymes**
Citrulline ureidase
Citrulline hydrolase [5]

**CAS Reg. No.**
59088-17-4

---

# 2 REACTION AND SPECIFICITY

**Catalysed reaction**
L-Citrulline $+ 2 H_2O \rightarrow$
$\rightarrow$ L-ornithine $+ CO_2 + NH_3$

**Reaction type**
Carboxylic acid amide hydrolysis

**Natural substrates**
L-Citrulline $+ H_2O$

**Substrate spectrum**
1  L-Citrulline $+ H_2O$

**Product spectrum**
1  L-Ornithine $+ CO_2 + NH_3$

---

**Inhibitor(s)**
Citrulline (high concentration) [3]; Sulfhydryl reagents (mercaptoethanol removes inactivation) [3]

**Cofactor(s)/prostethic group(s)**

**Metal compounds/salts**

---

**Turnover number (min$^{-1}$)**

**Specific activity** (U/mg)
  5.94 [5]

**K$_m$-value** (mM)
  1.06 (L-citrulline) [5]; 4.55 (L-citrulline) [3]

**pH-optimum**
  7.0–7.2 [3]; 7.5 [4]; 6.8 [5]

**pH-range**

**Temperature optimum** (°C)
  30 [3]

**Temperature range** (°C)

---

## 3 ENZYME STRUCTURE

**Molecular weight**
  94000 (Euglena gracilis, gel filtration) [3]

**Subunits**
  Trimer (3 × 31000, Euglena gracilis) [3]

**Glycoprotein/Lipoprotein**
  –

---

## 4 ISOLATION/PREPARATION

**Source organism**
  Tetrahymena pyriformis [5]; Tetrahymena thermophila [2]; Euglena gracilis
  [3, 4]; Phytomonas [1]; Leptomonas [1]; Heptomonas [1]; Streptococcus sp.
  [3]; Pseudomonas sp. [3]

**Source tissue**
  Cell [3]

**Localisation in source**
  Mitochondria [3]

**Purification**
  Tetrahymena thermophila [2]; Euglena gracilis [3]

**Crystallization**
  –

**Cloned**
  –

**Renaturated**

–

# 5 STABILITY

**pH**
6.0–7.0 [3]

**Temperature (°C)**
25–37 [2]; 40 [3]; 60 (inactivated after 10 minutes) [3]

**Oxidation**

**Organic solvent**

**General stability information**

**Storage**
Purified enzyme, –10°C, several days [2]

# 6 CROSSREFERENCES TO STRUCTURE DATABANKS

**PIR/MIPS code**

**Brookhaven code**

# 7 LITERATURE REFERENCES

[1] Camargo, E.P., Silva, S., Roitman, I., De Souza, W., Jankevicius, J.V., Dollet, M.: J.
Protozool., 34 (4) , 439–441 (1987)
[2] Eichler, W. in "Methods Enzym. Anal.", 3rd Ed. (Bergmeyer, H.U., Ed.) Vol.8, 412–418
(1985)
[3] Park, B.-S., Hirotani, A., Nakano, Y., Kitaoka, S.: Agric. Biol.Chem., 49 (7) , 2205–2206
(1985)
[4] Park, B.-S., Hirotani, A., Nakano, Y., Kitaoka, S.: Agric. Biol. Chem., 47 (11) ,
2561–2567 (1983)
[5] Hill, D.L., Chambers, P.: Biochim. Biophys. Acta, 148, 435–447 (1967)

## 1 NOMENCLATURE

**EC number**
3.5.1.21

**Systematic name**
N-Acetyl-beta-alanine amidohydrolase

**Recommended name**
N-Acetyl-beta-alanine deacetylase

**Synonymes**

**CAS Reg. No.**
37289-04-6

## 2 REACTION AND SPECIFICITY

**Catalysed reaction**
N-Acetyl-beta-alanine + $H_2O \rightarrow$
$\rightarrow$ acetate + beta-alanine

**Reaction type**
Carboxylic acid amide hydrolysis

**Natural substrates**
N-Acetyl-beta-alanine + $H_2O$ [1]

**Substrate spectrum**
1 N-Acetyl-beta-alanine + $H_2O$ [1]
2 N-Acetyl-taurine + $H_2O$ (to less extent) [1]

**Product spectrum**
1 Acetate + beta-alanine
2 Acetate + taurine

**Inhibitor(s)**
p-Chloromercuribenzoate [1]; Monoiodoacetate [1]; $Hg^{2+}$ [1]; $Ag^+$ [1];
$Cu^{2+}$ [1]

**Cofactor(s)/prostethic group(s)**

**Metal compounds/salts**

**Turnover number** (min$^{-1}$)

**Specific activity** (U/mg)
   11.9 (N-acetyl-beta-alanine) [1]; 4.4 (N-acetyl-taurine) [1]

**K$_m$-value** (mM)
   2.5 (acetyl-beta-alanine) [1]

**pH-optimum**
   7.6 [1]

**pH-range**

**Temperature optimum** (°C)

**Temperature range** (°C)

---

## 3 ENZYME STRUCTURE

**Molecular weight**

**Subunits**

**Glycoprotein/Lipoprotein**
   –

---

## 4 ISOLATION/PREPARATION

**Source organism**
   Hog [1]; Rat [1]

**Source tissue**
   Kidney [1]

**Localisation in source**

**Purification**
   Hog [1]

**Crystallization**
   –

**Cloned**
   –

**Renaturated**
   –

## 5 STABILITY

pH

**Temperature (°C)**
   60 (2 minutes , 20% loss of activity) [1]

Oxidation

Organic solvent

General stability information

Storage

## 6 CROSSREFERENCES TO STRUCTURE DATABANKS

PIR/MIPS code

Brookhaven code

## 7 LITERATURE REFERENCES

[1] Fujimoto, D., Koyama, T., Tamiya, N.: Biochim. Biophys. Acta, 167, 407–413 (1968)

## 1 NOMENCLATURE

**EC number**
3.5.1.22

**Systematic name**
Pantothenate amidohydrolase

**Recommended name**
Pantothenase

**Synonymes**
Pantothenate hydrolase

**CAS Reg. No.**
9076-90-8

## 2 REACTION AND SPECIFICITY

**Catalysed reaction**
Pantothenate + $H_2O$ →
→ pantoate + beta-alanine

**Reaction type**
Carboxylic acid amide hydrolysis

**Natural substrates**
Pantothenate + $H_2O$

**Substrate spectrum**
1 D-Pantothenate + $H_2O$ (r) [1, 2, 3, 5]
2 Hydroxypantothenate + $H_2O$ [3]
3 Pantoyl-gamma-aminobutyrate + $H_2O$ [3]

**Product spectrum**
1 Pantoate + beta-alanine
2 Hydroxypantoate + beta-alanine [3]
3 Pantoate + gamma-aminobutyrate [3]

**Inhibitor(s)**
Oxalate [8]; Oxalacetate [8]; Oxamate [8]; 2-Oxomalonate [8];
3-Oxoglutarate [8]; 3-Aminophenylboronic acid [5]; Phenylmethylsulfonylfluoride [8]

**Cofactor(s)/prostethic group(s)**

**Metal compounds/salts**
  More (no dissociable metal ion) [11]

---

**Turnover number** (min$^{-1}$)

**Specific activity** (U/mg)
  More [11]

**K$_m$-value** (mM)
  5 (pantothenate) [11]; More [2, 8]

**pH-optimum**
  7.4 [11]

**pH-range**

**Temperature optimum** (°C)
  28 [11]

**Temperature range** (°C)

---

## 3 ENZYME STRUCTURE

**Molecular weight**
  100000 (gel filtration, Pseudomonas fluorescens) [9]

**Subunits**
  Dimer (Pseudomonas fluorescens; 2 × 50000) [9]

**Glycoprotein/Lipoprotein**
  −

---

## 4 ISOLATION/PREPARATION

**Source organism**
  Pseudomonas P-2 (inducible) [11]; Pseudomonas fluorescens [1, 9, 8, 7, 6, 3]; Lactobacillus plantarum [4]; Pediococcus acidilactici [4]

**Source tissue**
  Cells [11]

**Localisation in source**

**Purification**
  Pseudomonas P-2 [11]; Pseudomonas fluorescens [9, 6, 3]

**Crystallization**
  −

**Cloned**
–

**Renaturated**
–

## 5 STABILITY

**pH**
    5.5–10 [9]; 5.5–9.5 [3]

**Temperature (°C)**
    28 (unstable above, reactivated when temperature is lowered again) [9, 10]

**Oxidation**

**Organic solvent**

**General stability information**
    Pantothenate and oxalate stabilize [9]; 3-Oxobutyrate, 2-oxomalonate, oxa-
    late and oxalacetate protect against thermal inactivation [7]; Partial reac-
    tivating of thermal inactivated pantothenase by oxalate, oxalacetate,
    pyruvate [6]

**Storage**
    Purified enzyme loses some activity on freezing [3]; Frozen enzyme, –20°C,
    2 years [3]

## 6 CROSSREFERENCES TO STRUCTURE DATABANKS

**PIR/MIPS code**

**Brookhaven code**

## 7 LITERATURE REFERENCES

[1] Airas, R.K.: Methods Enzymol., 122, 33–35 (1986) (Review)
[2] Airas, R.K.: Anal. Biochem., 134, 122–125 (1983)
[3] Airas, R.K.: Methods Enzymol., 62, 267–275 (1979) (Review)
[4] Solberg, O., Hegna, I.K.: Methods Enzymol., 62, 201–205 (1979) (Review)
[5] Airas, R.K.: Biochemistry, 17 (23), 4932–4938 (1978)
[6] Airas, R.K.: Biochim. Biophys. Acta, 452, 201–208 (1976)
[7] Airas, R.K.: Biochim. Biophys. Acta, 452, 193–200 (1976)
[8] Airas, R.K.: Biochem. J., 157, 415–421 (1974)
[9] Airas, R.K., Hietanen, E.A., Nurmikko, V.T.: Biochem. J., 157, 409–413 (1976)
[10] Airas, R.K.: Biochem. J., 130, 111–119 (1972)
[11] Nurmikko, V., Salo, E., Hakola, H., Mähinen, U., Snell, E.E.: Biochemistry, 5 (2),
      399–402 (1966)

## 1 NOMENCLATURE

**EC number**
3.5.1.23

**Systematic name**
N-Acylsphingosine amidohydrolase

**Recommended name**
Acylsphingosine deacylase

**Synonymes**
Ceramidase (N-acylsphingosine serves as substrate)
Glycosphingolipid ceramide deacylase (acidic glycosphingolipids serve as substrates) [1]

**CAS Reg. No.**
37289-06-8

---

## 2 REACTION AND SPECIFICITY

**Catalysed reaction**
N-Acylsphingosine + $H_2O$ →
→ a fatty acid anion + sphingosine

**Reaction type**
Carboxylic acid amide hydrolysis

**Natural substrates**

**Substrate spectrum**
1 N-Acylsphingosine + $H_2O$ (r) [2]
2 Ganglioside (GM1, GM2, GM3) + $H_2O$ [1]
3 More (not: N-acetylsphingosine, N-lignoceryldihydrosphingosine, sphingomyelin, cerebroside [2]) [2, 1]

**Product spectrum**
1 Sphingosine + fatty acid
2 Lyso-ganglioside (corresponding) + fatty acid [1]
3 ?

---

**Inhibitor(s)**
Sphingosine [2]; Fatty acids [2]; $Mn^{2+}$ [1]; $Cu^{2+}$ [1], $Sn^{2+}$ [1]; $Hg^{2+}$ [1]; EDTA [1]; $Zn^{2+}$ [1]

---

Enzyme Handbook © Springer-Verlag Berlin Heidelberg 1991
Duplication, reproduction and storage in data banks are only
allowed with the prior permission of the publishers

**Cofactor(s)/prostethic group(s)**

**Metal compounds/salts**
   $Ba^{2+}$ (10 mM, slight activation) [1]; $Mg^{2+}$ (10 mM, slight activation) [1]

---

**Turnover number (min$^{-1}$)**

**Specific activity (U/mg)**

**K$_m$-value (mM)**
   0.3 (N-oleyldi-$^3$H-sphingosine) [2]

**pH-optimum**
   4.8 [2]; 5.8 [1]

**pH-range**

**Temperature optimum (°C)**

**Temperature range (°C)**

---

**3 ENZYME STRUCTURE**

**Molecular weight**

**Subunits**

**Glycoprotein/Lipoprotein**
   −

---

**4 ISOLATION/PREPARATION**

**Source organism**
   Rat [2]; Nocardia spec. [1]

**Source tissue**
   Brain [2]

**Localisation in source**
   Membrane [1]

**Purification**
   Rat [2]

**Crystallization**
   −

**Cloned**
   −

Renaturated
–

---

# 5 STABILITY

pH

Temperature (°C)
  45 (30 minutes) [1]

Oxidation

Organic solvent

General stability information

Storage
  Purified enzyme, –20°C, 1 year [2]

---

# 6 CROSSREFERENCES TO STRUCTURE DATABANKS

PIR/MIPS code

Brookhaven code

---

# 7 LITERATURE REFERENCES

[1] Hirabayashi, Y., Kimura, M., Matsumoto, M., Yamamoto, K., Kadowaki, S., Tochikura, T.: J. Biochem., 103, 1–4 (1988)
[2] Gatt, S.: J. Biol. Chem., 241 (16) , 3724–3730 (1966)

# 1 NOMENCLATURE

**EC number**
3.5.1.24

**Systematic name**
3Alpha, 7alpha, 12alpha-trihydroxy-5beta-cholan-24-oylglycine amidohydrolase

**Recommended name**
Cholylglycine hydrolase

**Synonymes**
Glycocholase [1]
Bile salt hydrolase [2]
Cholylglycine hydrolase [3]

**CAS Reg. No.**
37289-07-9

---

# 2 REACTION AND SPECIFICITY

**Catalysed reaction**
3Alpha, 7alpha, 12alpha-trihydroxy-5beta-cholan-24-oylglycine + $H_2O \rightarrow$
$\rightarrow$ 3alpha, 7alpha, 12alpha-trihydroxy-5 beta-cholanate + glycine;
Cleavage of bile acid conjugates [1]

**Reaction type**
Carboxylic acid amide hydrolysis

**Natural substrates**
3Alpha, 7alpha, 12alpha-trihydroxy-5beta-cholan-24-oylglycine (cholyl-glycine) + $H_2O$

**Substrate spectrum**
1 3Alpha, 7alpha, 12alpha-trihydroxy-5beta-cholan-24-oylglycine + $H_2O$
2 Chenodeoxycholic acid + $H_2O$ [2, 3]
3 Deoxycholic acid + $H_2O$ [2, 3]
4 More (taurine- and glycine conjugates of cholic acid, not: lithocholic acid conjugates [2], $C_{24}$-bile acid conjugates with a tertiary amide group [3]) [2, 3]

**Product spectrum**

1  3Alpha, 7alpha, 12alpha-trihydroxy-5beta-cholanate + glycine
2  ?
3  ?
4  More (corresponding cholanates, chenodeoxycholanates) [2, 3]

**Inhibitor(s)**

Iodoacetate [1]; p-Chloromercuribenzoate [1]; $Hg^{2+}$ [1]; $Cu^{2+}$ [1]; $Zn^{2+}$ [1]; Cholic acid (product inhibition) [1]

**Cofactor(s)/prostethic group(s)**

**Metal compounds/salts**

**Turnover number** (min$^{-1}$)
20000 [2]

**Specific activity** (U/mg)
More [1, 2]

**$K_m$-value** (mM)
8 (glycocholic acid) [1]; 0.35 (glycocholic acid) [2]; More [1, 2]

**pH-optimum**
5.6–5.8 [1]; 4.2–4.5 [2]

**pH-range**
4.8–7.0 [1]

**Temperature optimum** (°C)
37 [1, 2]

**Temperature range** (°C)

# 3 ENZYME STRUCTURE

**Molecular weight**
250000 (gel filtration, Bacteroides fragilis) [2]

**Subunits**
Octamer (8 × 32500, Bacteroides fragilis, SDS-PAGE) [2]

**Glycoprotein/Lipoprotein**
–

## 4 ISOLATION/PREPARATION

**Source organism**
   Clostridium perfringens [1]; Bacteroides fragilis [2]

**Source tissue**
   Cell [1]; Spheroblast [2]

**Localisation in source**

**Purification**
   Clostridium perfringens [1]; Bacteroides fragilis [2]

**Crystallization**
   –

**Cloned**
   –

**Renaturated**
   –

## 5 STABILITY

pH

Temperature (°C)

Oxidation

Organic solvent

General stability information

**Storage**
   8 weeks at –20°C, pH 7.0 [1]

## 6 CROSSREFERENCES TO STRUCTURE DATABANKS

**PIR/MIPS code**

**Brookhaven code**

## 7 LITERATURE REFERENCES

[1] Nair, P.P., Gordon, M., Reback, J.: J. Biol. Chem., 242 (1) , 7–11 (1967)
[2] Stellwag, E.J., Hylemon, P.B.: Biochim. Biophys. Acta, 452, 165–176 (1976)
[3] Batta, A.K., Salen, G., Shefer, S.: J. Biol. Chem., 259 (24) , 15035–15039 (1984)

## 1 NOMENCLATURE

**EC number**
3.5.1.25

**Systematic name**
N-Acetyl-D-glucosamine-6-phosphate amidohydrolase

**Recommended name**
N-Acetylglucosamine-6-phosphate deacetylase

**Synonymes**
Acetylglucosamine phosphate deacetylase
Acetylaminodeoxyglucosephosphate acetylhydrolase
Deacetylase, acetylglucosaminephosphate
2-Acetamido-2-deoxy-D-glucose-6-phosphate amidohydrolase [4]

**CAS Reg. No.**
9027-50-3

---

## 2 REACTION AND SPECIFICITY

**Catalysed reaction**
N-Acetyl-D-glucosamine 6-phosphate + $H_2O$ →
→ D-glucosamine 6-phosphate + acetate

**Reaction type**
Carboxylic acid amide hydrolysis

**Natural substrates**
D-Acetyl-D-glucosamine 6-phosphate + $H_2O$

**Substrate spectrum**
1 N-Acetyl-D-glucosamine 6-phosphate + $H_2O$

**Product spectrum**
1 D-Glucosamine 6-phosphate + acetate

---

**Inhibitor(s)**
$Mn^{2+}$ [1]; $Ni^{2+}$ [1]; $Cu^{2+}$ [1]; $Co^{2+}$ (above 5 mM) [1]; Glucose 6-phosphate
[1]; p-Chloromercuribenzoate [1]; Glucosamine 6-phosphate (competitive)
[1–4, 6]; Fructose 6-phosphate [4, 6]

**Cofactor(s)/prostethic group(s)**

Enzyme Handbook © Springer-Verlag Berlin Heidelberg 1991
Duplication, reproduction and storage in data banks are only
allowed with the prior permission of the publishers

**Metal compounds/salts**
$Co^{2+}$ (below 5 mM: stimulation) [1]

---

**Turnover number** ($min^{-1}$)

**Specific activity** (U/mg)
56.0 [4, 6]

**$K_m$-value** (mM)
0.3 [1]; 0.8 [4]; 1.0 [5]; 0.4

**pH-optimum**
8.5 [4, 6]

**pH-range**
6.5–9.5 [4]

**Temperature optimum** (°C)

**Temperature range** (°C)

---

## 3 ENZYME STRUCTURE

**Molecular weight**
345000 (rat, gel filtration) [1]

**Subunits**

**Glycoprotein/Lipoprotein**
–

---

## 4 ISOLATION/PREPARATION

**Source organism**
Bacteria [1]; Yeast [1]; Mammalia [1]; Candida albicans [2 , 3]; Rat [1];
E. coli [4, 6]

**Source tissue**
Liver [1]

**Localisation in source**

**Purification**
Rat [1]; E. coli [4, 6]

**Crystallization**
–

**Cloned**
–

**Renaturated**
–

## 5 STABILITY

**pH**

**Temperature (°C)**
4 [5]

**Oxidation**

**Organic solvent**

**General stability information**

**Storage**
4°C [5]

## 6 CROSSREFERENCES TO STRUCTURE DATABANKS

**PIR/MIPS code**
S06989 (Escherichia coli)

**Brookhaven code**

## 7 LITERATURE REFERENCES

[1] Campbell, P., Laurent, T.C., Roden, L.: Anal. Biochem. , 166, 134–141 (1987)
[2] Corner, B.E., Poulter, R.T.M., Shepherd, M.G., Sullivan, P.A.: J. Gen. Microbiol., 132, 15–19 (1986)
[3] Gopal, P., Sullivan, P.A., Shepherd, M.G.: J. Gen. Microbiol., 128, 2319–2326 (1982)
[4] White, R.J., Pasternak, C.A.: Biochem. J., 105, 121–125 (1967)
[5] Gopal, P.K., Sullivan, P.A., Shepherd, M.G.: Anal. Biochem., 115, 30–33 (1981)
[6] White, R.J., Pasternak, C.A.: Methods Enzymol., 41, 497–502 (1975)

## 1 NOMENCLATURE

**EC number**
   3.5.1.26

**Systematic name**
   N⁴-(Beta-N-acetyl-D-glucosaminyl)-L-asparagine amidohydrolase

**Recommended name**
   N⁴-(Beta-N-acetylglucosaminyl)-L-asparaginase

**Synonymes**
   Aspartylglucosylamine deaspartylase
   Aspartylglucosylaminase
   Aspartylglucosaminidase
   Aspartylglycosylamine amidohydrolase
   N-Aspartyl-beta-glucosaminidase
   Glucosylamidase
   Beta-aspartylglucosylamine amidohydrolase
   Aspartylglucosylaminase

**CAS Reg. No.**
   9075-24-5

---

## 2 REACTION AND SPECIFICITY

**Catalysed reaction**
   N⁴-(Beta-N-acetyl-D-glucosaminyl)-L-asparagine + $H_2O$ →
   → N-acetyl-beta-glucosaminylamine + L-aspartate

**Reaction type**
   Carboxylic acid amide hydrolysis

**Natural substrates**
   Glycoproteins + $H_2O$ [4]

**Substrate spectrum**
   1 N⁴-(Beta-N-acetyl-D-glucosaminyl)-L-asparagine + $H_2O$
   2 Aspartylglycosylamines + $H_2O$ [7]
   3 Asparagine + $H_2O$ [7]

**Product spectrum**
   1 N-Acetyl-beta-glucosaminylamine + L-aspartate
   2 Glycosylamines + aspartate [7]
   3 $NH_4^+$ + aspartate [7]

## Inhibitor(s)

3-Hydroxybutanone [8]; Aspartylcyclohexylamine [7]; Aspartylaniline [7]; Aspartic acid (1mM) [6]; 5-Diazo-4-oxo-L-norvaline [5]; Dithiothreitol [1]; $MnCl_2$ [1]; $MgCl_2$ [1]; $FeCl_3$ [1]; $FeSO_4$ [1]; $CuSO_4$ [1]; $Cu^{2+}$ [2, 4, 6]; $Zn^{2+}$ [2, 6]; $Ni^{2+}$ [2, 4, 6]; $Mn^{2+}$ [2, 6]; N-Acetylcysteine [3, 8]; p-Chloromercuribenzoate [4]

## Cofactor(s)/prostethic group(s)

## Metal compounds/salts

---

## Turnover number (min⁻¹)

## Specific activity (U/mg)

20.0 (human) [1]; 0.722 [2, 3]; More [4, 5, 6]

## $K_m$-value (mM)

1.25 (aspartylglucosamine) [1]; 0.59 (N-acetylhexosamine) [4]; 0.77 (Asn-GlcNAc, ovalbumin glycopeptide) [6]; More [2, 5]

## pH-optimum

6.1 (human) [1]; 7.6 (rat) [4]; 7.7–9.0 (human) [2]; 5.1 (hen) [5]; 5.5 (pig) [6]

## pH-range

5.0–7.5 [1]; 5.0–9.5 [2]; 5.0–8.5 [4]; 3–9 [5]; More [6]

## Temperature optimum (°C)

## Temperature range (°C)

---

## 3 ENZYME STRUCTURE

## Molecular weight

63000 (gel filtration, human liver) [8]
70000 (gel filtration, SDS-PAGE, pig) [6]
80000 (human, gel filtration) [1]
84000 (human, SDS-PAGE) [1]
101000 (hen, gel filtration, sucrose density gradient) [5]

## Subunits

Monomer (pig) [1, 6]

## Glycoprotein/Lipoprotein

–

## 4 ISOLATION/PREPARATION

**Source organism**
Human (liver and seminal fluid enzyme not identical [2]) [1, 2]; Rat [1, 2, 4]; Pig [1, 2]; Hen [1, 2, 5]; Hog [6]

**Source tissue**
Liver [1, 4]; Oviduct [1, 5]; Ovine epidymis [1]; Plasma [1, 7]; Kidney [1, 6, 7]; Seminal fluid [1]; Brain [3]

**Localisation in source**
Lysosomes [1, 2, 4]

**Purification**
Hog [6]; Human [1, 2]; Rat [4]; Hen [5]

**Crystallization**
–

**Cloned**
–

**Renaturated**
–

## 5 STABILITY

**pH**
6 (above) [6]; 5.0 (67% loss of activity after 2 hours at 37°C) [6]

**Temperature (°C)**
60 (1 hour, 10% loss of activity) [1]; 37 (at least 16 hours at pH 7.6) [4]

**Oxidation**

**Organic solvent**

**General stability information**
Instability towards $(NH_4)SO_4$ [2, 3]; Substrate (stabilizes) [6]

**Storage**
1.5 years, −20°C [4]

## 6 CROSSREFERENCES TO STRUCTURE DATABANKS

**PIR/MIPS code**
S04228 (20K chain, rat, fragment); S04229 (24K chain, rat, fragment)

**Brookhaven code**

# 7 LITERATURE REFERENCES

[1] McGovern, M.M., Aula, P., Desnick, J.: J. Biol. Chem., 258 (17), 10743–10747 (1983)
[2] Dugal, B., Stromme, J.: Biochem. J., 165, 497–502 (1977)
[3] Dugal, B.: Biochem. J., 163, 9–14 (1977)
[4] Mahadevan, S., Tappel, A.L.: J. Biol. Chem., 242 (20), 4568–4576 (1967)
[5] Tarentino, A.L., Maley, F.: Arch. Biochem. Biophys., 130, 295–303 (1969)
[6] Kohno, M., Yamashina, I.: Biochim. Biophys. Acta, 258, 600–617 (1972)
[7] Tanaka, M., Kohno, M., Yamashina, J.: J. Biochem., 73, 1285–1289 (1973)
[8] Dugal, B.: Biochem. J., 171, 799–802 (1978)

## 1 NOMENCLATURE

**EC number**
3.5.1.27

**Systematic name**
N-Formyl-L-methionylaminoacyl-tRNA amidohydrolase

**Recommended name**
N-Formylmethionylaminoacyl-tRNA deformylase

**Synonymes**

**CAS Reg. No.**
37289-08-0

## 2 REACTION AND SPECIFICITY

**Catalysed reaction**
N-Formyl-L-methionylaminoacyl-tRNA $+ H_2O \rightarrow$
$\rightarrow$ formate $+$ L-methionylaminoacyl-tRNA

**Reaction type**
Carboxylic acid amide hydrolysis

**Natural substrates**
N-Formylmethionylaminoacyl-tRNA $+ H_2O$ [1]

**Substrate spectrum**
1 N-Formylmethionylaminoacyl-tRNA $+ H_2O$ [1]
2 N-Formylmethionylpuromycin $+ H_2O$ [1]
3 Peptides (N-terminal formylmethionine) $+ H_2O$ [1]

**Product spectrum**
1 Methionylaminoacyl-tRNA $+$ formate [1]
2 Methionylpuromycin $+$ formate [1]
3 Peptides (N-terminal methionine) $+$ formate [1]

**Inhibitor(s)**

**Cofactor(s)/prostethic group(s)**

**Metal compounds/salts**
$SO_4^{2-}$ (activates) [1]; $SO_3^{2-}$ (activates) [1]

Turnover number (min⁻¹)

Specific activity (U/mg)

$K_m$-value (mM)

pH-optimum

pH-range

Temperature optimum (°C)

Temperature range (°C)

---

## 3 ENZYME STRUCTURE

**Molecular weight**

**Subunits**

**Glycoprotein/Lipoprotein**
  –

---

## 4 ISOLATION/PREPARATION

**Source organism**
  Escherichia coli [1]

**Source tissue**

**Localisation in source**

**Purification**
  Escherichia coli (partial) [1]

**Crystallization**
  –

**Cloned**
  –

**Renaturated**
  –

---

## 5 STABILITY

**pH**

**Temperature (°C)**

Oxidation

Organic solvent

General stability information

**Storage**
   Liquid nitrogen, partially purified, 8 weeks [1]

---

# 6 CROSSREFERENCES TO STRUCTURE DATABANKS

PIR/MIPS code

Brookhaven code

---

# 7 LITERATURE REFERENCES

[1] Livingston, D.M., Leder, P.: Biochemistry, 8 (1) , 435–442 (1969)

Enzyme Handbook © Springer-Verlag Berlin Heidelberg 1991
Duplication, reproduction and storage in data banks are only
allowed with the prior permission of the publishers

3

## 1 NOMENCLATURE

**EC number**
3.5.1.28

**Systematic name**
Mucopeptide amidohydrolase

**Recommended name**
N-Acetylmuramoyl-L-alanine amidase

**Synonymes**
Acetylmuramyl-L-alanine amidase
N-Acetylmuramyl-L-alanine amidase
N-Acylmuramyl-L-alanine amidase
Acetylmuramoyl-alanine amidase
N-Acetylmuramic acid L-alanine amidase
Acetylmuramyl-alanine amidase
N-Acetylmuramylalanine amidase
Murein hydrolase
N-Acetylmuramoyl-L-alanine amidase Type I
N-Acetylmuramoyl-L-alanine amidase Type II

**CAS Reg. No.**
9013-25-6

## 2 REACTION AND SPECIFICITY

**Catalysed reaction**
Hydrolyses the link between N-acetylmuramoyl residues and L-amino acid residues in certain cell wall glycopeptides (preferentially: D-lactyl-L-Ala); Endopeptidase activity

**Reaction type**
Carboxylic acid amide hydrolysis

**Natural substrates**
Peptidoglycan + $H_2O$ (hydrolyses the link between N-acetylmuramoyl residues and L-amino acid residues in certain cell wall glycopeptides, preferentially: D-lactyl-L-Ala)

### Substrate spectrum

1 Peptidoglycan (bacterial cell wall) + $H_2O$
2 Peptidoglycan monomer + $H_2O$ [7]
3 Peptidoglycan dimer + $H_2O$ [5, 7]
4 N-Acetylamuramyl-pentapeptide + $H_2O$
5 N-Acetylmuramyl-tripeptide + $H_2O$ [6, 13]
6 More [5, 6, 13]

### Product spectrum

1 Glycan strands + peptides
2 Disaccharide (GlcNAc-MurNAc) + pentapeptide (corresponding)
3 Tetrasaccharide (GlcNAc-MurNAc-GlcNAc-MurNAc) + pentapeptide (corresponding)
4 N-Acetylmuramic acid + pentapeptide (corresponding)
5 N-Acetylmuramic acid + tripeptide (corresponding)
6 ?

---

### Inhibitor(s)

EDTA [16]; EGTA [3]; Mercaptoethanol [3]; Dithiothreitol [3]; Muramic acid [13]; N-Acetylmuramic acid [13]; N-Acetylglucosamine [13]; Teichonic acid glycan complex [14]; Lipoteichonic acid [12]; Phosphatidylglycerol (polycationic molecules reactivate) [1, 2]; Phospholipids [8]; p-Chloromercuribenzoate [3, 10]; More (high ionic strength) [11]; $Zn^{2+}$ [7, 12, 13]; $Cu^{2+}$ [5, 12]; $Hg^{2+}$ [7, 12]

### Cofactor(s)/prostethic group(s)

### Metal compounds/salts

$Mg^{2+}$ (stimulates) [7, 10, 15]; $K^+$ [15]; $Ca^{2+}$ [14]; $Mn^{2+}$ [14]

---

### Turnover number (min⁻¹)

### Specific activity (U/mg)

0.447 [13]; 0.18 [15]; 4.16 [14]; 23.62 [9]; More [2, 3, 5, 6, 7, 8]

### $K_m$-value (mM)

1.3 (penicillin murein) [16]; 50 (N-acetylmuramyl-tripeptide) [13]; 17 (peptidoglycan monomer) [7]; 8 (peptidoglycan monomer) [7]; 0.04 (N-acetylmuramoyl-tripeptide) [6]; More [5]

### pH-optimum

7.3 [16]; 6.8 [15]; 8.3 [13]; 6.9 [12]; 8.8 [11]; 9 [7]; 5.5 [5]; 8.6 [4]; 6.0 [3]

### pH-range

5.8–8.5 [10]; 5–7 [3, 5]

### Temperature optimum (°C)

43 [13]

### Temperature range (°C)

## 3 ENZYME STRUCTURE

**Molecular weight**
18500–21000 (gel electrophoresis, Streptomyces globisporus [5], bacteriophage T7 [10], Bacillus megaterium [15]) [5, 10, 15]
14000 (E. coli) [8]
35000–39000 (gel electrophoresis, E. coli [6], Diplococcus pneumoniae [12]) [6, 12, 13]
41000 (gel electrophoresis, E. coli) [2]
51000 (gel electrophoresis, Bacillus subtilis) [11, 14]
82000 (gel filtration, human) [3]
110000 (Bacillus subtilis, dimer or artificial aggregate) [9]
220000 (Bacillus subtilis, tetramer or artificial aggregate) [9]

**Subunits**
Monomer (1 × 82000, human) [3]
Monomer (1 × 185000, Streptomyces globisporus) [5]
Monomer (1 × 51000, Bacillus subtilis) [11, 14]

**Glycoprotein/Lipoprotein**
–

---

## 4 ISOLATION/PREPARATION

**Source organism**
Human [3, 7]; Mouse [7]; Rabbit [7]; Bovine [7]; Sheep [7]; Streptomyces sp. [17]; Streptomyces globisporus [5]; Staphylococcus aureus [16]; Bacillus subtilis (168 [9]) [9 , 11, 14, 16]; Bacillus cereus [16]; Bacillus megaterium [15]; E. coli K12 [1, 2, 6, 8, 13]; Proteus vulgaris [13]; Brucella abortus [13]; Diplococcus pneumoniae [12]; Myxobacter AL-1 [18]; Bacteriophage T7 [10]

**Source tissue**
Serum [3, 7]; Cell [14, 15, 16]

**Localisation in source**
Soluble [14, 15, 16]

**Purification**
E. coli [2, 6, 13]; Bacillus subtilis W-23 [14]; Bacillus subtilis ATCC 6051 [11]; Bacillus subtilis 168 [4, 9]; Bacillus megaterium [15]; Diplococcus pneumoniae [12]; Streptomyces globisporus [5]; Bacteriophage T7 [10]

**Crystallization**
–

**Cloned**
–

Renaturated

–

# 5 STABILITY

pH

Temperature (°C)
37 (inactivated after 60 minutes) [16]; 50 (inactivated after 5 minutes) [10];
50 (stable up to) [5]; 80 (inactivated after 10 minutes) [5]

Oxidation

Organic solvent

General stability information
Chitin stabilizes [10]

Storage
Purified enzyme at –20°C, 1 months [11, 12]; Crude extract, –20°C, 2 years
[16]; Several months at 4°C, protein concentration higher than 1 mg/ml [14]

# 6 CROSSREFERENCES TO STRUCTURE DATABANKS

PIR/MIPS code
MUBPA7 (bacteriophage T7); MUBPCP (bacteriophage CP-1); S07506
(bacteriophage T3); A25634 (Streptococcus pneumoniae)

Brookhaven code

# 7 LITERATURE REFERENCES

[1] Vanderwinkel, E., De Vliieghere, M., Charles, P., Baptist, V.: Biochim. Biophys. Acta,
    913, 238–244 (1987)
[2] Vanderwinkel, E., De Vlieghere, M.: Biochim. Biophys. Acta, 838, 54–59 (1985)
[3] Mollner, S., Braun, V.: Arch. Microbiol., 140, 171–177 (1984)
[4] Rogers, H.J., Taylor, C., Rayter, S., Ward, J.B.: J. Gen. Microbiol., 130, 2395–2402
    (1984)
[5] Kawata, S., Takemura, T., Takase, Y., Yokogawa, K.: Agric. Biol. Chem., 48 (2),
    261–269 (1984)
[6] Parquet, C., Flouret, B., Leduc, M., Hirota, Y., Van Heijenoort, J.: Eur. J. Biochem.,
    133, 371–377 (1983)
[7] Valinger, Z., Ladesic, B., Tomasic, J.: Biochim. Biophys. Acta, 701, 63–71 (1982)
[8] Vanderwinkel, E., De Vlieghere, M., De Tanhoffer De Volcsey, L.: Biochim. Biophys.
    Acta, 663, 46–57 (1984)
[9] Niwano, M., Fujita, H.: Agric. Biol. Chem., 44 (9), 2129–2133 (1980)

[10] Kleppe, G., Jensen, H.B., Pryme, J.F.: Eur. J. Biochem., 76, 317–326 (1977)
[11] Lindsay, B., Glaser, L.: J. Bacteriol., 127 (2) , 803–811 (1976)
[12] Höltje, J.V., Tomasz, A.: J. Biol. Chem., 251 (14) , 4199–4207 (1976)
[13] Van Heijenoort, J., Parquet, C., Flouret, B., Van Heijenoort, Y.: Eur. J. Biochem., 58, 611–619 (1975)
[14] Herbold, D.R., Glaser, L.: J. Biol. Chem., 250 (5) , 1676–1682 (1975)
[15] Chan, L., Glaser, L.: J. Biol. Chem., 247 (17) , 5391–5397 (1972)
[16] Singer, H.J., Wise, E.M., Park, J.T.: J. Bacteriol., 112 (2) , 932–939 (1972)
[17] Ghuysen, J.-M., Dierickx, L., Coyette, J., Leyh-Bouille, M., Guinand, M., Campbell, J.N.: Biochemistry, 8 (1) , 213–222 (1969)
[18] Jackson, R.L., Wolfe, R.S.: J. Biol. Chem., 243, 879–888 (1968)

## 1 NOMENCLATURE

**EC number**
3.5.1.29

**Systematic name**
2-(Acetamidomethylene)succinate amidohydrolase (deaminating, decarboxylating)

**Recommended name**
2-(Acetamidomethylene)succinate hydrolase

**Synonymes**
Alpha-(N-acetylaminomethylene)succinic acid hydrolase

**CAS Reg. No.**
37289-09-1

## 2 REACTION AND SPECIFICITY

**Catalysed reaction**
2-(Acetamidomethylene)succinate + 2 $H_2O$ →
→ acetate + succinate semialdehyde + $NH_3$ + $CO_2$

**Reaction type**
Carboxylic acid amide hydrolysis

**Natural substrates**
2-(Acetamidomethylene)succinate + $H_2O$ [1]

**Substrate spectrum**
1 2-(Acetamidomethylene)succinate + $H_2O$ (ir) [1]

**Product spectrum**
1 Succinate semialdehyde + acetate + $NH_3$ + $CO_2$ [1]

**Inhibitor(s)**
Sulfite [1]; $HgCl_2$ [1]; p-Chloromercuribenzoate [1]; Phosphate [1]; Pyrophosphate [1]; Dicarboxylic acids [1]; Tricarboxylic acids [1]; Itaconic acid [1]

**Cofactor(s)/prostethic group(s)**

**Metal compounds/salts**

**Turnover number** (min⁻¹)
100 [1]

**Specific activity** (U/mg)
579 [1]

**$K_m$-value** (mM)
0.003–0.069 (depending on buffer, 2-(acetamidomethylene)succinate) [1]

**pH-optimum**
6.5–7.5 (depending on buffer, 2-(acetamidomethylene)succinate) [1]

**pH-range**
6.0–10.0 [1]

**Temperature optimum** (°C)

**Temperature range** (°C)

---

## 3 ENZYME STRUCTURE

**Molecular weight**
30000–35000 (Pseudomonas MA-1, gel filtration) [1]
33500–34300 (Pseudomonas MA-1, gel electrophoresis) [1]

**Subunits**
Monomer (gel electrophoresis, Pseudomonas MA-1) [1]

**Glycoprotein/Lipoprotein**
–

---

## 4 ISOLATION/PREPARATION

**Source organism**
Pseudomonas MA-1 [1]

**Source tissue**

**Localisation in source**

**Purification**
Pseudomonas MA-1 [1, 2]

**Crystallization**
–

**Cloned**

–

**Renaturated**

–

## 5 STABILITY

**pH**
6.0–10.0 (room temperature) [1]

**Temperature (°C)**
70 (1 min, pH 8) [1]

**Oxidation**

**Organic solvent**

**General stability information**

**Storage**
2 days (room temperature, pH 7–8) [2]

## 6 CROSSREFERENCES TO STRUCTURE DATABANKS

**PIR/MIPS code**

**Brookhaven code**

## 7 LITERATURE REFERENCES

[1] Huynh, M.S., Snell, E.E.: J.Biol. Chem., 260 (4) , 2379–2383 (1985)
[2] Nyns, E.J., Zach, D., Snell, E.E.: J. Biol. Chem., 244 (10) , 2601–2605 (1969)

## 1 NOMENCLATURE

**EC number**
3.5.1.30

**Systematic name**
5-Aminopentanamide amidohydrolase

**Recommended name**
5-Aminopentanamidase

**Synonymes**
5-Aminovaleramidase
5-Aminonorvaleramidase

**CAS Reg. No.**
9054-60-8

## 2 REACTION AND SPECIFICITY

**Catalysed reaction**
5-Aminopentanamide + $H_2O \rightarrow$
$\rightarrow$ 5-aminopentanoate + $NH_3$

**Reaction type**
Carboxylic acid amide hydrolysis

**Natural substrates**
Amides (4–6 carbon atoms and an omega-amino group) + $H_2O$ [1]

**Substrate spectrum**
1 Amides (4–6 carbon atoms and an omega-amino group) + $H_2O$ [1]

**Product spectrum**
1 Corresponding carboxylic acids (4–6 carbon atoms and an
omega-amino group) + $NH_3$ [1]

**Inhibitor(s)**
$Zn^{2+}$ [2]; $Cd^{2+}$ [2]; $Hg^{2+}$ [2]; $Cu^{2+}$ [2]; $Sn^{2+}$ [2]; $Pb^{2+}$ [2]; $Co^{2+}$ [2]; $Mg^{2+}$;
$Ca^{2+}$; $Ba^{2+}$ [2]; $Fe^{2+}$ [2]; Arsenite [1]; p-Hydroxymercuribenzoate [1]

**Cofactor(s)/prostethic group(s)**

**Metal compounds/salts**

Turnover number (min$^{-1}$)

Specific activity (U/mg)
    130 [1]

K$_m$-value (mM)
    4.1 (gamma-aminobutyramide) [1]; 2.0 (delta-aminovaleramide) [1]; 12.0
    (epsilon-aminocaproamide) [1]

pH-optimum
    7.0–9.0 (delta-aminovaleramide) [1]; 7.5–8.5 (delta-aminovaleramide) [2]

pH-range
    10.5 (not active above) [1]

Temperature optimum (°C)

Temperature range (°C)

---

## 3 ENZYME STRUCTURE

Molecular weight
    67000 (Pseudomonas putida, gel filtration, density gradient centrifugation)
    [2]

Subunits

Glycoprotein/Lipoprotein
    –

---

## 4 ISOLATION/PREPARATION

Source organism
    Pseudomonas putida [1]

Source tissue

Localisation in source

Purification
    Pseudomonas putida [1, 2]

Crystallization
    –

Cloned
    –

Renaturated

–

## 5 STABILITY

pH
    9 (not stable above) [1]

Temperature (°C)

Oxidation

Organic solvent

General stability information
    Photoinactivation [1]

Storage
    EDTA and dithioerythritol stabilize [1]

## 6 CROSSREFERENCES TO STRUCTURE DATABANKS

PIR/MIPS code

Brookhaven code

## 7 LITERATURE REFERENCES

[1] Reitz, M.S., Rodwell, V.W.: Methods Enzymol., 17, 158–165 (1971)
[2] Reitz, M.S., Rodwell, V.M.: J. Biol. Chem., 245 (12) , 3091–3096 (1970)

## 1 NOMENCLATURE

**EC number**
3.5.1.31

**Systematic name**
N-Formyl-L-methionine amidohydrolase

**Recommended name**
Formylmethionine deformylase

**Synonymes**

**CAS Reg. No.**
9032-86-4

## 2 REACTION AND SPECIFICITY

**Catalysed reaction**
N-Formyl-L-methionine + $H_2O$ →
→ formate + L-methionine

**Reaction type**
Carboxylic acid amide hydrolysis

**Natural substrates**
N-Formyl-L-methionine + $H_2O$ [1]

**Substrate spectrum**
1 N-Formyl-L-methionine + $H_2O$ [1]

**Product spectrum**
1 L-Methionine + formate [1]

**Inhibitor(s)**
o-Phenantroline [1]; Dithiothreitol [1]; EDTA [1]

**Cofactor(s)/prostethic group(s)**

**Metal compounds/salts**

**Turnover number** (min$^{-1}$)

**Specific activity** (U/mg)

$K_m$-value (mM)
  3–12 (N-formyl-L-methionine) [2]; 3.8 (N-formyl-L-methionine) [3]

pH-optimum
  7.5 [1]; 7.4 [3]

pH-range
  5.5–10.0 [1]

Temperature optimum (°C)

Temperature range (°C)

---

## 3 ENZYME STRUCTURE

Molecular weight

Subunits

Glycoprotein/Lipoprotein
  –

---

## 4 ISOLATION/PREPARATION

Source organism
  Bacteria [1]; Plants [1]; Mammals [2]; Euglena gracilis [3]

Source tissue

Localisation in source
  Cytoplasm [1]

Purification

Crystallization
  –

Cloned
  –

Renaturated
  –

---

## 5 STABILITY

pH

Temperature (°C)

Oxidation

Organic solvent

General stability information

Storage

---

# 6 CROSSREFERENCES TO STRUCTURE DATABANKS

PIR/MIPS code

Brookhaven code

---

# 7 LITERATURE REFERENCES

[1] Ackerman, S.K., Douglas, S.D.: Biochem. J., 182, 885–887 (1979)
[2] Grisolia, S., Reglero, A., Rivas, J.: Biochem. Biophys. Res. Commun., 77 (1) , 237–244 (1977)
[3] Aronson, J.N., Lugay, J.C.: Biochem. Biophys. Res. Commun., 34 (3) , 311–314 (1969)

# 1 NOMENCLATURE

**EC number**
   3.5.1.32

**Systematic name**
   N-Benzoylamino-acid amidohydrolase

**Recommended name**
   Hippurate hydrolase

**Synonymes**
   Hydrolase, hippurate

**CAS Reg. No.**
   37278-43-6

# 2 REACTION AND SPECIFICITY

**Catalysed reaction**
   Hippurate + $H_2O$ →
   → benzoate + glycine

**Reaction type**
   Carboxylic acid amide hydrolysis

**Natural substrates**
   Hippurate + $H_2O$ [3, 4]

**Substrate spectrum**
   1 Hippurate + $H_2O$
   2 N-Benzoylamino acids + $H_2O$
   3 More [1, 5]

**Product spectrum**
   1 Benzoate + glycine
   2 Benzoate + amino acid
   3 ?

**Inhibitor(s)**
   $HgCl_2$ [2]; p-Chloromercuribenzoate [2]; 8-Hydroxychinoline [2]; EDTA [2];
   Iodoacetic acid [2, 5]; $FeSO_4$ [2]; $CuSO_4$ [2]; $AgNO_3$ [2]; $Ag^+$ [5]; $Hg^{2+}$ [5];
   $Cu^{2+}$ [5]; $Ni^{2+}$ [5]; $Mn^{2+}$ [5]; $Fe^{2+}$ [5]; $Fe^{3+}$ [5]; $Zn^{2+}$ [5]; $Co^{2+}$ [5];
   o-Phenanthroline [5]; Alpha, alpha-dipyridyl [5]

## Cofactor(s)/prostethic group(s)

## Metal compounds/salts

---

## Turnover number (min⁻¹)

## Specific activity (U/mg)
More [1, 5]; 7.4 [2]

## $K_m$-value (mM)
0.72 (N-benzoylglycine) [5]; 0.87 (N-benzoyl-L-alanine,
N-benzoyl-L-aminobutyric acid) [5]; More [2]

## pH-optimum
7.5 [2]; 7.0–8.0 [5]

## pH-range
5–9.5 [2]

## Temperature optimum (°C)
60 [2]; 50 [5]

## Temperature range (°C)
20–70 [2]

---

## 3 ENZYME STRUCTURE

## Molecular weight
170000 (Pseudomonas putida, gel filtration) [5]
42000 (Pseudomonas putida, SDS-PAGE, subunit) [5]

## Subunits
Tetramer (Pseudomonas putida, 4 × 42000) [5]

## Glycoprotein/Lipoprotein
–

---

## 4 ISOLATION/PREPARATION

## Source organism
Fusarium semitectum [1, 2]; Pseudomonas [3]; Bacillus [3, 4]; Pseudomonas putida [5]; Enterobacteriaceae [3]

## Source tissue

## Localisation in source

**Purification**
Fusarium semitectum [1]; Pseudomonas putida [5]

**Crystallization**
[5]

**Cloned**
–

**Renaturated**
–

---

# 5 STABILITY

**pH**
6.0–8.0 (at pH 6.0 and 8.0, 10°C, 20 hours) [5]

**Temperature (°C)**
60 (in absence of substrate complete loss of activity after 15 minutes) [2];
50 (stable against heating up to 50°C, 30 minutes) [5]

**Oxidation**

**Organic solvent**

**General stability information**

**Storage**

---

# 6 CROSSREFERENCES TO STRUCTURE DATABANKS

**PIR/MIPS code**

**Brookhaven code**

---

# 7 LITERATURE REFERENCES

[1] Röhr, M.: Monatsh. Chem., 99, 2255–2277 (1968)
[2] Röhr, M.: Monatsh. Chem., 99, 2278–2290 (1968)
[3] Zolg, W., Ottow, J.C.G.: Experientia, 29 (12) , 1573–1574 (1973)
[4] Ottow, J.C.G.: J. Appl. Bacteriol., 37, 15–30 (1974)
[5] Miyagawa, E., Yano, Y., Hamakado, T., Kido, Y., Nishimoto, K., Motoki, Y.: Agric. Biol. Chem., 49 (10) , 2881–2886 (1985)

---

## 1 NOMENCLATURE

**EC number**
3.5.1.33

**Systematic name**
N-Acetyl-D-glucosamine amidohydrolase

**Recommended name**
N-Acetylglucosamine deacetylase

**Synonymes**
Acetylaminodeoxyglucose acetylhydrolase
N-Acetyl-D-glucosaminyl N-deacetylase

**CAS Reg. No.**
9012-32-2

## 2 REACTION AND SPECIFICITY

**Catalysed reaction**
N-Acetyl-D-glucosamine + $H_2O$ →
→ D-glucosamine + acetate

**Reaction type**
Carboxylic acid amide hydrolysis

**Natural substrates**
N-Acetyl-D-glucosamine + $H_2O$ [2]
Peptidoglycan + $H_2O$ [1]

**Substrate spectrum**
1 N-Acetyl-D-glucosamine + $H_2O$ [2]
2 N-Acetylgalactosamine + $H_2O$ [2]
3 Peptidoglycan + $H_2O$ [1]

**Product spectrum**
1 D-Glucosamine + acetate [2]
2 Galactosamine + acetate [2]
3 Peptidoglycan with N-unsubstituted glucosamine residues + acetate [1]

**Inhibitor(s)**
Versene [2]

**Cofactor(s)/prostethic group(s)**

**Metal compounds/salts**
$Co^{2+}$ [1]

**Turnover number** (min$^{-1}$)

**Specific activity** (U/mg)

**$K_m$-value** (mM)
120 (N-acetylglucosamine) [2]

**pH-optimum**
8.0–8.5 (N-acetylglucosamine) [2]; 7.0 (peptidoglycan) [1]

**pH-range**
5.2–10.5 (N-acetylglucosamine) [2]

**Temperature optimum** (°C)

**Temperature range** (°C)

## 3 ENZYME STRUCTURE

**Molecular weight**

**Subunits**

**Glycoprotein/Lipoprotein**
–

## 4 ISOLATION/PREPARATION

**Source organism**
Bacteria [2]

**Source tissue**

**Localisation in source**

**Purification**

**Crystallization**
–

**Cloned**
–

**Renaturated**
–

## 5 STABILITY

pH

Temperature (°C)

Oxidation

Organic solvent

General stability information

Storage

---

## 6 CROSSREFERENCES TO STRUCTURE DATABANKS

PIR/MIPS code

Brookhaven code

---

## 7 LITERATURE REFERENCES

[1] Araki, Y., Fukuoka, S., Oba, S., Ito, E.: Biochem. Biophys. Res. Commun., 45 (3) ,
    751–758 (1971)
[2] Roseman, S.: J. Biol. Chem., 226, 115–124 (1957)

# 1 NOMENCLATURE

**EC number**
3.5.1.35

**Systematic name**
D-Glutamine amidohydrolase

**Recommended name**
D-Glutaminase

**Synonymes**

**CAS Reg. No.**
37289-12-6

# 2 REACTION AND SPECIFICITY

**Catalysed reaction**
D-Glutamine + $H_2O$ →
→ D-glutamate + $NH_3$

**Reaction type**
Carboxylic acid amide hydrolysis

**Natural substrates**
D-Glutamine + $H_2O$ [1]

**Substrate spectrum**
1 D-Glutamine + $H_2O$ [1]

**Product spectrum**
1 D-Glutamate + $NH_3$ [1]

**Inhibitor(s)**

**Cofactor(s)/prostethic group(s)**

**Metal compounds/salts**

**Turnover number** (min$^{-1}$)

**Specific activity** (U/mg)

**$K_m$-value** (mM)

pH-optimum

pH-range

Temperature optimum (°C)

Temperature range (°C)

---

## 3 ENZYME STRUCTURE

Molecular weight

Subunits

Glycoprotein/Lipoprotein

–

---

## 4 ISOLATION/PREPARATION

Source organism
  Blastocladiella emersonii [1]; Blastocladiella britannica [1]

Source tissue

Localisation in source

Purification

Crystallization

–

Cloned

–

Renaturated

–

---

## 5 STABILITY

pH

Temperature (°C)

Oxidation

Organic solvent

General stability information

Storage

---

# 6 CROSSREFERENCES TO STRUCTURE DATABANKS

PIR/MIPS code

Brookhaven code

---

# 7 LITERATURE REFERENCES

[1] Domnas, A., Cantino, E.C.: Phytochemistry, 4, 273–284 (1965)

Enzyme Handbook © Springer-Verlag Berlin Heidelberg 1991
Duplication, reproduction and storage in data banks are only
allowed with the prior permission of the publishers

## 1 NOMENCLATURE

**EC number**
3.5.1.36

**Systematic name**
N-Methyl-2-oxoglutaramate methylamidohydrolase

**Recommended name**
N-Methyl-2-oxoglutaramate hydrolase

**Synonymes**
5-Hydroxy-N-methyl-pyroglutamate synthase

**CAS Reg. No.**
9073-53-4

## 2 REACTION AND SPECIFICITY

**Catalysed reaction**
N-Methyl-2-oxoglutaramate + $H_2O$ →
→ 2-oxoglutarate + methylamine

**Reaction type**
Carboxylic acid amide hydrolysis

**Natural substrates**
N-Methyl-2-oxoglutaramate + $H_2O$ [2]
Alpha-ketoglutarate + methylamine + $H_2O$ [2]

**Substrate spectrum**
1 Delta-amides of alpha-ketoglutarate + $H_2O$ (r) [2]
2 Delta-esters of alpha-ketoglutarate + $H_2O$ (r) [2]

**Product spectrum**
1 Alpha-ketoglutarate + corresponding amine [2]
2 Alpha-ketoglutarate + corresponding alcohol [2]

**Inhibitor(s)**
p-Hydroxymercuribenzenesulfonate [3]; p-Hydroxymercuribenzoate [3];
Iodoacetate [3]; Iodoacetamide [3]; N-Ethylmaleimide [3];
N-Methylmaleimide [3]; Oxalacetate [3]; 5,5'-Dithiobis-(2-nitrobenzoate) [3]

**Cofactor(s)/prostethic group(s)**

Metal compounds/salts

---

**Turnover number** (min$^{-1}$)
  1070 [3]

**Specific activity** (U/mg)
  11.9 [3]

**$K_m$-value** (mM)
  46 (methylamine) [4]; 38 (alpha-ketoglutarate) [4]; 3 (alpha-ketoglutaramate) [3]; 0.0045 (ethyl-alpha-ketoglutarate) [2]; 0.003 (propyl-alpha-ketoglutarate) [2]; 0.005 (benzyl-alpha-ketoglutarate) [2]

**pH-optimum**
  8.0 (alpha-ketoglutarate) [4]

**pH-range**

**Temperature optimum** (°C)

**Temperature range** (°C)

---

## 3 ENZYME STRUCTURE

**Molecular weight**
  90000 (Pseudomonas M.A., gel chromatography) [3]

**Subunits**

**Glycoprotein/Lipoprotein**
  –

---

## 4 ISOLATION/PREPARATION

**Source organism**
  Pseudomonas M.A. [4]

**Source tissue**

**Localisation in source**

**Purification**
  Pseudomonas M.A. [3]

**Crystallization**
  –

**Cloned**

−

**Renaturated**

−

# 5 STABILITY

**pH**

**Temperature (°C)**

**Oxidation**

**Organic solvent**
Organic solvents (accelerate hydrolysis reactions) [1]

**General stability information**

**Storage**
20 mM potassium phosphate, 5 mM alpha-ketoglutarate, below 0°C , 6 months [3]

# 6 CROSSREFERENCES TO STRUCTURE DATABANKS

**PIR/MIPS code**

**Brookhaven code**

# 7 LITERATURE REFERENCES

[1] Hersh, L.B.: J. Biol. Chem., 246 (24) , 7804–7809 (1971)
[2] Hersh, L.B.: J. Biol. Chem., 246 (22) , 6803–6806 (1971)
[3] Hersh, L.B.: J. Biol. Chem., 245 (14) , 3526–3535 (1970)
[4] Hersh, L.B., Tsai, L., Stadtman, E.R.: J. Biol. Chem., 244 (17) , 4677–4683 (1969)

## 1 NOMENCLATURE

**EC number**
3.5.1.38

**Systematic name**
L-Glutamate(L-asparagine)amidohydrolase

**Recommended name**
Glutamin-(asparagin-)ase

**Synonymes**

**CAS Reg. No.**
39335-03-0

---

## 2 REACTION AND SPECIFICITY

**Catalysed reaction**
L-Glutamine + $H_2O$ →
→ L-glutamate + $NH_3$

**Reaction type**
Carboxylic acid amide hydrolysis

**Natural substrates**
L-Glutamine + $H_2O$ [1]
L-Asparagine + $H_2O$ [1]

**Substrate spectrum**
1 L-Glutamine + $H_2O$ [1]
2 L-Asparagine + $H_2O$ [1]
3 D-Glutamine + $H_2O$ (slowly) [3]
4 D-Asparagine + $H_2O$ (slowly) [3]

**Product spectrum**
1 L-Glutamate + $NH_3$ [1]
2 L-Aspartate + $NH_3$ [1]
3 D-Glutamate + $NH_3$ [3]
4 D-Aspartate + $NH_3$ [3]

## Inhibitor(s)

6-Diazo-5-oxo-L-norleucine [1] [9]; L-Methionine sulfoximine [1]; Azaserine [1]; Acivicin [1]; p-Chloromercuribenzoate [3]; Ammonium sulfate [6]; $Cu^{2+}$ [7]; $Ni^{2+}$ [7]; $NaAsO_2$ [7]

## Cofactor(s)/prostethic group(s)

## Metal compounds/salts

## Turnover number (min$^{-1}$)

## Specific activity (U/mg)

150 [1]; 104 [3]; 165 [5]; 160 [6]; 190 [9]

## $K_m$-value (mM)

0.41 (L-glutamine) [1]; 0.55 (L-asparagine) [1]; 0.15 (L-asparagine) [3]; 0.22 (L-glutamine) [3]; 0.046 (L-glutamine) [6]; 0.044 (L-asparagine) [6]; 0.3 (L-aparagine) [7]; 0.6 (L-glutamine) [7]; 0.006 (L-glutamine) [9]; 0.005 (L-asparagine) [9]

## pH-optimum

9.5 (L-glutamine, L-asparagine) [3]; 8.0 (L-glutamine) [6]; 10.0 (L-asparagine) [6]; 8.7 (L-asparagine, L-glutamine) [7]; 7.5 (L-glutamine) [9]

## pH-range

4 (above, L-glutamine, L-asparagine) [7]

## Temperature optimum (°C)

37–55 (L-glutamine, L-asparagine) [7]

## Temperature range (°C)

## 3 ENZYME STRUCTURE

## Molecular weight

156000 (gel filtration, Pseudomonas acidovorans) [3]
137000–143000 (sedimentation equilibrium, Pseudomonas 7A) [5]
93000–97000 (gel filtration, Acinetobacter glutaminasificans) [8]

## Subunits

Tetramer (4 × 39000, gel electrophoresis, Pseudomonas acidovorans) [3]
Tetramer (4 × 34000, amino acid analysis, Pseudomonas 7A) [5]
Tetramer (4 × 33000, gel electrophoresis, Acinetobacter glutaminasificans) [8]

## Glycoprotein/Lipoprotein

–

## 4 ISOLATION/PREPARATION

**Source organism**
Acinetobacter glutaminasificans [1, 8]; Pseudomonas 7A [2, 5]; Pseudomonas acidovorans [3]; Alcaligenes eutrophus [3]; Citrobacter freundii [4];
Tilachlidium humicola [7]; Verticillium malthoasei [7]; Microascus des-
mosporus [7]

**Source tissue**

**Localisation in source**

**Purification**
Pseudomonas acidovorans [3]; Citrobacter freundii [4]; Pseudomonas 7A
[6]; Acinetobacter glutaminasificans [9]

**Crystallization**
[2]

**Cloned**
–

**Renaturated**
[9]

---

## 5 STABILITY

**pH**

**Temperature (°C)**

**Oxidation**

**Organic solvent**

**General stability information**

**Storage**
Lyophilized, 4°C [1]; 4°C, 3–4 months [3]

---

## 6 CROSSREFERENCES TO STRUCTURE DATABANKS

**PIR/MIPS code**
A28063 (Acinetobacter calcoaceticus)

**Brookhaven code**

# 7 LITERATURE REFERENCES

[1] Steckel, J.: Biochem. Pharmacol., 32 (6) , 971–977 (1983)
[2] Wlodawer, A., Roberts, J., Holcenberg, J.S.: J. Mol. Biol., 112, 515–519 (1977)
[3] Davidson, L., Brear, D.R., Wingard, P., Hawkins, J., Kitto, G.B.: J. Bacteriol., 129 (3) ,
    1379–1386 (1977)
[4] Davidson, L., Burkom, M., Ahn, S., Chang, L.C., Kitto, B.: Biochim. Biophys. Acta, 480,
    282–294 (1977)
[5] Holcenberg, J.S., Teller, D.C.: J. Biol. Chem., 251 (17) , 5375–5380 (1976)
[6] Roberts, J.: J. Biol. Chem., 251 (7) , 2119–2123 (1976)
[7] Imada, A., Igarasi, S.: J. Takeda Res. Lab., 32 (2) , 140–151 (1973)
[8] Holcenberg, J.S., Teller, D.C., Roberts, J., Dolowy, W.C.: J. Biol. Chem., 247 (23) ,
    7750–7758 (1972)
[9] Roberts, J., Holcenberg, J.S., Dolowy, W.C.: J. Biol. Chem., 247 (1) , 84–90 (1972)

# 1 NOMENCLATURE

**EC number**
3.5.1.39

**Systematic name**
N-Methylhexanamide amidohydrolase

**Recommended name**
Alkylamidase

**Synonymes**

**CAS Reg. No.**
62213-19-8

---

# 2 REACTION AND SPECIFICITY

**Catalysed reaction**
N-Methylhexanamide + $H_2O$ →
→ hexanoate + methylamine

**Reaction type**
Carboxylic acid amide hydrolysis

**Natural substrates**

**Substrate spectrum**
1 N-Methylhexanamide + $H_2O$
2 N-Amines (mono-substituted) + $H_2O$
3 N, N-Amines (disubstituted) + $H_2O$
4 Amides (primary) + $H_2O$
5 More (not: short chain substrates) [1]

**Product spectrum**
1 Hexanoate + methylamine
2 ?
3 ?
4 ?
5 ?

---

**Inhibitor(s)**
Paraoxon [1]

---

Cofactor(s)/prostethic group(s)

Metal compounds/salts

Turnover number (min$^{-1}$)
  5.4 [1]; 551 [1]

Specific activity (U/mg)
  0.123 [1]

$K_m$-value (mM)
  More [1]

pH-optimum
  9.0 [1]

pH-range

Temperature optimum (°C)

Temperature range (°C)

# 3 ENZYME STRUCTURE

Molecular weight
  230000–250000 (sheep, gel filtration) [1]

Subunits
  Monomer [1]

Glycoprotein/Lipoprotein
  –

# 4 ISOLATION/PREPARATION

Source organism
  Sheep [1]

Source tissue
  Liver [1]

Localisation in source
  Microsomes [1]

Purification
  Sheep [1]

Crystallization
–

Cloned
–

Renaturated
–

---

## 5 STABILITY

pH

Temperature (°C)

Oxidation

Organic solvent
  Acetone (labile to) [1]; Ethanol (labile to) [1]

General stability information

Storage

---

## 6 CROSSREFERENCES TO STRUCTURE DATABANKS

PIR/MIPS code

Brookhaven code

---

## 7 LITERATURE REFERENCES

[1] Chen, P.R.S., Dauterman, W.C.: Biochim. Biophys. Acta, 250, 216–223 (1971)

---

## 1 NOMENCLATURE

**EC number**
3.5.1.40

**Systematic name**
Benzoylagmatine amidohydrolase

**Recommended name**
Acylagmatine amidase

**Synonymes**
Acrylagmatine amidohydrolase
Acylagmatine deacylase
Acylagmatine amidohydrolase

**CAS Reg. No.**
39419-74-4

## 2 REACTION AND SPECIFICITY

**Catalysed reaction**
Benzoylagmatine + $H_2O$ →
→ benzoate + agmatine

**Reaction type**
Carboxylic acid amide hydrolysis

**Natural substrates**
Benzoylagmatine + $H_2O$

**Substrate spectrum**
1 Benzoylagmatine + $H_2O$
2 Bleomycin $B_2$ + $H_2O$ [2]
3 Acetylagmatine + $H_2O$ [2]
4 Propanoylagmatine + $H_2O$

**Product spectrum**
1 Benzoate + agmatine
2 Bleomycinic acid + agmatine
3 Acetate + agmatine
4 Propanoate + agmatine

**Inhibitor(s)**
p-Chloromercuribenzoate (mercaptoethanol and cysteine protect) [1]; $Ni^{2+}$ [1]; $Zn^{2+}$ (10 mM) [1]; $Co^{2+}$ (10 mM) [1]; Benzoal-Arg-$NH_2$ [1]; Benzoyl-Arg-ethyl ester [1]

**Cofactor(s)/prostethic group(s)**

**Metal compounds/salts**

---

**Turnover number** (min$^{-1}$)

**Specific activity** (U/mg)
0.029 [2]; 0.168 [2]; 0.242 [2]; 0.279 [2]; 2.0 [1]

**$K_m$-value** (mM)
200 (propionylagmatine) [2]; 22 (benzoylagmatine) [2]; 80 (bleomycin $B_2$) [2]; 133 (acetylagmatine) [2]

**pH-optimum**
8.0 (phosphate buffer) [2]; 7.5 (Tris-HCl buffer) [2]

**pH-range**

**Temperature optimum** (°C)

**Temperature range** (°C)

---

## 3 ENZYME STRUCTURE

**Molecular weight**
65000 (gel filtration, Fusarium) [1]

**Subunits**

**Glycoprotein/Lipoprotein**
–

---

## 4 ISOLATION/PREPARATION

**Source organism**
Fusarium aguioides [1, 2]

**Source tissue**
Cell [2]

**Localisation in source**

Purification

Crystallization

–

Cloned

–

Renaturated

–

---

# 5 STABILITY

pH

Temperature (°C)

Oxidation

Organic solvent

General stability information

Storage

---

# 6 CROSSREFERENCES TO STRUCTURE DATABANKS

PIR/MIPS code

Brookhaven code

---

# 7 LITERATURE REFERENCES

[1] Takahashi, Y., Shirai, T., Ishii, S.: J. Biochem., 77, 823–830 (1975)
[2] Umezawa, H., Takahashi, Y., Fujii, A., Saino, T., Shirai, T., Takita, T.: J. Antibiot., 26, 117–119 (1975)

# 1 NOMENCLATURE

**EC number**
3.5.1.41

**Systematic name**
Chitin amidohydrolase

**Recommended name**
Chitin deacetylase

**Synonymes**

**CAS Reg. No.**
56379-60-3

---

# 2 REACTION AND SPECIFICITY

**Catalysed reaction**
Chitin + $H_2O$ →
→ chitosan + acetate (hydrolyses the N-acetamido groups of N-acetyl-D-glucosamine residues in chitin)

**Reaction type**
Carboxylic acid amide hydrolysis

**Natural substrates**
Chitin + $H_2O$ (nascent, tandem action of chitin synthetase E.C. 2.4.1.16 and chitin deacetylase, hydrolyses the N-acetamido groups of N-acetyl-D-glucosamine residues in chitin) [1]

**Substrate spectrum**
1 Chitin + $H_2O$ (hydrolyses the N-acetamido groups of N-acetyl-D-glucosamine residues in chitin)
2 Glycol chitin + $H_2O$ [4]
3 N-Acetylchitooligoses + $H_2O$ [4]
4 More (not: peptidoglycan [5], N-acetylated heparin [5], N-acetylgalactosamine polymer [5], N-acetylglucosamine [1], N-acetylglucosamine dimer [1]) [1, 5]

**Product spectrum**
1 Chitosan + acetate
2 Glycol chitosan + acetate [4]
3 ?
4 ?

---

**Inhibitor(s)**
Acetate (Mucor rouxii) [4, 5]; Glycol chitosan [4]; $Co^{2+}$ [4]; $Mn^{2+}$ [4]; $Na^+$ [4]; EDTA [4]

**Cofactor(s)/prostethic group(s)**

**Metal compounds/salts**

---

**Turnover number (min$^{-1}$)**

**Specific activity (U/mg)**
17.8 [4]

**$K_m$-value (mM)**
2.6 (glycol chitin) [4]

**pH-optimum**
5.5 [4, 5]; 8.5 [3]

**pH-range**
4.5–8.5 [4]

**Temperature optimum (°C)**

**Temperature range (°C)**

---

## 3 ENZYME STRUCTURE

**Molecular weight**

**Subunits**

**Glycoprotein/Lipoprotein**
–

---

## 4 ISOLATION/PREPARATION

**Source organism**
Mucor rouxii [4, 5]; Mucor miehei [3]; Colletotrichum lindemuthianum [3]

**Source tissue**
Cell [3, 5]

**Localisation in source**
Soluble [4]; Cytoplasm [4]; Wall (mycelial) [2]

**Purification**

**Crystallization**

–

**Cloned**

–

**Renaturated**

–

---

## 5 STABILITY

**pH**

**Temperature (°C)**

**Oxidation**

**Organic solvent**

**General stability information**

**Storage**
Purified enzyme, frozen, –18°C, several months [4]

---

## 6 CROSSREFERENCES TO STRUCTURE DATABANKS

**PIR/MIPS code**

**Brookhaven code**

---

## 7 LITERATURE REFERENCES

[1] Davis, L.L., Bartnicki-Garcia, S.: Biochemistry, 23, 1065–1073 (1984)
[2] Davis, L.L., Bartnicki-Garcia, S.: J. Gen. Microbiol., 130, 2095–2102 (1984)
[3] Kauss, H., Jeblick, W., Young, D.H.: Plant Sci. Lett., 28, 231–236 (1982/83)
[4] Araki, Y., Ito, E.: Eur. J. Biochem., 55, 71–78 (1975)
[5] Araki, Y., Ito, E.: Biochem. Biophys. Res. Commun., 56 (3) , 669–675 (1974)

## 1 NOMENCLATURE

**EC number**
3.5.1.42

**Systematic name**
Nicotinamide-nucleotide amidohydrolase

**Recommended name**
Nicotinamide-nucleotide amidase

**Synonymes**
NMN deamidase
Nicotinamide mononucleotide deamidase
Nicotinamide mononucleotide amidohydrolase

**CAS Reg. No.**
37355-58-1

---

## 2 REACTION AND SPECIFICITY

**Catalysed reaction**
Beta-nicotinamide D-ribonucleotide + $H_2O$ →
→ beta-nicotinate D-ribonucleotide + $NH_3$

**Reaction type**
Carboxylic acid amide hydrolysis

**Natural substrates**
Nicotinamide-ribonucleoside + $H_2O$ [1]
Nicotinamide-ribonucleotide + $H_2O$ [2–8]

**Substrate spectrum**
1 Nicotinamide-ribonucleoside + $H_2O$ [1]
2 Nicotinamide-ribonucleotide + $H_2O$ [2–8]

**Product spectrum**
1 Nicotinic acid ribonucleoside + $NH_3$ [1]
2 Nicotinic acid ribonucleotide + $NH_3$ [2–8]

---

**Inhibitor(s)**
$HgCl_2$ [1]; p-Hydroxymercuribenzoate [1]; $Cu^{2+}$ [7]

**Cofactor(s)/prostethic group(s)**

---

**Metal compounds/salts**

---

**Turnover number** (min$^{-1}$)

**Specific activity** (U/mg)
  22.7 [1]; 2.23 [7]

**K$_m$-value** (mM)
  3.6 (nicotinamide-ribonucleoside) [1]; 0.105 (nicoinamide-ribonucleotide) [7]

**pH-optimum**
  5–5.5 [1]; 9 [3]; 8.7 [5]; 5.6 [6]; 6.8–7.2 [7]

**pH-range**

**Temperature optimum** (°C)
  45 [1]; 64 [7]

**Temperature range** (°C)

---

**3 ENZYME STRUCTURE**

**Molecular weight**
  43000 (gel filtration, Vibrio cholerae [2], Salmonella typhimurium [4]) [2, 4]
  33000 (gel filtration, E. coli) [3]
  26300 (gel filtration, Azotobacter vinelandii) [7]

**Subunits**

**Glycoprotein/Lipoprotein**
  –

---

**4 ISOLATION/PREPARATION**

**Source organism**
  Aspergillus niger [1]; Vibrio cholerae [2]; E. coli [3]: Salmonella typhimurium [4, 5]; Propionibacterium shermanii [6, 8]; Azotobacter vinelandii [7]

**Source tissue**
  Cell

**Localisation in source**
  Cytoplasm (soluble)

## Purification

Aspergillus niger [1]; Vibrio cholerae [2]; E. coli [3]; Salmonella typhimurium [4]; Propionibacterium shermanii [6 , 8]; Azotobacter vinelandii [7]

## Crystallization
–

## Cloned
–

## Renaturated
–

## 5 STABILITY

### pH
5–6 [1]; 7.5 [7]

### Temperature (°C)
37 (30 minutes) [1]

### Oxidation

### Organic solvent

### General stability information

### Storage
–15°C (12% glycerol) [6]; 4°C (24% glycerol) [6]; pH 7.5 (4°C, Tris-HCl, 1 week) [7]; –22°C (1 month) [8]

## 6 CROSSREFERENCES TO STRUCTURE DATABANKS

### PIR/MIPS code

### Brookhaven code

## 7 LITERATURE REFERENCES

[1] Kuwahara, M., Ishida, Y., Okatani, M.: J. Ferment. Technol., 61 (1) , 61–66 (1983)
[2] Foster, J.W., Brestel, C.: J. Bacteriol., 149 (1) , 368–371 (1982)
[3] Hillyard, D., Rechsteiner, M., Manlapaz-Ramos, P., Imperial, J.S., Cruz, L.J., Olivera, B.M.: J. Biol. Chem., 256 (16) , 8491–8497 (1981)
[4] Foster, J.W.: J. Bacteriol., 145 (2) , 1002–1009 (1981)
[5] Kinney, D. M., Foster, J.W., Moat, A.G.: J. Bacteriol., 140 (3) , 607–611 (1979)
[6] Friedmann, H.C., Garstki, C.: Biochem. Biophys. Res. Commun., 50 (1) , 54–58 (1973)
[7] Imai, T.: J. Biochem., 73, 139–153 (1973)
[8] Friedmann, H.C.: Methods Enzymol., 18, Issue Pt.B, 192–197 (1971)

# 1 NOMENCLATURE

**EC number**
3.5.1.43

**Systematic name**
Peptidyl-L-glutamine amidohydrolase

**Recommended name**
Peptidyl-glutaminase

**Synonymes**
Peptidoglutaminase I
Peptideglutaminase
Peptidoglutaminase

**CAS Reg. No.**
37228-70-9

---

# 2 REACTION AND SPECIFICITY

**Catalysed reaction**
Alpha-N-peptidyl-L-glutamine + $H_2O$ →
→ alpha-N-peptidyl-L-glutamate + $NH_3$

**Reaction type**
Carboxylic acid amide hydrolysis

**Natural substrates**
Peptide-bound glutamine + $H_2O$ [1, 3, 4]

**Substrate spectrum**
1 Alpha-N-peptidyl-L-glutamine + $H_2O$ [1, 3, 5]
2 Acetamide + $H_2O$ (slight activity) [1]
3 Propionamide + $H_2O$ (slight activity) [1]
4 More (specific for hydrolysis of gamma-amide of glutamine substituted at alpha-amino group, e.g. glycyl-L-glutamine, N-acetyl-L-glutamine, L-leucylglycyl-L-glutamine)

**Product spectrum**
1 Alpha-N-peptidyl-L-glutamate + $NH_3$ [1, 3, 5]
2 Acetic acid + $NH_3$ [1]
3 Propionic acid + $NH_3$ [1]
4 ?

---

**Inhibitor(s)**
N-Bromosuccinimide [3]; Sodium lauryl sulfate [3]; $Cu^{2+}$ (slight) [5]; $Hg^{2+}$ (slight) [5]; $AgNO_3$ [3]

**Cofactor(s)/prostethic group(s)**

**Metal compounds/salts**

---

**Turnover number** (min$^{-1}$)

**Specific activity** (U/mg)
303 [1, 4]

**$K_m$-value** (mM)
0.11 (carbobenzoxy-L-glutamine) [3]; More [3]

**pH-optimum**
8.0 [1, 3]

**pH-range**

**Temperature optimum** (°C)
55 [1, 3]

**Temperature range** (°C)

---

## 3 ENZYME STRUCTURE

**Molecular weight**
89000–99000 (depending on method, Bacillus circulans) [2, 4]

**Subunits**
Dimer (2 × 42000–49000, depending on method, Bacillus circulans) [2]

**Glycoprotein/Lipoprotein**
–

---

## 4 ISOLATION/PREPARATION

**Source organism**
Bacillus circulans [1, 3–5]

**Source tissue**
Cell

**Localisation in source**
Cytoplasm (soluble)

**Purification**
  Bacillus circulans [1, 4, 5]

**Crystallization**
  −

**Cloned**
  −

**Renaturated**
  −

## 5 STABILITY

**pH**
  5–12.5 [2]

**Temperature (°C)**

**Oxidation**

**Organic solvent**

**General stability information**

**Storage**

## 6 CROSSREFERENCES TO STRUCTURE DATABANKS

**PIR/MIPS code**

**Brookhaven code**

## 7 LITERATURE REFERENCES

[1] Kikuchi, M., Sakaguchi, K.: Methods Enzymol., 45 (Proteolytic Enzymes Pt.B) ,
     485–492 (1976)
[2] Kikuchi, M., Sakaguchi, K.: Biochim. Biophys. Acta, 427, 285–294 (1976)
[3] Kikuchi, M., Sakaguchi, K.: Agric. Biol. Chem., 37 (8) , 1813–1821 (1973)
[4] Kikuchi, M., Sakaguchi, K.: Agric. Biol. Chem., 37 (4) , 827–835 (1973)
[5] Kikuchi, M., Hayashida, H., Nakano, E., Sakaguchi, K.: Biochemistry, 10 (7) ,
     1222–1229 (1971)

# 1 NOMENCLATURE

**EC number**
3.5.1.44

**Systematic name**
Protein-L-glutamine amidohydrolase

**Recommended name**
Protein-glutamine glutaminase

**Synonymes**
Peptidoglutaminase II
Glutaminyl-peptide glutaminase
Peptidylglutaminase II

**CAS Reg. No.**
62213-11-0

---

# 2 REACTION AND SPECIFICITY

**Catalysed reaction**
Protein L-glutamine + $H_2O$ →
→ protein L-glutamate + $NH_3$ (specific on derivatives substituted at both
alpha-amino- and alpha-carboxyl-group of glutamine [1, 3, 5], inactive
against high MW substrates [6]) [1, 3, 5, 6]

**Reaction type**
Carboxylic acid amide hydrolysis

**Natural substrates**
Protein L-glutamine + $H_2O$ (specific on derivatives substituted at both al-
pha-amino- and alpha-carboxyl-group of glutamine [1, 3, 5], inactive
against high MW substrates [6]) [1, 3, 5, 6]

**Substrate spectrum**
1 Protein L-glutamine + $H_2O$ (specific on derivatives substituted at both
alpha-amino- and alpha-carboxyl-group of glutamine [1, 3, 5], inactive
against high MW substrates [6]) [1, 3, 5, 6]

**Product spectrum**
1 Protein L-glutamate + $NH_3$

---

Enzyme Handbook © Springer-Verlag Berlin Heidelberg 1991
Duplication, reproduction and storage in data banks are only
allowed with the prior permission of the publishers

**Inhibitor(s)**
N-Bromosuccinate [3]; Sodium lauryl sulfate [3]; $Cu^{2+}$ (slight) [5]; $Hg^{2+}$ (slight) [5]; $AgNO_3$ [3]

**Cofactor(s)/prostethic group(s)**

**Metal compounds/salts**

---

**Turnover number (min$^{-1}$)**

**Specific activity (U/mg)**
78.0 [1, 4]

**$K_m$-value (mM)**
0.12 (tert-amyloxy carbonyl-L-glutaminyl-L-prolin) [3]; More [3]

**pH-optimum**
8.0 [1, 3]

**pH-range**

**Temperature optimum (°C)**
50 [1, 3]

**Temperature range (°C)**

---

## 3 ENZYME STRUCTURE

**Molecular weight**
104700–125000 (depending on method, Bacillus circulans) [2]

**Subunits**
Dimer (2 × 51000–70000, depending on method, Bacillus circulans) [2]

**Glycoprotein/Lipoprotein**
–

---

## 4 ISOLATION/PREPARATION

**Source organism**
Bacillus circulans [1–6]

**Source tissue**
Cells

**Localisation in source**
Cytoplasm (soluble)

## Purification
Bacillus circulans [1, 4, 5]

## Crystallization
–

## Cloned
–

## Renaturated
–

---

## 5 STABILITY

**pH**
6–11 [2]

**Temperature (°C)**

**Oxidation**

**Organic solvent**

**General stability information**

**Storage**

---

## 6 CROSSREFERENCES TO STRUCTURE DATABANKS

**PIR/MIPS code**

**Brookhaven code**

---

## 7 LITERATURE REFERENCES

[1] Kikuchi, M., Sakaguchi, K.: Methods Enzymol., 45 (Proteolytic Enzymes Pt.B) ,
     485–492 (1976)
[2] Kikuchi, M., Sakaguchi, K.: Biochim. Biophys. Acta, 427, 285–294 (1976)
[3] Kikuchi, M., Sakaguchi, K.: Agric. Biol. Chem., 37 (8) , 1813–1821 (1973)
[4] Kikuchi, M., Sakaguchi, K.: Agric. Biol. Chem., 37 (4) , 827–835 (1973)
[5] Kikuchi, M., Hayashida, H., Nakano, E., Sakaguchi, K.: Biochemistry, 10 (7) ,
     1222–1229 (1971)
[6] O'Shaughnessy, A.J., Gill, B.P., Headon, D.R.: Biochem. Soc. Trans., 13 (2) , 498
     (1985)

# 1 NOMENCLATURE

**EC number**
3.5.1.45

**Systematic name**
Urea amidohydrolase (ATP-hydrolysing)

**Recommended name**
Urease (ATP-hydrolysing)

**Synonymes**
Urea amidolyase [3]
ATP: urea amidolyase [2, 6]
UALase

**CAS Reg. No.**
72561-06-9

---

# 2 REACTION AND SPECIFICITY

**Catalysed reaction**
ATP + urea + $H_2O$ →
→ ADP + orthophosphate + $CO_2$ + 2 $NH_3$

**Reaction type**
Carboxylic acid amide hydrolysis

**Natural substrates**
Urea + ATP + $H_2O$

**Substrate spectrum**
1 Urea + ATP + $H_2O$
2 N-Methylurea + ATP + $H_2O$ [3]
3 Cyanamide + ATP + $H_2O$ [3]
4 Biuret + ATP + $H_2O$ [3]
5 Hydroxyurea + ATP + $H_2O$ [3]
6 Formamide + ATP + $H_2O$ [3]
7 Acetamide + ATP + $H_2O$ [3]
8 Propionamide + ATP + $H_2O$ [3]

---

## Product spectrum
1 $CO_2$ + $NH_3$ + ADP + orthophosphate
2 ?
3 ?
4 ?
5 ?
6 ?
7 ?
8 ?

## Inhibitor(s)
$BaCl_2$ (5 mM) [3]; $CoCl_2$ (5 mM) [3]; $CaCl_2$ (5 mM) [3]; ADP [3];
p-Chloromercuribenzoate (0.02 mM) [3]; $HgCl_2$ (0.01 mM) [3]; Iodoacetate
(10 mM) [3]; Acetohydroxamic acid [1]; Avidin [3]; Fluoride (1mM) [3];
Sulfhydryl reagents (dithiothreitol and mercaptoethanol protect) [3]

## Cofactor(s)/prostethic group(s)
Biotin [4]

## Metal compounds/salts
$Mg^{2+}$ [3]; $Mn^{2+}$ [3]; $K^+$ [3]; $Rb^+$ [3]; $Cs^+$ [3]; Bicarbonate (requires catalytic
amounts of bicarbonate)

## Turnover number (min$^{-1}$)

## Specific activity (U/mg)
0.8–1.2 [3]

## $K_m$-value (mM)
0.1 (urea) [3]; 0.25 (ATP) [3]

## pH-optimum
7.8–7.9 [3]

## pH-range

## Temperature optimum (°C)

## Temperature range (°C)

# 3 ENZYME STRUCTURE

## Molecular weight

## Subunits

## Glycoprotein/Lipoprotein
–

## 4 ISOLATION/PREPARATION

### Source organism

Candida utilis (inducible enzyme) [3]; Candida flareri [3]; Saccharomyces cerevisiae [3, 7, 8]; Chlorella ellipsoidea [2, 3, 6]; Chlorella pyrenoidosa [2, 3]; Chlorella vulgaris (var. viridis) [2]; Chlorella fusca (var. vacuolata) [2]; Chlamydomonas reinhardtii [2, 3]; Ankistrodesmus braunii [2]; Asterococcus superbus [2]; Dunaliella primolecta [2]; Nannochloris coccoides [2]; Scenedesmus brasilensis [2]; Scenedesmus obliquus [2]; Stichococcus bacillaris [2]; Geotrichum candidum [5]; Ureaplasma urealyticum [1]

### Source tissue

Cell [3]; Spores [5]

### Localisation in source

Cytoplasm [1]

### Purification

### Crystallization

–

### Cloned

–

### Renaturated

–

## 5 STABILITY

### pH

### Temperature (°C)

45 (50% loss of activity after 10 minutes) [3]; 55 (inactivated after 10 minutes) [3]

### Oxidation

### Organic solvent

### General stability information

### Storage

## 6 CROSSREFERENCES TO STRUCTURE DATABANKS

**PIR/MIPS code**

**Brookhaven code**

## 7 LITERATURE REFERENCES

[1] Romano, N., Tolone, G., Ajello, F., La Licata, R.: J. Bacteriol., 144 (2) , 830–832 (1980)
[2] Leftley, J.W., Syrett, P.J.: J. Gen. Microbiol., 77, 109–115 (1973)
[3] Roon, R.J., Levenberg, B.: J. Biol. Chem., 247 (13) , 4107–4113 (1972)
[4] Roon, R.J., Levenberg, B.: J. Biol. Chem., 245, 4593–4595 (1970)
[5] Shorer, J., Zelmanowicz, I., Barash, I.: Phytochemistry, II, 595–605 (1953)
[6] Thompson, J.F., Muenster, A.-M.E.: Biochem. Biophys. Res. Commun., 43, 1049–1005 (1971)
[7] Whitney, P.A., Cooper, T.G.: J. Biol. Chem., 247, 1349–1353 (1972)
[8] Whitney, P.A., Cooper, T.G.: Biochem. Biophys. Res. Commun., 49, 45–51 (1972)

## 1 NOMENCLATURE

**EC number**
3.5.1.46

**Systematic name**
N-(6-Aminohexanoyl)-6-aminohexanoate amidohydrolase

**Recommended name**
6-Aminohexanoate-dimer hydrolase

**Synonymes**
6-Aminohexanoic acid oligomer hydrolase

**CAS Reg. No.**

## 2 REACTION AND SPECIFICITY

**Catalysed reaction**
N-(6-Aminohexanoyl)-6-aminohexanoate + $H_2O$ →
→ 2 6-aminohexanoate

**Reaction type**
Carboxylic acid amide hydrolysis

**Natural substrates**

**Substrate spectrum**
1 N-(6-Aminohexanoyl)-6-aminohexanoate + $H_2O$
2 6-Aminohexanoate + $H_2O$ (oligomers of, up to 6 residues, residues are removed sequentially from the N-terminus)
3 More (not: cylic dimer, 6-N-carbobenzoxy-6-aminohexanoic acid oligomers, dipeptides, tripeptides, oligopeptides, casein, linear amides, cyclic amides, cyclic diamides) [1]

**Product spectrum**
1 6-Aminohexanoate
2 6-Aminohexanoate
3 ?

**Inhibitor(s)**
Diisopropylphosphofluoride [2]; p-Chloromercuribenzoate [2]

**Cofactor(s)/prostethic group(s)**

**Metal compounds/salts**

**Turnover number** (min$^{-1}$)
    144 [2]; 120 [2]

**Specific activity** (U/mg)
    1.05 [2]

**K$_m$-value** (mM)
    5.9 (6-aminohexanoic acid dimer) [2]; 6.2 (6-amino-hexanoic acid trimer) [2]

**pH-optimum**
    8–9 [2]

**pH-range**

**Temperature optimum** (°C)
    40 (1 hour reaction) [2]

**Temperature range** (°C)

---

## 3 ENZYME STRUCTURE

**Molecular weight**
    84000 (Flavobacterium sp., gel filtration) [2]

**Subunits**
    Dimer (2 × 42000, Flavobacterium sp.) [2]

**Glycoprotein/Lipoprotein**
    –

---

## 4 ISOLATION/PREPARATION

**Source organism**
    Flavobacterium sp. [1, 2]; Corynebacterium auranticum [3]

**Source tissue**
    Cell [2]

**Localisation in source**

**Purification**
    Flavobacterium sp. [2]

**Crystallization**
    –

**Cloned**
    –

Renaturated

–

---

## 5 STABILITY

**pH**
  6.5–9.5 [2]

**Temperature** (°C)
  35 (1 hour) [1]; 55 (inactivated) [2]

**Oxidation**

**Organic solvent**

**General stability information**

**Storage**

---

## 6 CROSSREFERENCES TO STRUCTURE DATABANKS

**PIR/MIPS code**
  A22644 (EII, Flavobacterium sp.); A29516 (EII, Flavobacterium sp. KI72, plasmid pOAD2); B22644 (EII', Flavobacterium sp.); B29516 (EII', Flavobacterium sp. KI72, plasmid pOAD2)

**Brookhaven code**

---

## 7 LITERATURE REFERENCES

[1] Kinoshita, S.: Hakkokogaku Kaishi, 60 (5) , 363–375 (1982)
[2] Kinoshita, S., Terada, T., Taniguchi, T., Takene, Y., Masuda, S., Matsunaga, N., Okada, H.: Eur. J. Biochem., 116, 547–551 (1981)
[3] Fukumara, T.: J. Biochem., 59, 537–544 (1959)

## 1 NOMENCLATURE

**EC number**
3.5.1.47

**Systematic name**
N-Acetyl-LL-2, 6-diaminoheptanedioate amidohydrolase

**Recommended name**
Acetyl diaminopimelate deacetylase

**Synonymes**
N-Acetyl-L-diaminopimelic acid deacylase [1]
N-Acetyl-LL-diaminopimelate deacylase [2]

**CAS Reg. No.**

## 2 REACTION AND SPECIFICITY

**Catalysed reaction**
N-Acetyl-LL-2, 6-diaminoheptanedioate + $H_2O \rightarrow$
$\rightarrow$ acetate + LL-2, 6-diaminoheptanedioate

**Reaction type**
Carboxylic acid amide hydrolysis

**Natural substrates**
N-Acetyl-L-diaminopimelic acid + $H_2O$ [1, 2]

**Substrate spectrum**
1 N-Acetyl-L-diaminopimelic acid + $H_2O$ [1, 2]

**Product spectrum**
1 Diaminopimelic acid + acetate [1, 2]

**Inhibitor(s)**

**Cofactor(s)/prostethic group(s)**

**Metal compounds/salts**

**Turnover number** (min$^{-1}$)

**Specific activity** (U/mg)
60 (wilde type) [1]; 80 (strain 1–16) [1]

$K_m$-value (mM)

pH-optimum

pH-range

Temperature optimum (°C)

Temperature range (°C)

## 3 ENZYME STRUCTURE

Molecular weight

Subunits

Glycoprotein/Lipoprotein

–

## 4 ISOLATION/PREPARATION

Source organism
   Bacillus megaterium [1]

Source tissue
   Cell wall [1]

Localisation in source
   Cell wall [1]

Purification
   Bacillus megaterium [1]

Crystallization
   –

Cloned
   –

Renaturated
   –

## 5 STABILITY

pH

Temperature (°C)

Oxidation

Organic solvent

General stability information

Storage

## 6 CROSSREFERENCES TO STRUCTURE DATABANKS

PIR/MIPS code

Brookhaven code

## 7 LITERATURE REFERENCES

[1] Sundharadas, G., Gilvarg, C.: J. Biol. Chem., 242, 3983–3988 (1967)
[2] Saleh, F., White, P.J.: J. Gen. Microbiol., 115, 95–100 (1979)

Enzyme Handbook © Springer-Verlag Berlin Heidelberg 1991
Duplication, reproduction and storage in data banks are only
allowed with the prior permission of the publishers

## 1 NOMENCLATURE

**EC number**
3.5.1.48

**Systematic name**
$N^1$-Acetylspermidine amidohydrolase

**Recommended name**
Acetylspermidine deacetylase

**Synonymes**
Deacetylase, acetylspermidine
$N^8$-Monoacetylspermidine deacetylase
$N^8$-Acetylspermidine deacetylase [5]

**CAS Reg. No.**
67339-07-5

## 2 REACTION AND SPECIFICITY

**Catalysed reaction**
$N^1$-Acetylspermidine + $H_2O$ →
→ acetate + spermidine

**Reaction type**
Carboxylic acid amide hydrolysis

**Natural substrates**
$N^8$-Acetylspermidine + $H_2O$ [8]

**Substrate spectrum**
1 $N^1$-Acetylspermidine + $H_2O$ (not deacetylated [6]) [2]
2 $N^8$-Acetylspermidine + $H_2O$ [1, 2, 6, 7]
3 $N^1$-Acetylspermidine + $H_2O$ [2]

**Product spectrum**
1 Acetate + spermidine
2 Acetate + spermidine
3 Acetate + spermidine

**Inhibitor(s)**
p-Chloromercuribenzoate [2]; Spermidine [2]; Spermine [2]; Putrescine [2];
$N^1$-Acetylspermidine (competitive inhibitor of $N^8$-acetylspermidine
deacetylation) [3]; Colchicine [3]; Phenacetin [3]; Neostigmine [3]; Acetyl-

penicillamine [3]; Acetylcysteine [3]; Acetylacetone [3]; Acetylornithine [3];
Acetlglycine [3]; Polyamines [4]; N-Ethylmaleimide [5]; EDTA [5]; Diethyl-
malonate [5]; Diisopropylfluorophosphate [5]; Neostigmine [5];
Echothiophate [5]; Na-butyrate [5]; Gamma-aminobutyric acid [5];
5-Aminovaleric acid [5]; N-Acetylputrescine [5]; N-Butyrylputrescine [5];
$CuCl_2$ [7]; 2-Mercaptoethanol [7]; Dithiothreitol [7]; Quinacrine [7];
Achriflavin [7]; 8-Hydroxyquinoline [7]; Pargyline [7]; $FeCl_2$ [7];
Iodoacetamide [7]; Semicarbazide [7]

**Cofactor(s)/prostethic group(s)**

**Metal compounds/salts**

---

**Turnover number** (min$^{-1}$)

**Specific activity** (U/mg)
0.00054 [2]

**$K_m$-value** (mM)
0.003 (N$^1$-acetylspermidine) [2]; 0.016 (N$^8$-acetylspermidine) [2]; More [3, 7]

**pH-optimum**
10.4 [2, 4]; 8.0–8.5 [7]

**pH-range**

**Temperature optimum** (°C)

**Temperature range** (°C)

---

# 3 ENZYME STRUCTURE

**Molecular weight**

**Subunits**

**Glycoprotein/Lipoprotein**
–

---

# 4 ISOLATION/PREPARATION

**Source organism**
Rat [1–6]; Human [7]

**Source tissue**
Liver [1–7]; Kidney [1, 7]; Lung [1, 7]; Heart [1, 7]; Brain [1, 7]; Muscle [1];
Spleen [1, 7]; Small intestine [7]; Pancreas [7]

Localisation in source
  Cytoplasm [1]

Purification
  Rat [2, 4]

Crystallization
  –

Cloned
  –

Renaturated
  –

## 5 STABILITY

pH

Temperature (°C)

Oxidation

Organic solvent

General stability information

Storage
  –20°C, glycerol and thioglycerol, 6 months [4]

## 6 CROSSREFERENCES TO STRUCTURE DATABANKS

PIR/MIPS code

Brookhaven code

## 7 LITERATURE REFERENCES

[1] Blankenship, J.: Arch. Biochem. Biophys., 189 (1) , 20–27 (1978)
[2] Libby, P.R.: Arch. Biochem. Biophys., 188 (2) , 360–363 (1978)
[3] Santacroce, M.J., Blankenship, J.: Proc. West. Pharmacol. Soc., 25, 113–118 (1982)
[4] Libby, P.: Methods Enzymol., 94, 329–331 (1983)
[5] Manneh, V.A., Blankenship, J.: Proc. West. Pharmacol. Soc., 28, 255–258 (1985)
[6] Marchant, P., Manneh, V.A., Blankenship, J.: Biochim. Biophys. Acta, 881, 297–299 (1986)
[7] Suzuki, O., Kumazawa, T., Seno, H., Matsumoto, T.: Med. Sci. Res., 15, 675–676 (1987)
[8] Blankenship, J., Marchant, P.E.: Proc. Soc. Exp. Biol. Med., 177, 180–187 (1984)

# 1 NOMENCLATURE

**EC number**
3.5.1.49

**Systematic name**
Formamide amidohydrolase

**Recommended name**
Formamidase

**Synonymes**

**CAS Reg. No.**

# 2 REACTION AND SPECIFICITY

**Catalysed reaction**
Formamide + $H_2O$ →
→ formate + $NH_3$

**Reaction type**
Carboxylic acid amide hydrolysis

**Natural substrates**

**Substrate spectrum**
1 Formamide + $H_2O$ [1, 2]
2 Aliphatic amides + $H_2O$ (also acts more slowly, on acetamide, propanamide and butanamide) [1, 3, 4, 7]

**Product spectrum**
1 Formate + $NH_3$
2 ?

**Inhibitor(s)**
Urea [5, 6]; $F^-$ [1]; p-Chloromercuribenzoate [1]; 5, 5-Dithiobis-2-nitrobenzoate [1]; Iodoacetamide [1]

**Cofactor(s)/prostethic group(s)**

**Metal compounds/salts**

**Turnover number** (min$^{-1}$)

**Specific activity** (U/mg)
    36 (formamide) [1]; 260 (acetamide) [1]; 698 (propionamide) [1]

**$K_m$-value** (mM)

**pH-optimum**

**pH-range**

**Temperature optimum** (°C)

**Temperature range** (°C)

---

## 3 ENZYME STRUCTURE

**Molecular weight**

**Subunits**

**Glycoprotein/Lipoprotein**
    −

---

## 4 ISOLATION/PREPARATION

**Source organism**
    Mycobacterium phlei [2]; Mycobacterium smegmatis [3]; Pseudomonas
    aeruginosa (enzyme also has transferase activity) [1]; Alcaligenes
    eutrophus [7]

**Source tissue**

**Localisation in source**

**Purification**

**Crystallization**
    −

**Cloned**
    −

**Renaturated**
    −

# 5 STABILITY

pH

Temperature (°C)

Oxidation

Organic solvent

General stability information

Storage

# 6 CROSSREFERENCES TO STRUCTURE DATABANKS

PIR/MIPS code

Brookhaven code

# 7 LITERATURE REFERENCES

[1] Clarke, P.H.: Adv. Microb. Physiol., 4, 179–222 (1970)
[2] Halpern, Y.S., Grossowicz, N.: Biochem. J., 65, 716 (1957)
[3] Kimura, T.: J. Biochem., Tokyo46, 1271 (1959)
[4] Draper, P.: J. Gen. Microbiol., 46, 111 (1967)
[5] Kelly, M., Kornberg, H.L.: Biochim. Biophys. Acta, 64, 190 (1962)
[6] Kelly, M., Kornberg, H.L.: Biochem. J., 93, 557 (1964)
[7] Friedrich, C.G., Mitrenga, G.: J. Gen. Microbiol., 125 , 367–374 (1981)

## 1 NOMENCLATURE

**EC number**
3.5.1.50

**Systematic name**
Pentanamide amidohydrolase

**Recommended name**
Pentanamidase

**Synonymes**
Valeramidase
Amidase, valer-

**CAS Reg. No.**
81032-50-0

## 2 REACTION AND SPECIFICITY

**Catalysed reaction**
Pentanamide + $H_2O \rightarrow$
$\rightarrow$ pentanoate + $NH_3$

**Reaction type**
Carboxylic acid amide hydrolysis

**Natural substrates**
Pentanamide + $H_2O$

**Substrate spectrum**
1 Pentanamide + $H_2O$
2 Aliphatic amides + $H_2O$ (acts, more slowly, on other short-chain aliphatic amides)

**Product spectrum**
1 Pentanoate + $NH_3$
2 Acid + $NH_3$

**Inhibitor(s)**

**Cofactor(s)/prostethic group(s)**

**Metal compounds/salts**

Turnover number (min$^{-1}$)

Specific activity (U/mg)

$K_m$-value (mM)

pH-optimum

pH-range

Temperature optimum (°C)

Temperature range (°C)

---

## 3 ENZYME STRUCTURE

Molecular weight

Subunits

Glycoprotein/Lipoprotein
  –

---

## 4 ISOLATION/PREPARATION

Source organism
  Alcaligenes eutrophus [1]

Source tissue

Localisation in source

Purification

Crystallization
  –

Cloned
  –

Renaturated
  –

---

## 5 STABILITY

pH

Temperature (°C)

Oxidation

Organic solvent

General stability information

Storage

---

# 6 CROSSREFERENCES TO STRUCTURE DATABANKS .

PIR/MIPS code

Brookhaven code

---

# 7 LITERATURE REFERENCES

[1] Friedrich, C.G., Mitrenga, G.: J. Gen. Microbiol., 125 , 367–374 (1981)

## 1 NOMENCLATURE

**EC number**
3.5.1.51

**Systematic name**
4-Acetamidobutanoyl-CoA amidohydrolase

**Recommended name**
4-Acetamidobutyryl-CoA deacetylase

**Synonymes**
Aminobutyryl-CoA thiolesterase
Deacetylase-thiolesterase

**CAS Reg. No.**

## 2 REACTION AND SPECIFICITY

**Catalysed reaction**
4-Acetamindobutanoyl-CoA + $H_2O$ →
→ acetate + 4-aminobutanoyl-CoA

**Reaction type**
Carboxylic acid amide hydrolysis

**Natural substrates**
4-Acetamidobutyryl-CoA + $H_2O$ [1]

**Substrate spectrum**
1  4-Acetamidobutanoyl-CoA + $H_2O$ [1]
2  4-Propionamidobutyryl-CoA + $H_2O$ [1]
3  4-Butyramidobutyryl-CoA + $H_2O$ [1]
4  4-Acetamidobutyryl pantetheine + $H_2O$ [1]
5  2-Acetamidoacetyl-CoA + $H_2O$ [1]
6  5-Acetamidovaleryl-CoA + $H_2O$ [1]
7  Acetyl-CoA + $H_2O$ (poor substrate) [1]
8  Propionyl-CoA + $H_2O$ [1]
9  Butyryl-CoA + $H_2O$ [1]
10  Valeryl-CoA + $H_2O$ [1]
11  Acetoacetyl-CoA + $H_2O$ [1]
12  3-Acetamidopropionyl-CoA + $H_2O$ [1]
13  Beta-alanyl-CoA + $H_2O$ [1]
14  DL-3-Aminobutyryl-CoA + $H_2O$ [1]
15  4-Aminobutanoyl-CoA + $H_2O$

## Product spectrum

1 Acetate + 4-aminobutanoyl-CoA
2 Propionate + 4-aminobutyryl-CoA
3 Butyrate + 4-aminobutyryl-CoA
4 Acetate + aminobutyryl pantetheine
5 Acetate + aminoacetyl-CoA
6 Acetate + aminovaleryl-CoA
7 ?
8 ?
9 ?
10 ?
11 ?
12 Acetate + 3-aminopropionyl-CoA
13 ?
14 ?
15 Aminobutanoate + coenzyme A

## Inhibitor(s)

$ZnCl_2$ (30% at 0.5 mM) [1]; $CaCl_2$ (about 10% at 0.5 mM) [1]; EDTA (26% at 62 mM) [1]; N-Ethylmaleimide (25% and 47% at 1 mM and 5 mM) [1]; More [1]

## Cofactor(s)/prostethic group(s)

## Metal compounds/salts

## Turnover number (min$^{-1}$)

## Specific activity (U/mg)

1.9 (thiolesterase activity) [1]; 1.29 (deacetylase activity) [1]

## $K_m$-value (mM)

0.005 (5-acetamidovaleryl CoA) [1]; 0.006 (4-acetamidobutyryl CoA); 0.005–0.012 (propionyl CoA) [1]; 0.007–0.020 (butyryl CoA) [1]; 0.007–0.026 (valeryl CoA) [1]; 0.016 (DL-3-aminobutytyl CoA) [1]; 0.02 (4-butyramidobutyryl CoA) [1]; 0.048 (3-acetamidopropionyl CoA) [1]; 0.05 (4-propionamidobutyryl CoA) [1]; 0.059 (beta-alanyl CoA) [1]; 0.077 (2-acetamidoacetyl CoA) [1]; 1 (acetoacetyl CoA) [1]; 1.5 (4-acetamidobutyryl pantetheine) [1]

## pH-optimum

7.0 [1]; 7.8 [1]; 8.7 [1]

## pH-range

7.6–8.3 [1]

## Temperature optimum (°C)

50 [1]

Temperature range (°C)
25–50 [1]

## 3 ENZYME STRUCTURE

### Molecular weight
275000 (Pseudomonas B4, gel filtration) [1]
257000 (Pseudomonas B4, gel filtration) [1]
292000 (Pseudomonas B4, gel filtration) [1]

### Subunits
Octamer (8 × 36500, native enzyme, Pseudomonas B4, gel ionophoresis)
[1]

### Glycoprotein/Lipoprotein
–

## 4 ISOLATION/PREPARATION

### Source organism
Pseudomonas B4 [1]

### Source tissue
Cell [1]

### Localisation in source

### Purification
Pseudomonas B4 [1]

### Crystallization
–

### Cloned
–

### Renaturated
–

## 5 STABILITY

### pH

### Temperature (°C)

### Oxidation

Organic solvent

General stability information
  Stable at protein concentration 1–3.8 mg/ml [1]

Storage
  –20°C or –75°C, 2 years [1]

---

# 6 CROSSREFERENCES TO STRUCTURE DATABANKS

PIR/MIPS code

Brookhaven code

---

# 7 LITERATURE REFERENCES

[1] Chsugi, M̄., Kahn, J., Hensley, C., Chew, S., Barker, H.A.: J. Biol. Chem., 256, 7642–7651 (1981)

# 1 NOMENCLATURE

**EC number**
  3.5.1.52

**Systematic name**
  N-Linked-glycopeptide-N⁴-(N-acetyl-beta-D-glucosaminyl)-L-asparagine
  amidohydrolase

**Recommended name**
  Peptide-N⁴-(N-acetyl-beta-glucosaminyl) asparagine amidase

**Synonymes**
  Glycopeptidase
  N-Oligosaccharide glycopeptidase [10]
  Jack-bean glycopeptidase [9]
  PNGase A [11]
  PNGase F [12]
  Glycopeptide N-glycosidase

**CAS Reg. No.**

# 2 REACTION AND SPECIFICITY

**Catalysed reaction**
  Asn-Asn(oligosaccharide)-Glu-Ser-Ser + $H_2O \rightarrow$
  $\rightarrow$ Asn-Asp-Glu-Ser-Ser + 1-amino-N-acetylglucosamine-oligosaccharide
  (step 1, catalysed by the enzyme [3]);
  1-Amino-N-acetylglucosamine-oligosaccharide + $H_2O \rightarrow$
  $\rightarrow$ N-acetylglucosamine-oligosaccharide + $NH_3$ (step 2, non enzymatically
  at acidic pH [3]); Hydrolysis of N⁴-(N-acetyl-beta-D-glucosaminyl)
  asparagine residues in which the glucosamine residue may be further
  glycosylated, to yield a (substituted) N-acetyl-beta-D-glucosaminylamine
  and a peptide containing an aspartic residue

**Reaction type**
  Carboxylic acid amide hydrolysis

**Natural substrates**

## Substrate spectrum

1  Glycopeptide (with 3–11 amino acid residues) + $H_2O$ [3, 6, 7]
2  Beta-aspartylglycosylamine (1-L-beta-aspartamido-2-acetamido-1,2-dideoxy-beta-D-glucose) + $H_2O$ [1]
3  Ovalbumin glycopeptide
    (Glu-Glu-Lys-Tyr-Asn(oligosaccharide)-Leu-Thr-Ser-Val) + $H_2O$ [1, 5]
4  Stem bromelian glycopeptide (Asn-Asn-(oligosaccharide)-Glu-Ser-Ser)
    + $H_2O$ [2, 3, 4]
5  Ovotransferrin glycopeptide (Gly-Leu-Ile-His-Asn(oligosaccharide)-Arg)
    + $H_2O$ [5]
6  Taka-amylase A (1.4-alpha-D-glucan glucanohydrolase, EC 3.2.1.1)
    + $H_2O$ [5, 8]
7  Transferrin (desialylated, human) + $H_2O$ [5]
8  Fetuin glycopeptide (Leu-Ala-Asn-(oligosaccharide)-AeCys-Ser) + $H_2O$
    [12]
9  More [1, 2, 3, 5, 9]

## Product spectrum

1  ?
2  Aspartic acid + $NH_3$ + N-acetylglucosamine [1]
3  Aspartic acid + $NH_3$ + N-acetylglucosamine [1, 5]
4  ?
5  ?
6  ?
7  ?
8  ?
9  ?

## Inhibitor(s)

$Cu^{2+}$ [3]; $Fe^{3+}$ [3]; $Zn^{2+}$ [3]; More (no effect: thiol inhibitors, iodoacetamide, N-ethylmaleimide, actinomycete protease inhibitors, leupeptins, chymostatin, pepstatin, $Mg^{2+}$, $Ca^{2+}$, $Mn^{2+}$, EDTA, L-cysteine, phenylmethyl-sulfonyl fluoride, gamma-D-gluconolactone) [3, 5]

## Cofactor(s)/prostethic group(s)

## Metal compounds/salts

## Turnover number (min⁻¹)

## Specific activity (U/mg)

0.00275 [1]; 0.019 (glycopeptidase group A) [5]; 0.0855 (glycopeptidase group B) [5]; 0.0623 (glycopeptidase group C) [5]; 5.128 [12]

**$K_m$-value (mM)**
  1 (beta-aspartylglycosylamine, pH 5.5) [1]; 4 (stem bromelain glycopeptide)
  [3]; 2.0 (stem bromelain undecapeptide, glycopeptidase group A) [5]; 2.3
  (stem bromelain undecapeptide, gycopeptidase group B) [5]; 4.0 (stem
  bromelain undecapeptide, gycopeptidase group C) [5]

**pH-optimum**
  5.5 [1]; 5.0 [2]; 5.2 [3, 4]; 6.0 (glycopeptidase group A) [5]; 5.0 (glycopeptidase group B/C) [5]; 5.3 (Taka-amylase A) [8]; 6.5 [9, 11]; 8.5 [12]

**pH-range**
  4–8.5 [1]; 2.5–7.5 [5]

**Temperature optimum (°C)**
  37 [12]

**Temperature range (°C)**

---

## 3 ENZYME STRUCTURE

**Molecular weight**
  79500 (Pisum sativum, HPLC) [11]

**Subunits**
? (x × 35500, Flavobacterium meningosepticum, SDS-PAGE) [12]

**Glycoprotein/Lipoprotein**
  –

---

## 4 ISOLATION/PREPARATION

**Source organism**
  Guinea pig [1]; Pig [1]; Rat [1]; Almonds [2, 3, 6, 7]; Jack-bean [9]; Pisum
  sativum [11]; Flavobacterium meningosepticum [12]; Lentil [11]; Wheat [11];
  More (overview) [11]

**Source tissue**
  Blood serum (mammalia) [1]; Tissues (mammalia) [1]; Liver (rat) [1]; Kidney
  (rat) [1]; Spleen [1]; Nuts (almond) [5, 8]; Meal (jack-bean) [9]; Split pea
  (Pisum sativum) [11]; Germ [11]; Cell (Flavobacterium) [12]

**Localisation in source**
  Lysosomes [1]

## Purification
Pig [1]; Almonds [5, 8]; Pisum sativum [11]; Flavobacterium meningosepticum [12]

## Crystallization
–

## Cloned
–

## Renaturated
–

---

## 5 STABILITY

### pH

### Temperature (°C)
37 (long periods in absence of detergent) [12]

### Oxidation

### Organic solvent

### General stability information

### Storage
Several weeks at 4°C [2]; 2 months at –20°C (activity loss less than 50%) [3]; Some months at 4°C [12]

---

## 6 CROSSREFERENCES TO STRUCTURE DATABANKS

### PIR/MIPS code
A35760 (precursor, Flavobacterium meningosepticum)

### Brookhaven code

---

## 7 LITERATURE REFERENCES

[1] Makino, M., Kojima, T., Ohgushi, T., Yamashina, I.: J. Biochem., 63, 186–192 (1968)
[2] Takahashi, N.: Biochem. Biophys. Res. Commun., 76, 1194–1201 (1977)
[3] Takahashi, N., Nishibe, H.: J. Biochem., 84, 1467–1473 (1978)
[4] Ishihara, H., Takahashi, N., Oguri, S., Tejima, S.: J. Biol. Chem., 254, 10715–10719 (1979)

[5] Takahashi, N., Nishibe, H.: Biochim. Biophys. Acta, 657, 457–467 (1981)
[6] Nishibe, H., Takahashi, N.: Biochim. Biophys. Acta, 661, 274–279 (1981)
[7] Takahashi, N., Shimizu, S., Yamada, K.: FEBS Lett., 146, 139–142 (1982)
[8] Takahashi, N., Toda, H., Nishibe, H., Yamamoto, K.: Biochim. Biophys. Acta, 707, 236–242 (1982)
[9] Sugiyama, K., Ishihara, H., Tejima, S., Takahashi, N.: Biochem. Biophys. Res. Commun., 112, 155–160 (1983)
[10] Tomiya, N., Kurono, M., Ishihara, H., Tejima, S., Endo, S., Arata, Y., Takahashi, N.: Anal. Biochem., 163, 489–499 (1987)
[11] Plummer Jr., T.H., Phelan, A.W., Tarentino, A.L.: Eur. J. Biochem., 163, 167–173 (1987)
[12] Tarentino, A.L., Plummer Jr., T.H.: Methods Enzymol., 138, 770–778 (1987)

# 1 NOMENCLATURE

**EC number**
3.5.2.1

**Systematic name**
Barbiturate amidohydrolase

**Recommended name**
Barbiturase

**Synonymes**

**CAS Reg. No.**
9025-16-5

# 2 REACTION AND SPECIFICITY

**Catalysed reaction**
Barbiturate + $H_2O$ →
→ malonate + urea

**Reaction type**
Carboxylic acid amide hydrolysis

**Natural substrates**
Barbiturate + $H_2O$

**Substrate spectrum**
1 Barbiturate + $H_2O$

**Product spectrum**
1 Malonate + urea

**Inhibitor(s)**

**Cofactor(s)/prostethic group(s)**

**Metal compounds/salts**

**Turnover number** (min$^{-1}$)

**Specific activity** (U/mg)
94.0 [2]

**$K_m$-value (mM)**
3.37 (barbiturate) [2]

**pH-optimum**
8–9 [2]

**pH-range**

**Temperature optimum (°C)**

**Temperature range (°C)**

---

## 3 ENZYME STRUCTURE

**Molecular weight**

**Subunits**

**Glycoprotein/Lipoprotein**
–

---

## 4 ISOLATION/PREPARATION

**Source organism**
Enterobacter aerogenes [1]; Mycobacterium [2]

**Source tissue**
Cell

**Localisation in source**

**Purification**
Mycobacterium (partial) [2]

**Crystallization**
–

**Cloned**
–

**Renaturated**
–

---

## 5 STABILITY

**pH**

**Temperature (°C)**

Oxidation

Organic solvent

General stability information

Storage

---

# 6 CROSSREFERENCES TO STRUCTURE DATABANKS

PIR/MIPS code

Brookhaven code

---

# 7 LITERATURE REFERENCES

[1] Patai, B.N., West, T.P.: FEMS Microbiol. Lett., 40, 33–36, (1987)
[2] Hayaishi, O., Kornberg, A.: J. Biol. Chem., 197, 717–732, (1952)

# 1 NOMENCLATURE

**EC number**
3.5.2.2

**Systematic name**
5, 6-Dihydroxypyridine amidohydrolase

**Recommended name**
Dihydropyrimidinase

**Synonymes**
Hydantoinase
Hydropyrimidine hydrase
Hydantoin peptidase
Pyrimidine hydrase
D-Hydantoinase

**CAS Reg. No.**
9030-74-4

---

# 2 REACTION AND SPECIFICITY

**Catalysed reaction**
5, 6-Dihydrouracil + $H_2O$ →
→ 3-ureidopropionate

**Reaction type**
Carboxylic acid amide hydrolysis

**Natural substrates**
Hydropyrimidine + $H_2O$
Dihydropyrimidine + $H_2O$
Hydantoins + $H_2O$

**Substrate spectrum**
1 5, 6-Dihydrouracil + $H_2O$
2 Hydrouracil + $H_2O$ (r) [19]
3 Hydrothymine + $H_2O$ (r) [19]
4 Hydantoin + $H_2O$ [19]
5 D-5-Hydantoins (aliphatic- and aromatic-5-monosubstituted) + $H_2O$
[2, 4, 13, 14, 16, 17]

---

6 Dihydropyrimidine + $H_2O$
7 Succinimides + $H_2O$ [21, 22]
8 More (not: 5, 5-disubstituted hydantoins [16], L-isomers of hydantoins [16], hydantoins having a charged group in amino acid moiety [14]) [14, 16]

## Product spectrum

1 3-Ureidopropionate
2 Carbamoyl-beta-alanine
3 Carbamoyl-beta-aminoisobutyric acid
4 Carbamoyl-glycine
5 Carbamoyl-D-amino acids (corresponding)
6 ?
7 ?
8 ?

## Inhibitor(s)

$Sn^{2+}$ [19]; $Mg^{2+}$ (5 mM) [19]; $Cu^{2+}$ [6]; Chelating agents (2, 6-dipicolinic acid, ortho-phenanthroline, 8-hydroxyquinoline, alpha, alpha-dipyridyl) [6, 9, 11, 12, 16]; p-Chloromercuribenzoate [6, 12, 16]; Sulfonamides [11]; L-Dihydroorotic acid [9]; p-Hydroxymercuribenzoate [6]

## Cofactor(s)/prostethic group(s)

## Metal compounds/salts

$Zn^{2+}$ (4 mol of $Zn^{2+}$ per mole active enzyme) [9, 11, 15]; $Mg^{2+}$ (activation) [6, 19]; $Mn^{2+}$ (activation) [5, 19]; $Fe^{2+}$ (activation) [6]; Divalent metal ions ($Zn^{2+}$, $Co^{2+}$, $Mn^{2+}$, reactivate dipicolinic acid-inhibited enzyme) [5, 11]

## Turnover number (min$^{-1}$)

## Specific activity (U/mg)

4.4 [19]; 30.3 [18]; 24.6 [6]; 13–18 [5]; 63.7 (Mn(II)-substituted dihydropyrimidase) [5]; More [3]

## $K_m$-value (mM)

830 (hydantoin) [19]; 280 (hydantoin) [16]; 350 (hydantoin) [12]; 1.7 (dihydrouracil) [12, 16]; 11 (dihydrouracil) [6]; 0.008 (dihydrouracil) [11]; 28 (S-(2-methylthioethyl)-hydantoin) [16]; 34 (S-(2-methylthioethyl)-hydantoin) [12]; 117.5 (hydrouracil) [19]; 2.1 (hydrothymine) [19]; 34 (5-isopropyl-hydantoin) [6]; 50 (5-phenylhydantoin) [12, 16]; More [11]

## pH-optimum

8.0–8.3 [12, 16]; 8.3–8.5 [12, 16]; 8.5 [19]; 8.8–9.0 [12, 16]; 9.0 [6]

## pH-range

6.0–9.5 [6]

**Temperature optimum (°C)**
55 [6, 12, 16]; 45 [12, 16]

**Temperature range (°C)**

## 3 ENZYME STRUCTURE

**Molecular weight**
190000 (gel filtration, Pseudomonas striata) [13, 16]
226000–230000 (gel filtration, Pseudomonas fluorescens [6], rat [10], bovine [11]) [6, 10, 11]
266000 (gel filtration, rat) [22]

**Subunits**
Tetramer (4 × 48000, Pseudomonas striata, gel electrophoresis) [13]
Tetramer (4 × 56500–60000, bovine [11], Pseudomonas fluorescens [11], gel electrophoresis) [6, 11]

**Glycoprotein/Lipoprotein**
–

## 4 ISOLATION/PREPARATION

**Source organism**
Calf [17, 19]; Bovine [5, 11, 15, 19]; Rat [10, 19, 21, 22]; Pigeon [19]; Guinea pig [21]; Rabbit [21]; Dog [21]; Mouse [21]; Wheat [18]; Pseudomonas striata [12–14, 16]; Pseudomonas fluorescens [6–8]; Corynebacterium sepedonicum [14]; Aerobacter cloacae [14]; Streptomyces griseus [14]; More [7, 8, 13, 21]

**Source tissue**
Liver [5, 10, 11, 15, 17, 19–22]; Kidney [21]; Wheat [18]; Cell [6–8, 12–14, 16]

**Localisation in source**
Soluble [11, 19]; Intracellular [7]

**Purification**
Pseudomonas striata [12, 16]; Pseudomonas fluorescens [16]; Bovine [11, 15]; Rat [22]

**Crystallization**
[12, 16]

## Cloned
–

## Renaturated
[1]

---

## 5 STABILITY

**pH**
6.0–7.0 [16]; 5.5–8.5 [6]

**Temperature (°C)**
60 (stable up to) [16]; 65 (unstable at) [16]; 40 (stable up to) [6, 7]

**Oxidation**

**Organic solvent**

**General stability information**

**Storage**
Purified enzyme, pH 7.5, room temperature or 4°C, several days [11]

---

## 6 CROSSREFERENCES TO STRUCTURE DATABANKS

**PIR/MIPS code**

**Brookhaven code**

---

## 7 LITERATURE REFERENCES

[1] Jacob, E., Henco, K., Marcinowski, S., Schenk, G.: (BASF A.–6) Ger. Offen. DE 3, 535, 987 (Cl. C12N15/00) (1987)
[2] Syldatk, C., Cotoras, D., Dombach, G., Groß, C., Kallwaß, H., Wagner, F.: Biotechnol. Lett., 9 (1) , 25–30 (1987)
[3] Morin, A., Hummel, W., Kula, M.-R.: J. Gen. Microbiol. , 133, 1201–1207 (1987)
[4] Nishida, Y., Nakamichi, K., Nabe, K., Tosa, T.: Enzyme Microb. Technol., 9, 721–725 (1987)
[5] Lee, M.L., Pettigrew, W.W., Sander, E.G., Nowak, T.: Arch. Biochem. Biophys., 259 (2) , 597–664 (1987)
[6] Morin, A., Hummel, W., Schütte, H., Kula, M.-R.: Biotechnol. Appl. Biochem., 8, 564–574 (1986)
[7] Morin, A., Hummel, W., Kula, M.-R.: Appl. Microbiol. Biotechnol., 25, 91–96 (1986)
[8] Morin, A., Hummel, W., Kula, M.-R.: Biotechnol. Lett., 8 (8) , 573–576 (1986)
[9] Lee, M.H., Cowling, R.A., Sander, E.G., Pettigrew, D. W.: Arch. Biochem. Biophys., 248 (1) , 368–378 (1986)

[10] Traut, T.W., Loechel, S.: Biochemistry, 23, 2533–2539 (1984)

[11] Brooks, K.P., Jones, E.A., Kim, B.-D., Sander, E.G.: Arch. Biochem. Biophys., 226 (2), 469–483 (1983)

[12] Takahashi, S.: Hakkokogaku Kaishi, 61, 139–151 (1983)

[13] Shimizu, S., Shimada, H., Takahashi, S., Ohashi, T., Tani, Y., Yamada, H.: Agric. Biol. Chem., 44 (9), 2233–2234 (1980)

[14] Takahashi, S., Ohashi, T., Kii, Y., Kumagai, H., Yamada, H.: J. Ferment. Technol., 57 (4), 328–332 (1979)

[15] Brooks, K.P., Kim, B.D., Sander, E.G.: Biochim. Biophys. Acta, 570, 213–214 (1979)

[16] Takahashi, S., Kii, Y., Kumagai, H., Yamada, H.: J. Ferment. Technol., 56 (5), 492–498 (1978)

[17] Cerere, F., Galli, G., Morisi, F.: FEBS Lett., 57 (2), 192–194 (1975)

[18] Mazús, B., Buchowicz, J.: Phytochemistry, 11, 77–82 (1972)

[19] Wallach, D.P., Grisolia, S.: J. Biol. Chem., 226, 277–288 (1956)

[20] Yamada, H., Takahashi, S., Kii, Y., Kumagai, H.: J. Ferment. Technol., 56, 484 (1978)

[21] Dudley, K.H., Buttler, T.C., Bius, D.L.: Drug Metab. Dispos., 2, 103–112 (1974)

[22] Maguire, J.H., Dudley, K.H.: Drug Metab. Dispos., 6, 601–605 (1978)

## 1 NOMENCLATURE

**EC number**
3.5.2.3

**Systematic name**
(S)-Dihydroorotate amidohydrolase

**Recommended name**
Dihydroorotase

**Synonymes**
Carbamoylaspartic dehydrase
Dihydroorotate hydrolase

**CAS Reg. No.**
9024-93-5

## 2 REACTION AND SPECIFICITY

**Catalysed reaction**
(S)-Dihydroorotate + $H_2O$ →
→ N-carbamoyl-L-aspartate

**Reaction type**
Carboxylic acid amide hydrolysis

**Natural substrates**
L-Dihydroorotate + $H_2O$

**Substrate spectrum**
1 L-Dihydroorotate + $H_2O$ (r)

**Product spectrum**
1 N-Carbamoylaspartate

**Inhibitor(s)**
Cysteine [3]; 1, 10-Phenanthroline [3, 15]; 8-Hydroxychinoline [3]; EDTA [7, 15]; 2-Mercaptoethanol [8]; Phosphate (more than 200 mM/l) [15]; $HgCl_2$ [16]; $AgNO_3$ [16]; $CuSO_4$ [16]; $ZnCl_2$ [16]; $CdCl_2$ [16]

**Cofactor(s)/prostethic group(s)**

**Metal compounds/salts**
Zinc (1 mole per subunit) [4, 8]; Zinc (2 moles per subunit) [5, 7]

Enzyme Handbook © Springer-Verlag Berlin Heidelberg 1991
Duplication, reproduction and storage in data banks are only
allowed with the prior permission of the publishers

**Turnover number** (min$^{-1}$)
11.760 [4]; 334 [3]

**Specific activity** (U/mg)
371 [4]; 278 [8]; 168 [5]; More [7, 10, 13, 16]

**K$_m$-value** (mM)
1.07 (N-carbamoylaspartate) [8]; 0.0756 (dihydroorotate) [8]; 0.015
(N-carbamoylaspartate) [6]; 0.028 (dihydroorotate) [6]; More [3, 4, 11, 15]

**pH-optimum**
8 (hydrolysis) [6]; 6 (cyclization) [6]; 4.4 (cyclization) [12]; 8.8 (hydrolysis)
[12]

**pH-range**
4.4–7.0 (cyclization) [12]; 7.0–9.2 (hydrolysis) [12]; 5. 6–6.2 (cyclization) [15];
7.5–9.0 (hydrolysis) [15]

**Temperature optimum** (°C)

**Temperature range** (°C)

---

## 3 ENZYME STRUCTURE

**Molecular weight**
75000–80900 (E.coli, gel filtration, equilibrium sedimentation centrifugation)
[4, 8]
95000–100000 (Clostridium oroticum, gel permeation chromatography, thin
layer gel filtration) [7, 15]
220000 (mammalian cells, covalently linked to EC 6.3.5.5 and EC 2.1.3.2, active fragment 82000) [3, 9, 12]

**Subunits**
Dimer (2 × 38000–55000, depending on method) [3, 4, 5, 7, 8, 9, 15]

**Glycoprotein/Lipoprotein**
–

---

## 4 ISOLATION/PREPARATION

**Source organism**
Leishmania donovani [1]; E.coli [4, 8]; Clostridium oroticum [5, 7, 11, 15];
Hamster [2, 9, 13]; Trypanosoma cruzi [10]; Mouse [12]; Rat [14]

## Source tissue
Cell [4, 5, 7, 8, 15]; Promastigote form [1, 10]; Amastigote form [1, 10]; Epimastigote form [10]; Kidney [13]; Liver [14, 16]; Organs (distribution in) [16]

## Localisation in source
Cytoplasm (soluble, subcellular localization) [16]

## Purification
E.coli [4, 8]; Clostridium oroticum [7, 11, 15]; Rat [14, 16]; Hamster [13]; Mouse [12]

## Crystallization
–

## Cloned
–

## Renaturated
–

---

## 5 STABILITY

### pH

### Temperature (°C)

### Oxidation
Sensible to air oxidation [4, 8]

### Organic solvent

### General stability information

### Storage
4°C (metal free, 10 mM carbamoyl-aspartate, Tris-phosphate buffer, pH 7) [4, 8]; Unstable –20°C [8], –196°C (several months) [13]

---

## 6 CROSSREFERENCES TO STRUCTURE DATABANKS

### PIR/MIPS code
DEECOO (Escherichia coli); A33917 (Chinese hamster, fragment); A27143 (Salmonella typhimurium); S00902 (URA4, yeast, Saccharomyces cerevisiae)

### Brookhaven code

# 7 LITERATURE REFERENCES

[1] Mukherjee, T., Ray, M., Bhaduri, A.: J. Biol. Chem., 263 (2), 708–713 (1988)

[2] Carrey, E.A., Hardie, D.G.: Eur. J. Biochem., 171, 583–588 (1988)

[3] Kelly, R.E., Mally, M.I., Evans, D.R.: J. Biol. Chem., 261 (13), 6073–6083 (1986)

[4] Washabaugh, M.W., Collins, K.D.: J. Biol. Chem., 261 (13), 5920–5929 (1986)

[5] Pettigrew. D.W., Metha, B.J., Bidigare, R.R., Choudhury, R.R., Scheffler, J.E., Sander, E.G.: Arch. Biochem. Biophys., 243 (2), 447–453 (1985)

[6] Bidigare, R.R., Sander, E.G., Pettigrew, D.W.: Biochim. Biophys. Acta, 831, 159–160 (1985)

[7] Pettigrew, D.W., Bidigare, R.R., Metha, B.J., Williams, M.I., Sander, E.G.: Biochem. J., 230, 101–108 (1985)

[8] Washabaugh, M.W., Collins, K.D.: J. Biol. Chem., 259 (5), 3293–3298 (1984)

[9] Davidson, J.N., Rumsby, P.C., Tamaren, J.: J. Biol. Chem., 256 (10), 5220–5225 (1981)

[10] Hammond, D.J., Gutteridge, W.E.: FEBS Lett., 118 (2), 259–262 (1980)

[11] Scheffler, J.E., Ma, J., Sander, E.G.: Biochem. Biophys. Res. Commun., 91 (2), 563–568 (1979)

[12] Christopherson, R.I., Jones, M.E.: J. Biol. Chem., 254 (24), 12506–12512 (1979)

[13] Coleman, P.F., Suttle, D.P., Stark, G.R.: Methods Enzymol., 121–134 (1978)

[14] Mori, M., Tatibana, M.: Methods Enzymol., 111–121 (1978)

[15] Taylor, W.H., Taylor, M.L., Balch, W.E., Gilchrist, P.S.: J. Bacteriol., 127 (2), 863–873 (1976)

[16] Kennedy, J.: Arch. Biochem. Biophys., 160, 358–365 (1974)

## 1 NOMENCLATURE

**EC number**
3.5.2.4

**Systematic name**
L-5-Carboxymethylhydantoin amidohydrolase

**Recommended name**
Carboxymethylhydantoinase

**Synonymes**
Hydantoin hydrolase [2]

**CAS Reg. No.**
9025-14-3

## 2 REACTION AND SPECIFICITY

**Catalysed reaction**
L-5-Carboxymethylhydantoin + $H_2O$ →
→ N-carbamoyl-L-aspartate

**Reaction type**
Carboxylic acid amide hydrolysis

**Natural substrates**

**Substrate spectrum**
1 L-5-Hydantoin (monosubstituted) + $H_2O$ (r) [1, 3, 5]
2 D, L-5-Hydantoin (monosubstituted) + $H_2O$ (r) [3, 7]
3 D, L-5-Hydanzoin (substituted, corresponding to aromatic amino acids) + $H_2O$ [2]
4 More [2]

**Product spectrum**
1 N-Carbamoyl-L-amino acids (corresponding) [1, 3, 5]
2 N-Carbamoyl-D, L-amino acid (corresponding) [3]
3 N-Carbamoyl-D, L-aromatic amino acids (corresponding) [2]
4 ?

**Inhibitor(s)**
EDTA [4]; $Cu^{2+}$ [3]; $Zn^{2+}$ [3]

**Cofactor(s)/prostethic group(s)**

**Metal compounds/salts**
   $Mn^{2+}$ [1, 4]; $Co^{2+}$ [1, 4]

**Turnover number** (min⁻¹)

**Specific activity** (U/mg)
   31 [2]; 1.23 [2]

$K_m$-**value** (mM)
   2.67 (L-5-benzylhydantoin) [2]

**pH-optimum**
   9.7 [2]; 8.5 [1]

**pH-range**
   9–10 [2]

**Temperature optimum** (°C)
   40 [2]; 45–55 [1]

**Temperature range** (°C)

## 3 ENZYME STRUCTURE

**Molecular weight**

**Subunits**

**Glycoprotein/Lipoprotein**
   –

## 4 ISOLATION/PREPARATION

**Source organism**
   Zymobacterium oroticum [4]; Arthrobacter sp. [3, 6]; Flavobacterium sp. [1, 3, 7]

**Source tissue**
   Cell [4]; Intact cells [1, 3]

**Localisation in source**
   Soluble [7]; Intracellular [7]

**Purification**

**Crystallization**
   –

Cloned

–

Renaturated

–

## 5 STABILITY

pH
   4–10 [1]

Temperature (°C)
   35 (stable below) [1]; 45 (unstable above) [1]

Oxidation

Organic solvent

General stability information

Storage

## 6 CROSSREFERENCES TO STRUCTURE DATABANKS

PIR/MIPS code

Brookhaven code

## 7 LITERATURE REFERENCES

[1] Yokozeki, K., Sano, K., Eguchi, C., Iwagami, H., Mitsugi, K.: Agric. Biol. Chem., 51 (3) , 729–736 (1987)
[2] Yokozeki, K., Hirose, Y., Kubota, K.: Agric. Biol. Chem. , 51 (3) , 737–746 (1987)
[3] Syldatk, C., Cotoras, D., Dombach, G., Groß, C., Kallaß, H., Wagner, F.: Biotechnol. Lett., 9 (1) , 25–30 (1987)
[4] Lieberman, J., Kronberg, A.: J. Biol. Chem., 207, 911–924 (1953)
[5] Tsugawa, R., Okumura, S., Ito, T., Katsuya, N.: Agric. Biol. Chem., 30, 27 (1966)
[6] Kitagawa, H., Miyoshi, T., Kato, M., Ikemi, M., Omine, H., Chiba, S.: Jpn. Kokai Tokkyo Koho, JP60241888 A2 JP84–99578, 8pp (1985)
[7] Nishida, Y., Nakamichi, K., Nabe, K., Tosa, T.: Enzyme Microb. Technol., 9, 721–725 (1987)

# 1 NOMENCLATURE

**EC number**
  3.5.2.5

**Systematic name**
  Allantoin amidohydrolase

**Recommended name**
  Allantoinase

**Synonymes**

**CAS Reg. No.**
  9025-20-1

---

# 2 REACTION AND SPECIFICITY

**Catalysed reaction**
  Allantoin + $H_2O$ →
  → allantoate

**Reaction type**
  Carboxylic acid amide hydrolysis

**Natural substrates**
  Allantoin + $H_2O$

**Substrate spectrum**
  1 Allantoin + $H_2O$

**Product spectrum**
  1 Allantoate

---

**Inhibitor(s)**
  $Hg^{2+}$ [3, 8]; $Cu^{2+}$ [7]; Dithiothreitol [4]; p-Chloromercuribenzoate [6, 10];
  p-Chloromercuribenzene-sulphonic acid [4]; N-Ethylmaleimide [6];
  2,2'-Dinitro-5, 5'-dithiodibenzoic acid (DTNB) [6]; Heavy metal ions [6];
  Thioglycollate [8]; Inhibitor protein (from Pseudomonas aeruginosa grown
  on citrate-nitrate) [9, 11]

**Cofactor(s)/prostethic group(s)**

**Metal compounds/salts**
  $Mn^{2+}$ (increases activity) [7]

---

Turnover number (min$^{-1}$)

Specific activity (U/mg)
   27.9 [2]; 989 [6]; 60 [9]; More [4, 5, 10, 12]

$K_m$-value (mM)
   13.3 (allantoin) [3]; 13.89 (allantoin) [15]; More [4, 6, 7, 8, 10]

pH-optimum
   7.8–8.0 [1]; 6.0–7.7 [3]; 8.4 [6]; 7.5 [7, 8, 10, 13, 15]

pH-range
   4–9 [7]

Temperature optimum (°C)
   35 [3]

Temperature range (°C)
   20–60 [6]; 30–80 [8, 13]

---

## 3 ENZYME STRUCTURE

Molecular weight
   200000 (frog, gel filtration, complex with EC 3.5.3.4) [2 , 5]
   125000 (pigeonpea, gel permeation) [3]
   50000 (soybean, gel permeation) [4]
   140000–150000 (Pseudomonas aeruginosa, gel filtration) [6]
   80000 (Pseudomonas aeruginosa, gel filtration) [9]

Subunits
   Tetramer (4 × 38000, Pseudomonas aeruginosa) [6]
   Tetramer (4 × 48000, frog, complex with EC 3.5.3.4 , 54000) [2]

Glycoprotein/Lipoprotein
   –

---

## 4 ISOLATION/PREPARATION

Source organism
   Soybean [1, 4]; Frog [2, 5, 16]; Pigeonpea (cajanus cajon) [3]; Pseudomonas aeruginosa [6, 9, 11]; Vigna radiata [7]; Dolichos biflorus [8];
   Lathyrus sativus [10]; Castor bean (Ricinus communis) [12, 14, 15]; Peanut
   [13]

Source tissue
   Leaves [1, 4]; Fruits [4]; Seedlings [7, 15]; Nitrogen fixing root nodules [3];
   Liver [2, 5, 16]

**Localisation in source**
Membrane bound [1, 3]

**Purification**
Frog [2, 5]; Pigeonpea [3]; Soybean [4]; Pseudomonas aeruginosa [6, 9]; Dolichos biflorus [8]

**Crystallization**
−

**Cloned**
−

**Renaturated**
−

## 5 STABILITY

**pH**

**Temperature** (°C)
70 [4]; 70 (10 minutes) [3]; More (plant enzyme cold labile) [7]

**Oxidation**

**Organic solvent**

**General stability information**

**Storage**
8°C, 3 weeks [3]; −20°C, 5 weeks [5]; 0–5°C, 2 weeks [5]; Frozen, 2 weeks, purified enzyme [10]

## 6 CROSSREFERENCES TO STRUCTURE DATABANKS

**PIR/MIPS code**

**Brookhaven code**

## 7 LITERATURE REFERENCES

[1] Costigan, S.A., Franceschi, V.R., Ku, M.S.B.: Plant Sci., 50, 179–187, (1987)
[2] Noguchi, T., Fujiwara, S., Hayashi, S.: J. Biol. Chem. , 261 (9) , 4221–4223, (1986)
[3] Amarjit, Singh, R.: Phytochemistry, 24 (3) , 415–418, (1985)
[4] Thomas, R.J., Meyers, S.T., Schrader, L.E.: Phytochemistry , 22 (5) , 1117–1120, (1983)

[5] Takada, Y., Noguchi, T.: J. Biol. Chem., 258 (8) , 4762–4764, (1983)
[6] Janssen, D.B., Smits, R.A.M.M., Van Der Drift, C.: Biochim. Biophys. Acta, 718, 212–219, (1982)
[7] Mary, A., Nirmala, J., Sastry, K.S.: Phytochemistry, 20 (12) , 2647–2650, (1981)
[8] Mary, A., Sastry, K.S.: Phytochemistry, 17, 397–399, (1978)
[9] De Windt, F.E., Van Der Drift, C.: Arch. Microbiol., 111, 117–122, (1976)
[10] Nirmala, J., Sastry, K.S.: Phytochemistry, 14, 1971–1973, (1975)
[11] Rijnierse, V.F.M., Van Der Drift, C.: Arch. Microbiol., 96, 319–328, (1974)
[12] Theimer, R.R., Beevers, H.: Plant Physiol., 47, 246–251, (1971)
[13] Singh, R., St. Angelo, A., Neucere, N.J.: Phytochemistry, 9, 1535–1538, (1970)
[14] St. Angelo, A.J., Ory, R.L.: Biochem. Biophys. Res. Commun., 40 (2) , 290–296, (1970)
[15] Ory, R.L., Gordon, C.V., Singh, R.: Phytochemistry, 8, 401–404, (1969)
[16] Visentin, L.P., Allen, J.M.: Science, 163, 1463–1464, (1969)

# 1 NOMENCLATURE

**EC number**
3.5.2.6

**Systematic name**
Beta-lactamhydrolase

**Recommended name**
Beta-lactamase

**Synonymes**
Cephalosporinase
Neutrapen
Penicillin beta-lactamase
Exopenicillinase
Ampicillinase
Penicillin amido-beta-lactamhydrolase
Penicillinase I, II (different in pH and temperature stability, molecular weight) [15]
Cephalosporin-beta-lactamase
Beta-lactamase AME I [6]
Beta-lactamase A, B, C (different in substrate specifity, amino acid sequence, mechanistic properties, metal ion requirement) [37, 38]
Beta-lactamase I-III (different in molecular weight, localization, metal ion requirement) [5, 12, 27]
E. C. 3.5.2.8 (formerly, a group of enzymes of varying specificity hydrolysing beta-lactams, some act more rapidly on penicillins, some more rapidly on cephalosporins. The latter were formerly listed as E. C. 3.5.2.8)
Penicillinase

**CAS Reg. No.**
9001-74-5

# 2 REACTION AND SPECIFICITY

**Catalysed reaction**
A beta-lactam + $H_2O \rightarrow$
$\rightarrow$ a substituted beta-amino acid

**Reaction type**
Carboxylic acid amide hydrolysis

## Natural substrates

Penicillins + $H_2O$
Cephalosporins + $H_2O$
Beta-lactams + $H_2O$

## Substrate spectrum

1 Beta-lactam + $H_2O$
2 Penicillin + $H_2O$
3 Penicillin (derivatives) + $H_2O$
4 Cephalosporin + $H_2O$
5 Cephalosporin (derivatives) + $H_2O$
6 Benzylpenicillin + $H_2O$
7 Ampicillin + $H_2O$
8 Carbenicillin + $H_2O$
9 Oxacillin + $H_2O$
10 Cloxacillin + $H_2O$
11 Methicillin + $H_2O$
12 Cephaloridine + $H_2O$
13 Cephaloglycine + $H_2O$
14 Cepholothin + $H_2O$
15 Cephalexin + $H_2O$
16 Cephazoline + $H_2O$
17 More [4, 6, 8, 19, 22–27, 31, 33–35]

## Product spectrum

1 Beta-amino acid (substituted)
2 ?
3 ?
4 ?
5 ?
6 ?
7 ?
8 ?
9 ?
10 ?
11 ?
12 ?
13 ?
14 ?
15 ?
16 ?
17 ?

## Inhibitor(s)

2-Benzylimidazole [36]; Benzylpenicillic acid [36]; $Ag^+$ [15]; $Cu^{2+}$ [5, 9, 22, 35]; $Zn^{2+}$ [9]; $Co^{2+}$ [9]; $Ca^{2+}$ [36]; $Fe^{2+}$ [36]; $Hg^{2+}$ [9, 15, 23, 35]; p-Chloromercuribenzoate [16, 17, 23, 27, 34, 35]; Cloxacillin [8, 16, 17, 34]; Iodine [9–11, 23, 26, 27, 34]; EDTA [27]; Methicillin [23]; N-Bromosuccinimide [5, 11, 30]; 2-Hydroxy-5-nitrobenzyl bromide [30]; Clavulanic acid [4, 16, 17]; Urea [15]; Beta-lactam-antibiotics (semi-synthetic) [8, 9]; Sulbactam [4]; More [8, 9, 16, 19, 34–36]

## Cofactor(s)/prostethic group(s)

## Metal compounds/salts

$Zn^{2+}$ (beta-lactamase II, penicillinase, beta-lactamase B) [15, 32, 35]

## Turnover number (min⁻¹)

2800–160000 (benzyl penicillin, overview) [35]; 210000 (lactamase I, benzyl-penicillin) [32]; 80000 (lactamase II, benzylpenicillin) [32]; 50600 (lactamase II, cephalosporin C) [32]; 28400 (benzylpenicillin) [11]

## Specific activity (U/mg)

1966 [24]; 33 [22]; 5666 [21, 23]; 5916 [21]; 190 [20]; 163 [33]; 134.3 [33]; 264 [31]; 49.7 [9]; 24.0 [8]; 2200 [4]; 980 [17]; 2430 [11]; 388 [10]; 5616.6 [27]; 5250 [25]

## $K_m$-value (mM)

More [4, 5, 8–11, 14, 15, 17, 20, 22–27, 31–36]

## pH-optimum

6.0–6.5 [36]; 7.0 [15, 34, 36]; 6.0–7.0 [27, 35]; 5.0–8.5 [35]; 7.5 [33]; 8.2 [23, 33]; 6.0 [32]; 8.0 [9, 10, 31]; 6–8 [22]; 6.5 [4, 11, 21]; 8.5 [8, 20]; 7.0 [17]

## pH-range

4.3–7.8 [21]; 6.5–10 [20]; 6–8 [2]

## Temperature optimum (°C)

30 [35]; 35–40 [35]; 45–55 [35]; 50 [20]; 30–35 [17]; 40–45 [11]; 45 [4, 10]; 40 [8, 9]

## Temperature range (°C)

---

## 3 ENZYME STRUCTURE

## Molecular weight

12400–17000 (gel filtration, SDS-PAGE, Bacillus sp. [15], Pseudomonas aeruginosa [16], Enterobacter sp. [23]) [15, 16, 23, 35]
20000–26000 (gel filtration, analytical ultracentrifugation, SDS-PAGE, Levinia malonatica, Streptomyces sp., E. coli, Bacillus cereus, Enterobacter cloacae) [13, 17, 24, 27, 32–34]

28000–35000 (SDS-PAGE, gel filtration, Proteus penneri, Pseudomonas aeruginosa, Streptomyces cacaoi, Bacillus sp., Bacillus licheniformis, Enterobacter sp., Staphylococci sp.) [4, 5, 9, 11, 12, 15, 21, 23, 25–27, 31, 32, 35, 36]
38000–41000 (SDS-PAGE, gel filtration, Proteus morganii, Citrobacter freundii, Enterobacter sp.) [8, 10, 20, 23]
49000 (SDS-PAGE, Enterobacter sp., Klebsiella aerogenes) [23, 33]
More [16, 19, 35]

**Subunits**
Monomer (SDS-PAGE)

**Glycoprotein/Lipoprotein**
Lipoprotein [21]

---

## 4 ISOLATION/PREPARATION

**Source organism**
Staphylococci [25, 28, 36]; Bacillus sp. [12, 15]; Bacillus cereus [3, 5, 14, 30, 32, 36]; Bacillus subtilis [36]; Bacillus licheniformis [21, 26]; E. coli [18, 24, 28, 34]; Enterobacter cloacae [33]; Klebsiella aerogenes [33]; Enterobacter sp. [23]; Pseudomonas aeruginosa [9, 16, 19, 31]; Streptomyces sp. [17, 22]; Streptomyces cacaoi [11]; Streptomyces antibioticus [6]; Streptomyces albus [1]; Proteus morganii [8, 20]; Proteus pennerei [4]; Levinia malonatica [13]; Citrobacter freundii [10]; Clostridium butyricum [7]; Alcaligenes eutrophus [2]; Bacteria [29]; More [19, 35]

**Source tissue**
Culture medium [7, 11, 15, 17, 26, 32, 36]; Cell [2, 5, 9, 10, 13, 26, 31]

**Localisation in source**
Extracellular [7, 11, 26, 27, 32, 35, 36]; Cell bound [5, 21, 27, 34]; Periplasm [26, 31]; More [35]

**Purification**
Bacillus cereus [3, 5, 27, 32, 36]; E. coli [24, 34]; Klebsiella aerogenes [33]; Enterobacter cloacae [33]; Pseudomonas aeruginosa [9, 16, 31]; Bacillus licheniformis [21, 26]; Staphylococcus aureus [25]; Enterobacter sp. [23]; Proteus morganii [8, 20]; Streptomyces sp. [17]; Streptomyces cacaoi [11]; Citrobacter freundii [10]; Proteus penneri [4]; Streptomyces albus [1]; More [35]

**Crystallization**
[1, 3, 27, 32, 36]

**Cloned**
–

Renaturated

–

---

## 5 STABILITY

**pH**
   3.0–10.0 [36]; 5.0–9.0 [27]; 6–8 [15]; 9–10 [15]

**Temperature (°C)**
   40 (stable up to) [15]; 45 (stable up to) [33]; 50 (unstable above) [11]; 60
   (unstable at) [17, 22, 23, 34]

**Oxidation**

**Organic solvent**

**General stability information**

**Storage**
   Purified enzyme, 20°C, 1 year [33]; Purified enzyme, 15°C, many years [26];
   Purified enzyme, 2°C, pH 7.0 [24]

---

## 6 CROSSREFERENCES TO STRUCTURE DATABANKS

**PIR/MIPS code**
   PNSAP (precursor, Staphylococcus aureus, PC-1); PNBSL (precursor,
   Bacillus licheniformis); PNBSU (I, precursor, Bacillus cereus); PNBS5B (I,
   precursor, Bacillus cereus, 5/B); PNBSLC (III, precursor, Bacillus cereus);
   PNECP (precursor, Escherichia coli, plasmids); PNBSU2 (II, precursor,
   Bacillus cereus); PNBS2S (II, precursor, Bacillus sp.); QKEC (precursor,
   Escherichia coli); PNKBM (precursor, Enterobacter cloacae, strain MNH1);
   PNKBQ (precursor, Enterobacter cloacae, strain Q908R, fragment); PNKBP
   (precursor, Enterobacter cloacae, strain P99); PNEBT (OXA2, precursor, Salmonella typhimurium); A32882 (Citrobacter freundii); A35001 (PSE-4, precur-
   sor, Staphylococcus aureus); A35257 (Neisseria gonorrhoeae); A35395 (2A,
   precursor, plasmid BWH77, Klebsiella pneumoniae); D26839 (OXA2, precursor, Escherichia coli); S03852 (OXA2, precursor, Salmonella typhimurium,
   plasmid R46); S06757 (Staphylococcus aureus); S08296 (precursor,
   Citrobacter freundii); A27028 (Citrobacter freundii); A24869 (precursor,
   Citrobacter freundii); S00464 (class A, Escherichia coli, plasmid p453);
   S03557 (OXA2, precursor, Escherichia coli); S02434 (SHV-2, Escherichia coli);
   A24469 (precursor, Klebsiella pneumoniae); S06264 (SHV-1, Klebsiella
   pneumoniae, fragment); S04649 (precursor, Rhodobacter capsulatus);
   A23600 (Staphylococcus aureus, PC-1); A32017 (II, precursor, Bacillus cereus,
   5/B/6); S03167 (precursor, Bacillus cereus); A28183 (Bacillus licheniformis,
   strain 749/C, fragment); S02714 (precursor, Streptomyces aureofaciens)

---

**Brookhaven code**
1BLM (Staphylococcus aureus)

# 7 LITERATURE REFERENCES

[1] Dideberg, O., Charlier, P., Wéry, J.-P., Dehottay, P., Dusart, J., Erpicum, T., Frére, J.-M., Ghuysen, J.-M.: Biochem. J., 245, 911–913 (1987)
[2] Sebo, P., Stastná, J.: Folia Microbiol., 32, 376–381 (1987)
[3] Sutton, B.J., Artymiuk, P.J., Cordero-Barboa, A.E., Little, C., Phillips, D.C., Waley, S.G.: Biochem. J., 248, 181–188 (1987)
[4] Grace, M.E., Gregory, F.M., Hung, P.P., Fu, K.P.: J. Antibiot., 7, 938–942 (1986)
[5] Conolly, A.K., Waley, S.G.: Biochemistry, 22, 4647–4651 (1983)
[6] Erne, A.M., Zahner, H., Werner, R.G.: FEMS Microbiol. Lett., 16, 117–121 (1983)
[7] Magot, M.: J. Gen. Microbiol., 127, 113–119 (1981)
[8] Toda, M., Inoue, M., Mitsuhashi, S.: J. Antibiot., 11, 1469–1475 (1981)
[9] Murata, T., Minami, S., Yasuda, K., Iyobe, S., Inoue, M., Mitsuhashi, S.: J. Antibiot., 9, 1164–1170 (1981)
[10] Tajima, M., Takenouchi, Y., Sugawara, S., Inoue, M., Mitsuhashi, S.: J. Gen. Microbiol., 121, 449–456 (1980)
[11] Ogawara, H., Mantoku, A., Shimada, S.: J. Biol. Chem., 256 (6), 2649–2655 (1981)
[12] Akiba, T., Horikoshi, K.: Agric. Biol. Chem., 44 (11), 2741–2742 (1980)
[13] Philippon, A., Paul, G., Barthelemy, M., Labia, R., Nevot, P.: FEMS Microbiol. Lett., 8, 191–194 (1980)
[14] Klemes, Y., Citri, N.: Biotechnol. Bioeng., 21, 897–905 (1979)
[15] Sunaga, T., Akiba, T., Horikoshi, K.: Agric. Biol. Chem., 43 (3), 477–480 (1979)
[16] Matthew, M.: FEMS Microbiol. Lett., 4, 241–244 (1978)
[17] Ogawara, H., Minagawa, T., Nishizaki, H.: J. Antibiot., 9, 923–925 (1978)
[18] Ambler, R.P., Scott, G.K.: Proc. Natl. Acad. Sci. USA, 75 (8), 3732–3736 (1978)
[19] Matthew, M., Sykes, R.B.: J. Bacteriol., 132 (1), 341–345 (1977)
[20] Fujii-Kuriyama, Y., Yamamoto, M., Sugawara, S.: J. Bacteriol., 131 (3), 726–734 (1977)
[21] Yamamoto, S., Lampen, J.O.: J. Biol. Chem., 251 (13), 4095–4101 (1976)
[22] Johnson, K., Duez, C., Frére, J.-M., Ghuysen, J.-M.: Methods Enzymol., 43, 687–698 (1975)
[23] Ross, G.W.: Methods Enzymol., 43, 678–687 (1975)
[24] Richmond, M.H.: Methods Enzymol., 43, 672–677 (1975)
[25] Richmond, M.H.: Methods Enzymol., 43, 664–672 (1975)
[26] Thatcher, D.R.: Methods Enzymol., 43, 652–664 (1975)
[27] Thatcher, D.R.: Methods Enzymol., 43, 640–652 (1975)
[28] Richmond, M.H.: Methods Enzymol., 43, 86–100 (1975)
[29] Ross, G.W., O'Callaghan, C.H.: Methods Enzymol., 43, 69–85 (1975)
[30] Ogawara, H., Umezawa, H.: Biochim. Biophys. Acta, 391, 435–447 (1975)
[31] Furth, A.J.: Biochim. Biophys. Acta, 377, 431–443 (1975)
[32] Davies, R.B., Abrahym, E.P.: Biochem. J., 143, 115–127 (1974)
[33] Ross, G.W., Boulton, M.G.: Biochim. Biophys. Acta, 309, 430–439 (1973)

[34] Ogawara, H., Maeda, K., Umezawa, H.: Biochim. Biophys. Acta, 289, 203–211 (1972)
[35] Citri, N. in "The Enzymes", 3rd Ed. (Boyer, P.D., Ed.) Vol.4, 23–46 (1971) (Review)
[36] Pollock, M.R. in "The Enzymes", 2nd Ed. (Boyer, P.D., Ed.) Vol.4, 269–278 (1960)
    (Review)
[37] Ambler, R.P.: Philos. Trans. R. Soc. Lond. B Biol. Sci., 289, 321–331 (1980)
[38] Jaurin, B., Grundstrom, T.: Proc. Natl. Acad. Sci. USA, 78, 4897–4901 (1981)

# 1 NOMENCLATURE

**EC number**
3.5.2.7

**Systematic name**
4-Imidazolone-5-propanoate amidohydrolase

**Recommended name**
Imidazolonepropionase

**Synonymes**

**CAS Reg. No.**
9024-91-3

# 2 REACTION AND SPECIFICITY

**Catalysed reaction**
4-Imidazolone-5-propanoate + $H_2O \rightarrow$
$\rightarrow$ N-formimino-L-glutamate

**Reaction type**
Carboxylic acid amide hydrolysis

**Natural substrates**
4-Imidazolone-5-propanoate + $H_2O$

**Substrate spectrum**
1 4-Imidazolone-5-propanoate + $H_2O$

**Product spectrum**
1 N-Formimino-L-glutamate

**Inhibitor(s)**
p-Chloromercuribenzoate [3, 7]

**Cofactor(s)/prostethic group(s)**

**Metal compounds/salts**

**Turnover number** (min$^{-1}$)

**Specific activity** (U/mg)

**K$_m$-value (mM)**
    0.2 (4-imidazolone-5-propanoate) [8]; 0.007 (4-imidazolone-5-propanoate)
    [3, 7]; 0.1 (4-imidazolone-5-propanoate) [6]

**pH-optimum**
    7.4 [3, 6, 7]

**pH-range**
    7–8 [3]

**Temperature optimum (°C)**

**Temperature range (°C)**

---

**3 ENZYME STRUCTURE**

**Molecular weight**

**Subunits**

**Glycoprotein/Lipoprotein**
    –

---

**4 ISOLATION/PREPARATION**

**Source organism**
    Bovine [8]; Hog [8]; Sheep [8]; Guinea pig [8]; Rat [3, 7, 8]; Pseudomonas
    sp. [8]; Pseudomonas fluorescens [8]; Pseudomonas putida [1]; Aerobacter
    aerogenes [8]; Clostridium tetanomorphum [8]; Clostridium cylindrosporum
    [8]; Salmonella typhimurium [6]; Bacillus subtilis [4–5]; Streptomyces
    coelicolor [2]

**Source tissue**
    Liver [3, 7, 8]; Cell [2, 4, 6, 8]

**Localisation in source**

**Purification**
    Pseudomonas fluorescens [8]; Rat [3, 7]

**Crystallization**
    –

**Cloned**
    –

**Renaturated**
    [1]

## 5 STABILITY

pH

Temperature (°C)

Oxidation

Organic solvent

General stability information
   $Mn^{2+}$ stabilizes during storage [6]

Storage
   Partially purified enzyme, −10°C, several weeks [8]; Purified enzyme, 10°C, 2 weeks [3]

---

## 6 CROSSREFERENCES TO STRUCTURE DATABANKS

PIR/MIPS code

Brookhaven code

---

## 7 LITERATURE REFERENCES

[1] Consevage, M.W., Porter, R.D., Phillips, A.T.: J. Bacteriol., 162 (1) , 138–146 (1985)
[2] Kendrick, K.E., Wheelis, M.L.: J. Gen. Microbiol., 128 , 2029–2040 (1982)
[3] Snyder, S.H.: Methods Enzymol., 17, Pt. B, 92–95 (1971)
[4] Magasanik, B., Kaminskas, E., Kimhi, Y.: Methods Enzymol., 17, Pt. B, 55–57 (1971)
[5] Hassall, H., Greenberg, D.M.: Methods Enzymol., 17, Pt. B, 89–91 (1971)
[6] Smith, G.R., Halpern, Y.S., Magasanik, B.: J. Biol. Chem., 246 (10) , 3320–3329 (1971)
[7] Snyder, S.H., Silva, O.L., Kies, M.W.: J. Biol. Chem., 236 (11) , 2996–2998 (1961)
[8] Rao, D.R., Greenberg, D.M.: J. Biol. Chem., 236 (6) , 1758–1763 (1961)

## 1 NOMENCLATURE

**EC number**
3.5.2.9

**Systematic name**
5-Oxo-L-proline amidohydrolase (ATP-hydrolysing)

**Recommended name**
5-Oxoprolinase (ATP-hydrolysing)

**Synonymes**
Pyroglutamase (ATP-hydrolysing)
Oxoprolinase
Pyroglutamase
5-Oxoprolinase
Pyroglutamate hydrolase
Pyroglutamic hydrolase
L-Pyroglutamate hydrolase
5-Oxo-L-prolinase
5-Oxo-L-prolinase-Components A (Beta-5) and B (F8)
More (Pseudomonas putida: component A exhibits 5-oxo-L-proline depen-
dent ATPase activity, component B is a catalyst that converts a phosphory-
lated form of 5-oxo-L-proline to glutamate) [2, 4]

**CAS Reg. No.**
9075-46-1

---

## 2 REACTION AND SPECIFICITY

**Catalysed reaction**
ATP + 5-oxo-L-proline + 2 $H_2O$ →
→ ADP + orthophosphate + L-glutamate

**Reaction type**
Carboxylic acid amide hydrolysis

**Natural substrates**
5-Oxo-L-proline + ATP + $H_2O$ [1, 15, 17]

## Substrate spectrum

1 ATP + 5-oxo-L-proline + $H_2O$ (r)
2 dATP + 5-oxo-L-proline + $H_2O$ [2, 11, 12]
3 ATP + 3-oxy-5-oxo-L-proline + $H_2O$ [12]
4 ATP + 4-oxy-5-oxo-L-proline + $H_2O$ [12]
5 ATP + piperidone carboxylate + $H_2O$ [12]
6 Nucleotide (ATP, ITP, GTP, UTP) + $H_2O$ [6, 7, 11]
7 ITP + 5-oxo-L-proline + $H_2O$ [6, 10]
8 GTP + 5-oxo-L-proline + $H_2O$ [10]
9 Alpha-hydroxyglutarate lactone + $H_2O$ [8]
10 More [5]

## Product spectrum

1 ADP + orthophosphate + L-glutamate
2 dADP + orthophosphate + L-glutamate
3 ADP + orthophosphate + 3-oxy-L-glutamate
4 ADP + orthophosphate + 4-oxy-L-glutamate
5 ADP + orthophosphate + 2-aminoadipate
6 Nucleoside diphosphate (ADP, IDP, GDP, UDP) + orthophosphate
7 IDP + orthophosphate + L-glutamate
8 GDP + orthophosphate + L-glutamate
9 ?
10 ?

## Inhibitor(s)

2-Imidazolidone-4-carboxylate [1, 10, 17];
L-2-Oxothiazolidine-4-carboxylate [1]; DL-3-Methyl-5-oxo-proline [1];
2-Piperidone-6-carboxylate [12]; 3-Oxy-5-oxo-proline [12];
4-Oxy-5-oxoproline [12]; L-Dihydroorotate [12]; ADP [10, 12];
Dithionitrobenzoic acid [6]; N-Ethylmaleimide [2, 6, 10, 12];
p-Chloromercuribenzoate [12]; p-Hydroxymercuribenzoate [2];
Iodoacetamide [2, 10, 12]; 5-p-Fluorosulfonylbenzoyl adenosine [6];
5-p-Fluorosulfonylbenzoyl inosine [6]; More [5]

## Cofactor(s)/prostethic group(s)

## Metal compounds/salts

$Mg^{2+}$ [6, 10, 12, 18]; $Mn^{2+}$ [6, 10, 18]; $K^+$ [2, 6, 10, 12, 18]; $Ca^{2+}$ [6]; $Co^{2+}$ [6]

## Turnover number (min$^{-1}$)

7 [14]

## Specific activity (U/mg)

0.122 [14]; 0.88 [12]; 1.45 [3, 6]; 4. 05 (component A) [2, 4]; 1.25 (component
A + B) [2, 4]

**K$_m$-value (mM)**
0.05 (5-oxo-L-proline) [12, 17]; 0.14 (5-oxo-L-proline) [16]; 0.031
(5-oxo-L-proline) [14]; 0.1 (ATP) [18]; 1.0 (ATP) [16]; 0.17 (ATP) [3, 6, 12, 14]

**pH-optimum**
7 [12]; 7.8 [18]; 7.8–8.0 [16]; 9.5 [10]; 9.7 [12]; 8.2 (component A) [2, 4]; 9.5
(component A + B) [2, 4]

**pH-range**
5.5–11.2 [12]

**Temperature optimum (°C)**

**Temperature range (°C)**

---

## 3 ENZYME STRUCTURE

**Molecular weight**
460000 (gel filtration, rat) [13]
230000 (zonal sedimentation, rat) [13]
115000 (gel electrophoresis, Pseudomonas putida, rat) [4, 13]
750000 (gel filtration, native, Pseudomonas putida, component A)
325000 (gel filtration, rat) [3, 6, 12]
51000 (gel electrophoresis, Pseudomonas putida) [2, 4]
64000 (gel electrophoresis, Pseudomonas putida, component B) [2, 4]
650000 (gel filtration, Pseudomonas putida, component B) [4]

**Subunits**
Dimer (2 × 142000, rat, gel electrophoresis) [3, 6]
Dimer (2 × 115000, rat, gel electrophoresis) [13]
Tetramer (4 × 115000, rat, gel electrophoresis) [13]
Dodecamer (6 heterodimers, (51000 + 64000), Pseudomonas putida, com-
ponent A, gel electrophoresis) [4]
Octamer (8 × 82000, Pseudomonas putida, gel electrophoresis, component
B) [4]

**Glycoprotein/Lipoprotein**
–

---

## 4 ISOLATION/PREPARATION

**Source organism**
Rat [3, 6–9, 12–14, 17, 18]; Mouse [1, 12, 15, 17]; Pig [12, 18]; Sheep [12, 18];
Human [18]; Rabbit [12]; Cow [12]; Dog [12]; Cat [12]; Wheat [10]; Pseudo-
monas sp. [16]; Pseudomonas putida [4, 2]; More [10]

### Source tissue
Kidney [3, 6–9, 12–14, 17, 18]; Spleen [18]; Liver [18]; Intestine [18]; Heart muscle [18]; Brain [18]; Germ [10]

### Localisation in source
Soluble [3, 6–9, 12–14, 17, 18]; Ciliary body [12, 18]; Choroid plexus [19]

### Purification
Rat [3, 6, 12–14, 18]; Pseudomonas sp. [16]; Pseudomonas putida [2, 4]

### Crystallization
–

### Cloned
–

### Renaturated
–

---

## 5 STABILITY

### pH

### Temperature (°C)
47 (stable in presence of 5-oxo-L-proline up to) [12]

### Oxidation

### Organic solvent

### General stability information
Dithiothreitol and 2-mercaptoethanol stabilize [2, 14]; 5-Oxo-L-proline stabilizes [2–4]

### Storage
Partially purified enzyme, 0°C, 2 months [16]; Partially purified enzyme, 15°C, several weeks [10]

---

## 6 CROSSREFERENCES TO STRUCTURE DATABANKS

### PIR/MIPS code

### Brookhaven code

## 7 LITERATURE REFERENCES

[1] Hsu, T., Meister, A.: Methods Enzymol., 113, 468–471 (1985)
[2] Seddon, A.P., Li, L., Meister, A.: Methods Enzymol., 113, 451–458 (1985) (Review)
[3] Meister, A., Griffith, O.W., Williamson, J.M.: Methods Enzymol., 113, 445–451 (1985) (Review)
[4] Seddon, A.P., Li, L., Meister, A.: J. Biol. Chem., 259 (13) , 8091–8094 (1984)
[5] Williamson, J.M., Meister, A.: J. Biol. Chem., 257 (20) , 12039–12042 (1982)
[6] Williamson, J.M., Meister, A.: J. Biol. Chem., 257 (15) , 9161–9172 (1982)
[7] Griffith, O.W., Meister, A.: J. Biol. Chem., 257 (8) , 4392–4397 (1982)
[8] Griffith, O.W., Meister, A.: J. Biol. Chem., 256 (19) , 9981–9985 (1981)
[9] Tsui, E., Yeung, A.: Experientia, 35, 1293–1295 (1979)
[10] Mazelis, M., Creveling, R.G.: Plant Physiol., 62, 798–801 (1978)
[11] Griffith, O.W., Meister, A.: Biochem. Biophys. Res. Commun., 70 (3) , 759–765 (1976)
[12] Van Der Werf, P., Griffith, O.W., Meister, A.: J. Biol. Chem., 250 (17) , 6686–6692 (1975)
[13] Wendel, A., Flügge, U.-J., Jenke, H.-S.: Hoppe-Seyler's Z. Physiol. Chem., 356, 881–885 (1975)
[14] Wendel, A., Flügge, U.-J.: Hoppe-Seyler's Z. Physiol. Chem., 356, 873–880 (1975)
[15] Van Der Werf, P., Stephani, R.A., Meister, A.: Proc. Natl. Acad. Sci. USA, 71 (4) , 1026–1029 (1974)
[16] Van Der Werf, P., Meister, A.: Biochem. Biophys. Res. Commun., 56 (1) , 90–96 (1974)
[17] Van Der Werf, P., Stephani, R.A., Orlowski, M., Meister, A.: Proc. Natl. Acad. Sci. USA, 70 (3) , 759–761 (1973)
[18] Van Der Werf, P., Orlowski, M., Meister, A.: Proc. Natl. Acad. Sci. USA, 68 (12) , 2982–2985 (1971)
[19] Tate, S.S., Ross, L.L., Meister, A.: Proc. Natl. Acad. Sci. USA, 70, 1447–1449 (1973)

# 1 NOMENCLATURE

**EC number**
3.5.2.10

**Systematic name**
Creatinine amidohydrolase

**Recommended name**
Creatininase

**Synonymes**
Creatinine hydrolase

**CAS Reg. No.**
9025-13-2

---

# 2 REACTION AND SPECIFICITY

**Catalysed reaction**
Creatinine + $H_2O$ →
→ creatine

**Reaction type**
Carboxylic acid amide hydrolysis

**Natural substrates**
Creatinine + $H_2O$

**Substrate spectrum**
1 Creatinine + $H_2O$ (r) [2, 3]
2 Glycocyamide + $H_2O$ [2]

**Product spectrum**
1 Creatine (r)
2 Glycocyamine

---

**Inhibitor(s)**
Azide [6]; Cyanide [6]; Heavy metal ions [2, 6]; $Cu^{2+}$ [2, 6]; $Mg^{2+}$ [2, 6]; EDTA [2, 6]; Sulfhydryl reagents [6]; N-Bromosuccinimide [2]; o-Phenanthroline [2]; More [2]

**Cofactor(s)/prostethic group(s)**

---

**Metal compounds/salts**
$Zn^{2+}$ (one gram atom $Zn^{2+}$ per mol of subunit) [2]; $Mn^{2+}$ (activates) [2];
$Co^{2+}$ (activates) [2]; $Mg^{2+}$ (activates) [2]; $Fe^{2+}$ (activates) [2]; $Ni^{2+}$ (activates) [2]

**Turnover number** (min$^{-1}$)

**Specific activity** (U/mg)
193 [6]; 488 [2]

**$K_m$-value** (mM)
125 (creatinine) [6]

**pH-optimum**
8.0 [1]; 8.7 (immobilized enzyme) [1]; 7.5–8.5 [4]; 7.5–8.0 [3]; 7–9 [2]; 8.3 [6]

**pH-range**

**Temperature optimum** (°C)

**Temperature range** (°C)

## 3 ENZYME STRUCTURE

**Molecular weight**
240000 (gel filtration, Arthrobacter ureafaciens) [6]
175000 (ultracentrifugal analysis, Pseudomonas putida) [2]
23000 (chemical analysis, Pseudomonas putida) [2]

**Subunits**
Octamer (8 × 31000, SDS-PAGE) [6]
Octamer (8 × 23000, SDS-PAGE) [2]

**Glycoprotein/Lipoprotein**
Glycoprotein (3.5% glucose) [2]

## 4 ISOLATION/PREPARATION

**Source organism**
Arthrobacter ureafaciens [6]; Pseudomonas putida [2, 3]; Pseudomonas aeruginosa [7]; Pseudomonas ovalis [8]

**Source tissue**
Cell [2, 3, 6–8]

**Localisation in source**

## Purification
Arthrobacter ureafaciens [6]; Pseudomonas putida [2]

## Crystallization
–

## Cloned
–

## Renaturated
–

## 5 STABILITY

### pH
5–10 [3]; 6–12 [2]; 6–10 (apoprotein) [2]

### Temperature (°C)
55 (15 minutes [6], 30 minutes, apoenzyme [2]) [2, 6]; 70 (30 minutes) [2]

### Oxidation
Unstable to photooxidation [2]

### Organic solvent

### General stability information

### Storage
Purified enzyme, –20°C, 6 months [5]; Lyophilized [3]

## 6 CROSSREFERENCES TO STRUCTURE DATABANKS

### PIR/MIPS code

### Brookhaven code

## 7 LITERATURE REFERENCES

[1] Guilbault, G.G., Chen, S.P., Kuan, S.S.: Anal. Lett., 13 (B18) , 1607–1624 (1980)
[2] Rikitake, K., Oka, I., Ando, M., Yoshimoto, T., Tsuru, D.: J. Biochem., 86, 1109–1117 (1979)
[3] Tsuru, D., Oka, I., Yoshimoto, T.: Agric. Biol. Chem., 40 (5) , 1011–1018 (1976)
[4] Wahlefeld, A.W., Holz, G., Bergmeyer, H.U.: Methods Enzymol., Vol.2, 1834–1938 (1974)
[5] Thompson, H., Rechnitz, G.A.: Anal. Chem., 46 (2) , 246–249 (1974)
[6] Kaplan, A., Szabo, L.L.: Mol. Cell. Biochem., 3 (1) , 17–25 (1974)
[7] Kopper, P.H., Beard, H.H.: Arch. Biochem. Biophys., 15, 195 (1947)
[8] Appleyard, G., Woods, D.D.: J. Gen. Microbiol., 14, 351 (1956)

# 1 NOMENCLATURE

**EC number**
3.5.2.11

**Systematic name**
L-Lysine-1, 6-lactam lactamhydrolase

**Recommended name**
L-Lysine-lactamase

**Synonymes**
L-Alpha-aminocaprolactam hydrolase
L-Lysinamidase

**CAS Reg. No.**

---

# 2 REACTION AND SPECIFICITY

**Catalysed reaction**
L-Lysine 1, 6-lactam + $H_2O$ →
→ L-lysine

**Reaction type**
Carboxylic acid amide hydrolysis

**Natural substrates**
L-Alpha-amino-epsilon-caprolactam + $H_2O$

**Substrate spectrum**
1 L-Alpha-amino-epsilon-caprolactam + $H_2O$
2 More (also hydrolyses L-lysinamide)

**Product spectrum**
1 L-Lysine
2 ?

---

**Inhibitor(s)**

**Cofactor(s)/prostethic group(s)**

**Metal compounds/salts**
$MnCl_2$ (activation) [1]; $MgCl_2$ (activation) [1]

---

**Turnover number** (min$^{-1}$)

---

**Specific activity** (U/mg)
244 [1]

**K$_m$-value** (mM)
2.6 (L-alpha-amino-epsilon-caprolactam) [1]

**pH-optimum**
9.0 [1]

**pH-range**

**Temperature optimum** (°C)

**Temperature range** (°C)

---

## 3 ENZYME STRUCTURE

**Molecular weight**
185000 (Cryptococcus laurentii, gel filtration) [1]

**Subunits**

**Glycoprotein/Lipoprotein**
–

---

## 4 ISOLATION/PREPARATION

**Source organism**
Cryptococcus laurentii [1]

**Source tissue**
Cell [1]

**Localisation in source**
Cytoplasm (soluble) [1]

**Purification**
Cryptococcus laurentii [1]

**Crystallization**
–

**Cloned**
–

**Renaturated**
–

# 5 STABILITY

pH

Temperature (°C)

Oxidation

Organic solvent

General stability information

Storage

---

# 6 CROSSREFERENCES TO STRUCTURE DATABANKS

PIR/MIPS code

Brookhaven code

---

# 7 LITERATURE REFERENCES

[1] Fukumura, T., Talbot, G., Misono, H., Teramura, Y., Kato, K., Soda, K.: FEBS Lett., 89 (2) , 298–300, (1978)

## 1 NOMENCLATURE

**EC number**
   3.5.2.12

**Systematic name**
   1, 8-Diazacyclotetradecane-2, 9-dione lactamhydrolase

**Recommended name**
   6-Aminohexanoate-cyclic-dimer hydrolase

**Synonymes**

**CAS Reg. No.**

---

## 2 REACTION AND SPECIFICITY

**Catalysed reaction**
   1, 8-Diazacyclotetradecane-2, 9-dione + $H_2O$ →
   → N-(6-aminohexanoyl)-6-aminohexanoate

**Reaction type**
   Carboxylic acid amide hydrolysis

**Natural substrates**
   1, 8-Diazacyclotetradecane-2, 9-dione + $H_2O$

**Substrate spectrum**
   1  1, 8-Diazacyclotetradecane-2, 9-dione + $H_2O$

**Product spectrum**
   1  N-(6-Aminohexanoyl)-6-aminohexanoate

---

**Inhibitor(s)**
   Diisopropylphosphofluoridate [3]; p-Chloromercuribenzoate [3]

**Cofactor(s)/prostethic group(s)**

**Metal compounds/salts**

---

**Turnover number** (min$^{-1}$)
   480 [3]

**Specific activity** (U/mg)
   2.2 [1]; 2.25 [3]

---

$K_m$-value (mM)
   6 (6-aminohexanoic acid, cyclic dimer) [3]

pH-optimum
   7.3 [1, 3]

pH-range

Temperature optimum (°C)
   33 [1, 3]

Temperature range (°C)

---

## 3 ENZYME STRUCTURE

Molecular weight
   100000 (Flavobacterium sp., gel filtration) [1]
   110000 (Acromobacter guttatus, gel filtration) [3]

Subunits
   Dimer (2 × 55000, Flavobacterium sp. [1], Achromobacter sp.,
   SDS-electrophoresis) [1, 3]

Glycoprotein/Lipoprotein
   –

---

## 4 ISOLATION/PREPARATION

Source organism
   Flavobacterium sp. KI-72 [1, 2]; Acromobacter guttatus KI-72 [3]

Source tissue
   Cell

Localisation in source
   Cytoplasm (soluble)

Purification
   Acromobacter guttatus [3]; Flavobacterium sp. KI-72 [1]

Crystallization
   –

Cloned
   –

Renaturated
   –

## 5 STABILITY

**pH**
   5.5–8.5 (several days) [3]

**Temperature (°C)**
   5 (several days) [3]

**Oxidation**

**Organic solvent**

**General stability information**
   Phosphate (0.02–0.2 M stabilizes) [3]; Glycerol (5–25% stabilizes) [3]

**Storage**

---

## 6 CROSSREFERENCES TO STRUCTURE DATABANKS

**PIR/MIPS code**
   S06849 (Flavobacterium sp., fragment)

**Brookhaven code**

---

## 7 LITERATURE REFERENCES

[1] Kinoshita, S.: Hakkokogaku Kaishi, 60 (5) , 363–375, (1982)
[2] Negoro, S., Shinagawa, H., Nakata, A., Kinoshita, S., Hatozaki, T., Okada, H.: J. Bac-
    teriol., 143 (1) , 238–245, (1980)
[3] Kinoshita, S., Negoro, S., Muramatsu, M., Bisaria, V. S., Sawada, S., Okada, H.: Eur. J.
    Biochem., 80, 489–495, (1977)

## 1 NOMENCLATURE

**EC number**
3.5.3.1

**Systematic name**
L-Arginine amidinohydrolase

**Recommended name**
Arginase

**Synonymes**
Arginine amidinase
Canavanase
L-Arginase
Arginine transamidinase

**CAS Reg. No.**
9000-96-8

---

## 2 REACTION AND SPECIFICITY

**Catalysed reaction**
L-Arginine + $H_2O$ →
→ L-ornithine + urea

**Reaction type**
Amidine hydrolysis

**Natural substrates**
L-Arginine + $H_2O$ [10]

**Substrate spectrum**
1 L-Arginine + $H_2O$
2 Argininic acid + $H_2O$ [1]
3 L-Arginines (alpha-N-substituted) + $H_2O$
4 Canavanine + $H_2O$ [8, 11]

**Product spectrum**
1 L-Ornithine + urea
2 ?
3 ?
4 Urea + $NH_2OCH_2CH_2CH(NH_2)COOH$

---

## Inhibitor(s)

$Zn^{2+}$ [1]; $Ag^+$ [1]; $Hg^{2+}$ [1, 8]; Adenosine [20]; Inosine [20]; Citrate (buffer) [1]; Borate (buffer) [1, 8]; EDTA [7]; L-Canavanine [8]; DL-Homocysteine [8]; L-Homoarginine [8]; L-Lysine [1, 8, 9]; L-Proline [8]; L-Isoleucine [8]; L-Methionine [8]; L-Leucine [8]; L-Valine [8]; NaCl [8]; L-Glutamate [8]; $Fe^{2+}$ [8]; $Co^{2+}$ [8]; $Cd^{2+}$ [8]; $Ni^{2+}$ [8]; $Na_2AsO_4$ [9]; L-Tryptophan [14]; Ornithine (no effect [9]) [1, 8, 15]; Indospicine [15]; Urea [21]; Citrulline [15]

## Cofactor(s)/prosthetic group(s)

## Metal compounds/salts

$Mn^{2+}$ (metalloenzyme binding 4 moles $Mn^{2+}$ per mole enzyme [6], no effect on rat kidney enzyme [8]) [1, 6, 9, 12–17, 20, 23]; $Co^{2+}$ [1, 9, 23]

---

## Turnover number ($min^{-1}$)

132000 [1]

## Specific activity (U/mg)

5.8 [6]; 23.9 [8]; More [9–15, 19, 21, 24]

## $K_m$-value (mM)

13.5 [9]; 50 [11]; More [8, 12–16, 18, 19, 21–24]

## pH-optimum

10 (with $Mn^{2+}$) [4, 9]; 7 (with $Co^{2+}$ and $Ni^{2+}$) [1]; 10.5 [12]; 8.5 [13]; More [15, 17, 20, 23]

## pH-range

6.5–11 [4]; 6.5–12.5 [9]; More [16]

## Temperature optimum (°C)

60 [12]

## Temperature range (°C)

---

# 3 ENZYME STRUCTURE

## Molecular weight

138000 (sedimentation analysis) [3]
115000 (sedimentation analysis, amino acid analysis, ox) [6]
110000 (gel filtration, rabbit) [7]
138000 (gel filtration, Glycine hispida) [11]
More [12–17, 20–24]

## Subunits

Oligomer (rabbit [7], Pista pacifica [12], rat [14], human [15]) [7, 12, 14, 15]
Tetramer (4 × 30800, rat) [10]
Hexamer (6 × 36500, Iris hollandica) [17]
Tetramer (4 × 37000, human) [13]
More [18, 22]

## Glycoprotein/Lipoprotein

Glycoprotein (280 glucose residues, 27 fructose residues, 85 mannose
residues per molecule) [24]

## 4 ISOLATION/PREPARATION

### Source organism

Chicken [5]; Ox [6]; Rabbit [7]; Rat [8, 10, 14, 18]; Bacillus subtilis [9];
Glycine hispida (var. Cheepewa) [11]; Iris hollandica [17]; Pista pacifica
[12]; Artichoke [16]; Evernia prunastri [21, 24]; Canavalia ensiformis [19];
Neurospora crassa (multiple forms) [22]; Human (2 isoenzymes [23]) [13,
15, 23]; More [20, 23]

### Source tissue

Liver [5, 6, 7, 8, 15, 18, 20]; Small intestine [14]; Tuber [16]; Thallus [24];
Lymphocytes [13]; Granulocytes [13]; Erythrocytes [15]; Bulb [17]; Kidney
[8]

### Localisation in source

Mitochondria [8, 19]

### Purification

Ox [6]; Rabbit [7]; Rat [8, 14, 18]; Bacillus subtilis [9]; Glycine hispida [11];
Human [13, 15]; Artichoke [16]; Iris hollandica [17]; Canavalia ensiformis
[19]; Evernia prunastri [21, 24]; Neurospora crassa [22]

### Crystallization

[2, 9, 11]

### Cloned

—

### Renaturated

[18]

## 5 STABILITY

**pH**

6–9 (below pH 6.0 and above 9.0 rapid loss of activity) [1]; 9–12 (after 60 minutes incubation at 50°C no loss of activity) [9]

**Temperature (°C)**

60 (no loss of activity after 10 minutes) [8]; 0–4 (no loss of activity after 12 days) [8]; 55 (no loss of activity after 60 minutes below 55°C) [9]; More (metal ions protect from heat inactivation [1], $MnCl_2$ or L-threonine protect from heat inactivation [9], frozen: rapid loss of activity [8]) [1, 8, 9, 14, 23]

**Oxidation**

**Organic solvent**

**General stability information**

Unstable at high dilution (ornithine or glycine stabilize) [1]; At pH 2 or with SDS at pH 10 in absence of $Mn^{2+}$, inactivation due to dissociation into subunits [7]; L-Valine (25 mM, stabilizes during purification) [14]; More [8]

**Storage**

Aqueous solutions, pH 7, 4°C [1]; –10°C or 4°C (unstable) [14]; –20°C, for at least 3 months [20]

---

## 6 CROSSREFERENCES TO STRUCTURE DATABANKS

**PIR/MIPS code**

WZBYR (yeast, Saccharomyces cerevisiae); PX0031 (rat, fragment); S06118 (Agrobacterium tumefaciens, plasmid C58); A26370 (hepatic, human); S02132 (hepatic, human); B26370 (hepatic, rat); A28358 (hepatic, rat); A26702 (hepatic, rat)

**Brookhaven code**

---

## 7 LITERATURE REFERENCES

[1] Greenberg, D.M. in "The Enzymes", 2nd. Ed. (Boyer, P.D., Ed.) Vol.4, 257–267 (1960)
[2] Bach, S.J., Killip, J.D.: Biochim. Biophys. Acta, 29, 273 (1958)
[3] Greenberg, D.M., Bagot A.E., Roholt, O.A.: Arch. Biochem. Biophys., 62, 446–453 (1956)
[4] Roholt, O.A., Greenberg, D.M.: Arch. Biochem. Biophys. , 62, 444 (1956)
[5] Grazi, E., Magri, E.: Biochem. J., 126, 667–674 (1972)
[6] Harell, D., Sokolovsky, M.: Eur. J. Biochem., 25, 102–108 (1972)
[7] Vielle-Breitburd, F., Orth, G.: J. Biol. Chem., 247 (4) , 1227–1235 (1972)
[8] Kaysen, G.A., Strecker, H.J.: Biochem. J., 133, 779–788 (1973)

[9] Nakamura, N., Fujita, M., Kimura, K.: Agric. Biol. Chem. , 37 (12) , 2827–2833 (1973)
[10] Ratner, S.: Adv. Enzymol. Relat. Areas Mol. Biol., 39, 1–90 (1973)
[11] Dumitru, J.F.: Acta Vitaminol. Enzymol., 27, 207–210 (1973)
[12] O'Malley, K.L., Terwilliger, R.C.: Biochem. J., 143, 591–597 (1974)
[13] Reyero, C., Dorner, F.: Eur. J. Biochem., 56, 137–147 (1975)
[14] Fujimoto, M., Kameji, T., Kanaya, A., Hagihira, H.: J. Biochem., 79, 441–449 (1976)
[15] Berüter, J., Colombo, J., Bachmann, C.: Biochem. J., 175, 449–454 (1978)
[16] Wright, L.C., Brady, C.J., Hinde, R.W.: Phytochemistry, 20 (12) , 2641–2645 (1981)
[17] Boutin, J.: Eur. J. Biochem., 127, 237–243 (1982)
[18] Aguirre, R., Kasche, V.: Eur. J. Biochem., 130, 373–381 (1983)
[19] Downun, K.R., Rosenthal, G.A., Cohen, W.S.: Plant Physiol., 73, 965–968 (1983)
[20] Colombo, J., Konarska, L. in "Methods Enzym. Anal.", 3rd. Ed. (Bergmeyer, H.U., Ed.) Vol.4, 285–294 (1984)
[21] Martín-Falquina, Legaz, M.E.: Plant Physiol., 76, 1065–1069 (1984)
[22] Borkovich, K.A., Weiss, R.L.: J. Biol. Chem., 262 (15) , 7081–7086 (1987)
[23] Grody, W.W., Dizikes, G.J., Cederbaum, S.D.: Isozymes Curr. Top. Biol. Med. Res., 13, 181–214 (1987)
[24] Planelles, V., Legaz, M.E.: Plant Sci., 51, 9–16 (1987)

Enzyme Handbook © Springer-Verlag Berlin Heidelberg 1991
Duplication, reproduction and storage in data banks are only
allowed with the prior permission of the publishers

# 1 NOMENCLATURE

**EC number**
3.5.3.2

**Systematic name**
Guanidinoacetate amidinohydrolase

**Recommended name**
Glycocyaminase

**Synonymes**

**CAS Reg. No.**
9024-92-4

---

# 2 REACTION AND SPECIFICITY

**Catalysed reaction**
Guanidinoacetate + $H_2O$ →
→ glycine + urea

**Reaction type**
Amidine hydrolysis

**Natural substrates**
Guanidinoacetate + $H_2O$

**Substrate spectrum**
1 Guanidinoacetate + $H_2O$ (ir) [1, 2, 5]

**Product spectrum**
1 Glycine + urea (ir) [1, 2, 5]

---

**Inhibitor(s)**
$Cu^{2+}$ [1]; Diethylthiocarbamate [1]; $N_3Na$ [1]; KCN [1]; $Fe^{2+}$ [2]; $Co^{2+}$ [2]; $Ni^{2+}$ [2]; p-Chloromercuribenzoate [2, 8]; ATP [2]; ADP [2]; $Zn^{2+}$ [2]; Sulfhydryl reagents [2]; 1, 10-Phenanthroline [2]; $J^-$ [8]; Glycine [8]; n-Alkylamine [8]; Metal ions [8]

**Cofactor(s)/prostethic group(s)**

**Metal compounds/salts**
$Mn^{2+}$ [1, 2]; $Zn^{2+}$ [4]; $Co^{2+}$ [4]

---

**Turnover number** (min$^{-1}$)
 3245 [2]; 3610 [4]

**Specific activity** (U/mg)
 63.7 [2]; 190 [4]

**K$_m$-value** (mM)
 9.1 (guanidinoacetate) [2]; 0.0013 (Mn$^{2+}$) [2]; 15 (guanidinoacetate) [4]; 16 (guanidinoacetate) [8]

**pH-optimum**
 7.8 [1]; 8–8.5 [4]; 9.0–9.5 [2, 8]

**pH-range**
 9–10 [2]; 5.5–9.5 [4]

**Temperature optimum** (°C)
 30 [2, 4]

**Temperature range** (°C)

---

## 3 ENZYME STRUCTURE

**Molecular weight**
 160000 (Pseudomonas sp., gel filtration) [2]
 281000 (Flavobacterium sp., gel electrophoresis) [4]
 150000 (Corynebacterium sp., gel filtration) [8]

**Subunits**
 Tetramer (4 × 38000, Pseudomonas sp., SDS-PAGE) [2]
 Tetramer (4 × 70000, Flavobacterium sp., SDS-PAGE) [4]

**Glycoprotein/Lipoprotein**
 –

---

## 4 ISOLATION/PREPARATION

**Source organism**
 Pseudomonas ovalis [1]; Pseudomonas eisenbergii [1]; Pseudomonas putida [7]; Penicillium roqueforti [2]; Pseudomonas sp. ATCC 14676 [2, 3, 6, 9]; Flavobacterium sp. GE-1 [4]; Corynebacterium sp. [8]

**Source tissue**
 Cell

**Localisation in source**
 Cytoplasm

## Purification
Pseudomonas sp. ATCC 14676 [2, 8]; Corynebacterium sp. [8]; Flavobacterium sp. GE-1 [4]

## Crystallization
[2, 4]

## Cloned
–

## Renaturated

---

## 5 STABILITY

**pH**
6.0–10.5 [8]

**Temperature** (°C)
–15 [2]

**Oxidation**

**Organic solvent**

**General stability information**

**Storage**
–15°C [2]; 4°C (2-mercaptoethanol) [2]

---

## 6 CROSSREFERENCES TO STRUCTURE DATABANKS

**PIR/MIPS code**

**Brookhaven code**

---

## 7 LITERATURE REFERENCES

[1] Roche, J., Lacombe, G.: Biochim. Biophys. Acta, 6, 210–216 (1950)
[2] Yorifuji, T., Tamai, H., Usami, H.: Agric. Biol. Chem. , 41 (6) , 959–966 (1977)
[3] Tamai, H., Usami, H., Yorifuji, T.: Agric. Biol. Chem. , 42 (6) , 1295–1296 (1978)
[4] Yorifuji, T., Komaki, N., Oketani, K., Entani, E.: Agric. Biol. Chem., 43 (1) , 55–62 (1979)
[5] Shirokane, Y., Utsushikawa, M., Nakajima, M.: Clin. Chem., 33 (3) , 394–397 (1987)
[6] Yorifuji, T., Shiritani, Y.: Agric. Biol. Chem., 46 (1) , 317–318 (1982)
[7] Yorifuji, T., Kobayashi, T., Tabuchi, A., Shiritani, Y., Yonaha, K.: Agric. Biol. Chem., 47 (12) , 2825–2830 (1983)
[8] Shirokane, Y., Nakayima, M.: J. Ferment. Technol., 64 (1) , 29–36 (1986)
[9] Yorifuji, T., Shiritani, Y., Eguchi, S., Yonaha, K.: J. Appl. Biochem., 5, 375–381 (1983)

---

# 1 NOMENCLATURE

**EC number**
3.5.3.3

**Systematic name**
Creatine amidinohydrolase

**Recommended name**
Creatinase

**Synonymes**
Creatinase (Boeringer Toyo Jozo Co.)
Creatine amidohydrolase (Toyobo Co., Noda)
Creatine hydrolases (Eiken chemical Co.)

**CAS Reg. No.**
37340-58-2

---

# 2 REACTION AND SPECIFICITY

**Catalysed reaction**
Creatine + $H_2O \rightarrow$
$\rightarrow$ sarcosine + urea

**Reaction type**
Amidine hydrolysis

**Natural substrates**
Creatine + $H_2O$ (microbial decomposition of creatinine [12–15]) [1–15]

**Substrate spectrum**
1 Creatine + $H_2O$ (ir)

**Product spectrum**
1 Sarcosine + urea (ir)

---

**Inhibitor(s)**
NaN$_3$ [1]; KCN [1]; Diethylthiocarbamate [1]; N-Bromosuccinimide [11, 14];
Zn$^{2+}$ [11, 14]; Hg$^{2+}$ [14, 11]; Diisopropylfluorophosphate [14]; EDTA [14];
CoCl$_2$ [14]; FeCl$_2$ [14]; Pb-Acetate [14]; MgSO$_4$ [14]; Na-Tetrathionate [14];
Dithiothreitol [14]; Cu$^{2+}$ [1, 11, 14]; p-Chloromercuribenzoate [11, 14]

---

**Cofactor(s)/prostethic group(s)**

**Metal compounds/salts**
$Fe^{2+}$ (activates) [1]

---

**Turnover number** (min$^{-1}$)
105000 [10]

**Specific activity** (U/mg)
5.8 [6]; 11.3 [10]; 303 [10]; 15.3 [3]; 90 [7]; 0.042–0.101 [2]

**$K_m$-value** (mM)
17.2 (creatine) [11]; 1.33 (creatine) [14]

**pH-optimum**
7.0–9.0 [3]; 8.0 [9, 11, 14]; 6.7 [2]

**pH-range**
6.5–9.0 [11]

**Temperature optimum** (°C)
35–45 [3]; 30 [14]

**Temperature range** (°C)
35–45 [3]

---

## 3 ENZYME STRUCTURE

**Molecular weight**
51000 (Alcaligenes sp., SDS-gel electrophoresis) [10]
94000 (Pseudomonas putida, meniscus depletion method) [14, 15]

**Subunits**
Monomer (Alcaligenes sp.) [11]
Dimer (2 × 47000, Pseudomonas putida, SDS-gel electrophoresis) [14, 15]

**Glycoprotein/Lipoprotein**
–

---

## 4 ISOLATION/PREPARATION

**Source organism**
Acinetobacter CRH-1Q4D [2, 3]; Bacillus B-068 [6]; Alcaligenes AK-2 [7];
Flavobacterium [8]; Corynebacterium [8]; Pseudomonas sp. [9, 13]; Pseudomonas putida [13, 14, 15]; Alcaligenes sp. WS 51400 [10]; Pseudomonas
eisenbergii [1]; Alcaligenes denitrificans [12]; Arthrobacter spp J5, J9 [12]

**Source tissue**
   Cell

**Localisation in source**
   Cytoplasm [1]; Extracellular [11]

**Purification**
   Acinetobacter CRH-104D [2]; Bacillus B-068 [6]; Alcaligenes AK-2 [7];
   Flavobacterium [8]; Corynebacterium [8]; Pseudomonas sp. [9]; Alcaligenes
   sp. [10, 11]; Pseudomonas putida [14, 15]

**Crystallization**
   [14, 15]

**Cloned**
   –

**Renaturated**
   –

---

## 5 STABILITY

**pH**
   5–10 [9]; 5–9 [11]; 4 (labile) [11]

**Temperature (°C)**
   40 [9]; 45 (50% loss of activity after 30 minutes) [15]; 55 (completely
   destroyed) [16]

**Oxidation**

**Organic solvent**

**General stability information**

**Storage**
   At 30°C stable for 4 weeks, with gluconic acid [5]; Lyophilized storable [11]

---

## 6 CROSSREFERENCES TO STRUCTURE DATABANKS

**PIR/MIPS code**

**Brookhaven code**

# 7 LITERATURE REFERENCES

[1] Roche, J., Lacombe, G.: Biochim. Biophys. Acta, 6, 210–216 (1950)
[2] Kikuchi, T., Takenada, H., Aisui, S.: Jpn. Kokai Tokkyo Koho JP61/67485 A2 [86/674853], 6 Pp., 7 Apr. (1986)
[3] Kikuchi, T., Takenaka, H., Aisui, S.: Jpn. Kokai Tokkyo Koho JP61/67486 A2 [86/67486], 12 Pp., 7 Apr. (1986)
[4] Ito, K.: Jpn. Kokai Tokkyo Koho JP59/85290 A2 [84/85290], , 4 Pp., 17 May (1984)
[5] Nunokawa, Y.: Jpn. Kokai Tokkyo Koho JP57/36985 A2 [82/36985] , 4 Pp., 27 Feb. (1982)
[6] Ikuta, S., Matsuura, K., Misaki, H.: Ger. Offen. DE3024915, 19 Pp., 22 Jan. (1981)
[7] Nunokawa, Y.: Jpn. Kokai Tokkyo Koho JP55/34029 [80/34029], 6 Pp., 10 Mar. (1980)
[8] Suzuki, M., Saito, N.: Ger. Offen. DE2659878, 28 Pp., 10 Nov., Division Of Ger. Offen.2, 614, 114 (1977)
[9] Tsuru, D., Yoshimoto, T., Oka, I.: Japan Kokai JP52/61292 [77/61292], 5 Pp., 20 May (1977)
[10] Moellering, H., Beaucamp, K., Nelboek-Hochstetter, M. , Bergmeyer, H.U.: Ger. Offen. DE2122298, 16 Pp., 23 Nov. (1972)
[11] Matsuda, Y., Wakamutse, N., Inouye, Y., Uede, S., Hashimoto, Y., Asano, K., Nakamura, S.: Chem. Pharm. Bull., 34 (5) , 2155–2160 (1986)
[12] Min Kim, J., Shimizu, S., Yamada, H.: Agric. Biol. Chem., 50 (11) , 2811–2816 (1986)
[13] Shimizu, S., Min Kim , J., Shinmen, Y., Yamada, H.: Arch. Microbiol., 145, 322–328 (1986)
[14] Yoshimoto, T., Oka, I., Tsuru, D.: Arch. Biochem. Biophys., 177, 508–515 (1976)
[15] Yoshimoto, T., Oka, I., Tsuru, D.: J. Biochem., 79, 1381–1383 (1976)
[16] Kaplan, A., Naugler, A.: Mol. Cell. Biochem., 3 (3) , 9–15 (1974)

## 1 NOMENCLATURE

**EC number**
3.5.3.4

**Systematic name**
Allantoate amidinohydrolase

**Recommended name**
Allantoicase

**Synonymes**

**CAS Reg. No.**
9025-21-2

## 2 REACTION AND SPECIFICITY

**Catalysed reaction**
Allantoate + $H_2O$ →
→ (-)-ureidoglycolate + urea

**Reaction type**
Amidine hydrolysis

**Natural substrates**
Allantoate + $H_2O$ [3]

**Substrate spectrum**
1 Allantoate + $H_2O$
2 (+)-Ureidoglycolate + $H_2O$

**Product spectrum**
1 (-)-Ureidoglycolate + urea
2 Glyoxylate + urea

**Inhibitor(s)**
$Hg^{2+}$ [1]; $Cu^{2+}$ [1]; (-)Ureidoglycolate [1]; N-Carbamoyl-(R)-asparagine [3]; Glycolic acid [3]; Hydantoic acid [3]; Phosphate [7]; L-Asparagine [7]

**Cofactor(s)/prostethic group(s)**

**Metal compounds/salts**
$Mg^{2+}$ [2, 3]; $Cd^{2+}$ [2]; Bivalent cations [2]

Turnover number (min$^{-1}$)

Specific activity (U/mg)
177 [2]; More [8, 10]

K$_m$-value (mM)
More [1]; 10 [7]

pH-optimum
7.2 (Pseudomonas aeruginosa, Penicillium) [1]; 6.0 (Pseudomonas fluorescens) [1]; 8.0 [2]; 7.5 [7]

pH-range
4.5–9 [2]; 5.5–9 [7]

Temperature optimum (°C)
40 [7]

Temperature range (°C)
10–60 [7]

---

## 3 ENZYME STRUCTURE

Molecular weight
11000 (sedimentation diffusion equilibrium, 0.9-S-allantoicase, Pseudomonas aeruginosa) [3]
154000 (sedimentation, 10.8-S-allantoicase, Pseudomonas aeruginosa) [3]

Subunits

Glycoprotein/Lipoprotein
–

---

## 4 ISOLATION/PREPARATION

Source organism
Pseudomonas aeruginosa [1–4]; Pseudomonas fluorescens [1]; Penicillium
citreo-viride [1]; Penicillium notatum [1]; Fish [5]; Crustaceae [5];
Saccharomyces cerevisiae [6]; Streptomyces viridiviolaceus [7, 8];
Enterobacter cloacae [9]; Mackerel [5]; Frog (enzyme also has allantoinase
activity, E.C. 3.5.2.5) [10]; Mycobacterium intracellulare [11]; Mycobacterium
scrofulaceum [11]; Rhodopseudomonas capsulata [12]

Source tissue
Hepatopancreas [5]; Liver [5, 10]

Localisation in source
Peroxisomes (membrane) [5]

2

## Purification
Pseudomonas aeruginosa [2, 3]; Mackerel [5]; Frog [10], Streptomyces viridoviolaceus [8]

## Crystallization
–

## Cloned
–

## Renaturated
–

---

## 5 STABILITY

### pH

### Temperature (°C)
75 (in presence of $Mn^{2+}$ stable up to 75°C, in absence of $Mn^{2+}$ completely inactivated within 5 minutes) [2]; 50 (34% loss of activity after 30 minutes, 79% loss of activity after 90 minutes) [7]

### Oxidation

### Organic solvent

### General stability information
Freezing and thawing (stable) [7]

### Storage
–15°C, 6 months, 0.02M Tris-HCl buffer, pH 7.5 [7]

---

## 6 CROSSREFERENCES TO STRUCTURE DATABANKS

### PIR/MIPS code

### Brookhaven code

# 7 LITERATURE REFERENCES

[1] Tribjels, F., Vogels, G.D.: Biochim. Biophys. Acta, 118, 387–395 (1966)

[2] Tribjels, F., Vogels, G.D.: Biochim. Biophys. ACTA, 132, 115–126 (1967)

[3] 'S-Gravenmade, E.J., Van Der Drift, C., Vogels, G.D.: Biochim. Biophys. Acta, 251, 393–406 (1971)

[4] Rijnierse, V.F.M., Van Der Drift, C., Vogels, G.D.: Can. J. Microbiol., 23, 633–637 (1977)

[5] Nogucchi, T., Takada, Y., Fujiwara, S.: J. Biol. Chem., 254 (12), 5272–5275 (1979)

[6] Zacharski, C.A., Cooper, T.G.: J. Bacteriol., 135 (2), 490–497 (1978)

[7] Elzainy, T.A., Elawamry, Z.A.: Egypt. J. Bot., 23 (3), 137–144 (1980)

[8] Zeinab, A.E., Elzainy, T.A.: Egypt. J. Bot., 23 (3), 131–136 (1980)

[9] Machida, Y., Nakanishi, T.: Agric. Biol. Chem., 46 (8), 2135–2136 (1982)

[10] Takada, Y., Nogucchi, T.: J. Biol. Chem., 258 (8), 4762–4764 (1983)

[11] Falkinham, J.O., George, K.L., Parker, B.C., Gruft, H.: J. Bacteriol., 155 (1), 36–39 (1983)

[12] Kaspari, H., Busse, W.: Arch. Microbiol., 144, 67–70 (1986)

## 1 NOMENCLATURE

**EC number**
3.5.3.5

**Systematic name**
N-Formimino-L-aspartate iminohydrolase

**Recommended name**
Formiminoaspartate deiminase

**Synonymes**
Deiminase, formiminoaspartate

**CAS Reg. No.**
9025-07-4

## 2 REACTION AND SPECIFICITY

**Catalysed reaction**
N-Formimino-L-aspartate + $H_2O$ →
→ N-formyl-L-aspartate + $NH_3$

**Reaction type**
Amidine hydrolysis

**Natural substrates**
N-Formimino-L-aspartate + $H_2O$

**Substrate spectrum**
1 N-Formimino-L-aspartate + $H_2O$

**Product spectrum**
1 N-Formyl-L-aspartate + $NH_3$

**Inhibitor(s)**

**Cofactor(s)/prostethic group(s)**

**Metal compounds/salts**

**Turnover number** (min$^{-1}$)

**Specific activity** (U/mg)

**$K_m$-value** (mM)

pH-optimum

pH-range

Temperature optimum (°C)

Temperature range (°C)

---

# 3 ENZYME STRUCTURE

Molecular weight

Subunits

Glycoprotein/Lipoprotein
–

---

# 4 ISOLATION/PREPARATION

Source organism
  Bacteria [1]

Source tissue

Localisation in source

Purification

Crystallization
–

Cloned
–

Renaturated
–

---

# 5 STABILITY

pH

Temperature (°C)

Oxidation

Organic solvent

General stability information

Storage

# 6 CROSSREFERENCES TO STRUCTURE DATABANKS

PIR/MIPS code

Brookhaven code

# 7 LITERATURE REFERENCES

[1] Elödi, P., Szörenyi, E.: Acta Physiol. Acad. Sci. Hung., 9, 367–379 (1956)

# 1 NOMENCLATURE

**EC number**
3.5.3.6

**Systematic name**
L-Arginine iminohydrolase

**Recommended name**
Arginine deiminase

**Synonymes**
Arginine dihydrolase
Citrulline iminase
L-Arginine deiminase
Deiminase, arginine

**CAS Reg. No.**
9027-98-6

# 2 REACTION AND SPECIFICITY

**Catalysed reaction**
L-Arginine + $H_2O$ →
→ L-citrulline + $NH_3$

**Reaction type**
Amidine hydrolysis

**Natural substrates**
L-Arginine + $H_2O$ [3]

**Substrate spectrum**
1 L-Arginine + $H_2O$
2 Guanidine (derivatives) + $H_2O$ [8]
3 Canavanine + $H_2O$ [8]
4 More [8]

**Product spectrum**
1 L-Citrulline + $NH_3$
2 Ureido derivatives + $NH_3$
3 ?
4 ?

## Inhibitor(s)

L-Homoarginine [1]; L-Alpha-amino-beta-guanidinopropionic acid [1];
p-Chloromercuribenzoate [1]; $Ag^+$ [1]; $Hg^{2+}$ [1]; $Al^{3+}$ [1]; $Fe^{3+}$ [1]; $Fe^{2+}$ [1];
$Cu^{2+}$ [1]; $Zn^{2+}$ [1]; $Sn^{2+}$ [1]; Formamidinium ion [6]; D-Arginine [10];
L-Homoarginine [10]; Citrulline [10]; Ornithine [10];
5,5-Dithiobis-(2-nitrobenzoate) [10]; N-Ethylmaleimide [10]; Mersaryl [10];
L-Arginine-HCl [11]; L-Ornithine [11]; p-Hydroxymercuribenzoate [11];
L-Alpha-amino-gamma-guanidinobutyric acid [1]; More (competitive inhibitors) [8]

## Cofactor(s)/prostethic group(s)

## Metal compounds/salts

$Co^{2+}$ [10]; $Mg^{2+}$ [10]

---

## Turnover number (min⁻¹)

$\text{Turnover number (min}^{-1}\text{)}$

1740 [8]

## Specific activity (U/mg)

58.8 [1]; 57.05 [2]; More [4, 10, 11, 12, 13]

## $K_m$-value (mM)

0.166 (crude extract, L-arginine) [3]; 0.004 (L-arginine) [4];
More [1, 7, 8, 10, 11]

## pH-optimum

9.7–10.3 [10]; 6.0 [1, 11]

## pH-range

4.5–8 [1]; 8.5–11 [10]

## Temperature optimum (°C)

50 [1]; 30 [10]

## Temperature range (°C)

30–60 [1]; 10–40 [10]

---

## 3 ENZYME STRUCTURE

## Molecular weight

130000 (Pseudomonas putida, gel filtration) [1]
120000 (Pseudomonas putida, sedimentation equilibrium measurement)
[1, 2]
80000 (Mycoplasma arthriditis, gel filtration) [4]
More [7, 9, 10, 11, 13]

## Subunits
Dimer (2 × 44000, Mycoplasma arthriditis [6], 2 × 48000, Euglena gracilis [10], 2 × 54000, Pseudomonas putida, SDS-PAGE [1]) [1, 4, 6, 9, 10]

## Glycoprotein/Lipoprotein
–

---

# 4 ISOLATION/PREPARATION

## Source organism
Pseudomonas putida [1, 2, 9]; Mycoplasma hominis [3, 13]; Mycoplasma arthriditis (3 forms) [4–8]; Eulena gracilis [10]; Lactobacillus buchneri [11]; Clostridium perfringens [12]; Streptococcus mitis [14]

## Source tissue

## Localisation in source
Mitochondrial matrix [10]; Membrane [13]; Cytoplasm [13]; Cell wall [14]

## Purification
Pseudomonas putida [1, 2]; Mycoplasma arthriditis [4]; Euglena gracilis [10]; Lactobacillus buchneri [11]; Mycoplasma hominis [13]

## Crystallization
[1]

## Cloned
–

## Renaturated
–

---

# 5 STABILITY

## pH
2.5 (loss of activity) [7]

## Temperature (°C)
40 (up to, 15 minutes) [10]; 50 (39% loss of activity, 15 minutes) [10]

## Oxidation

## Organic solvent

## General stability information

## Storage

---

# 6 CROSSREFERENCES TO STRUCTURE DATABANKS

**PIR/MIPS code**
  S02138 (Pseudomonas aeruginosa)

**Brookhaven code**

# 7 LITERATURE REFERENCES

[1] Shibatani, T., Kakimoto, T., Chibata, I.: J. Biol. Chem., 250 (12), 4580–4583 (1975)
[2] Kakimoto, T., Shibatani, T., Chibata, I.: FEBS Lett., 19 (2), 166–168 (1971)
[3] Fenske, J.D., Kenny, G.E.: J. Bacteriol., 126 (1), 501–510 (1976)
[4] Weickann, J.L., Fahrney, D.E.: J. Biol. Chem., 252 (8), 2615–2620 (1977)
[5] Smith, D.W., Fahrney, D.E.: Biochem. Biophys. Res. Commun., 83 (1), 101–106 (1978)
[6] Weickmann, J.L., Himmel, M.E., Smith, D.W., Fahrney, D.E.: Biochem. Biophys. Res. Commun., 83 (1), 107–113 (1978)
[7] Weickmann, J.L., Himmel, M.E., Squire, P.G., Fahrney, D.E: J. Biol. Chem., 253 (17), 6010–6015 (1978)
[8] Smith, D.W., Ganaway, R.L., Fahrney, D.E.: J. Biol. Chem., 6016–6020 (1978)
[9] Shibatani, T., Kakimoto, T., Chibata, I.: FEBS Lett., 96 (2), 389–391 (1978)
[10] Park, B., Hirotani, A., Nakano, Y., Kitaoka, S.: Agric. Biol. Chem., 48 (2), 483–489 (1984)
[11] Manca De Nadra, M.C., Pesce De Ruiz Holgado, A.A., Oliver, G.: J. Appl. Biochem., 6, 184–187 (1984)
[12] Sacks, L.E.: Experientia, 41, 1435–1437 (1984)
[13] Lin, J.L.: J. Gen. Microbiol., 132, 1467–1474 (1986)
[14] Hiraoka, B.Y., Harada, M., Fukasawa, K., Mogi, M.: Curr. Microbiol., 15, 81–84 (1987)

## 1 NOMENCLATURE

**EC number**
   3.5.3.7

**Systematic name**
   4-Guanidinobutanoate amidinohydrolase

**Recommended name**
   Guanidinobutyrase

**Synonymes**
   Gamma-guanidobutyrase
   4-Guanidinobutyrate amidinobutyrase
   Gamma-guanidinobutyrate amidinohydrolase [3]
   G-Base [4, 5]
   GBH [6, 7]
   Guanidinobutyrate ureahydrolase [1]

**CAS Reg. No.**
   9013-69-8

---

## 2 REACTION AND SPECIFICITY

**Catalysed reaction**
   4-Guanidinobutanoate + $H_2O \rightarrow$
   $\rightarrow$ 4-aminobutanoate + urea

**Reaction type**
   Amidine hydrolysis

**Natural substrates**
   4-Guanidinobutyrate + $H_2O$ [3]

**Substrate spectrum**
   1 4-Guanidinobutyrate + $H_2O$ [3]
   2 5-Guanidinovalerate + $H_2O$ [5]
   3 6-Guanidinocaproate + $H_2O$ [5]
   4 Beta-guanidinopropionate + $H_2O$ [2, 3, 5, 6]
   5 L-Alpha-amino-gamma-guanidinobutyrate + $H_2O$ [3]
   6 L-/D-Arginine + $H_2O$ [2, 3, 5]
   7 L-Homoarginine + $H_2O$ [3]
   8 Agmatine + $H_2O$ [3, 5]

9 Creatine + $H_2O$ [4, 5]
10 N-Amidinoalanine + $H_2O$ [4, 5]
11 Guanidine + $H_2O$ [4]
12 3-Guanidino-n-butyrate + $H_2O$ [5]
13 Taurocyamine + $H_2O$ [5, 6]
14 D-/L-N-Acetylarginine + $H_2O$ [5]
15 Canavanine + $H_2O$ [5]
16 Methylguanidine + $H_2O$ [5]
17 Guanylurea + $H_2O$ [5]
18 More [1]

## Product spectrum

1 4-Aminobutyrate + urea [3]
2 5-Aminovalerate + urea [5]
3 6-Aminocaproate + urea [5]
4 3-Aminopropanoate + urea
5 2,4-Diaminobutyrate + urea
6 ?
7 ?
8 $NH_2(CH_2)_4NH_2$ + urea (i.e. diaminobutane + urea)
9 $CH_3$-NH-$CH_2$-COOH + urea
10 ?
11 ?
12 ?
13 ?
14 ?
15 $NH_2OCH_2$-CH($NH_2$)COOH + urea (asparagine + urea)
16 ?
17 ?
18 ?

## Inhibitor(s)

EDTA [5]; p-Chloromercuribenzoate (reactivation possible with 2-mercaptoethanol) [5]; Propionate (competitive) [5, 7]; n-Butyrate (competitive) [5, 7]; Trans-crotonate (competitive) [7]; 4-Aminobutyrate (competitive) [7]; More (no inhibition: D/L-lactate, DL-2-hydroxybutyrate, 4-hydroxybutyrate) [7]

## Cofactor(s)/prostethic group(s)

## Metal compounds/salts

$Mn^{2+}$ (activator [1, 3], restores activity of inactivated enzyme [5]) [1, 3, 5]; $K_3PO_4$ (stabilizes, pH 7.5–10) [5]; NaCl (stabilizes, pH 7.5–10) [5]; More (not activated by $Co^{2+}$ or $Cd^{2+}$, Helix pomatia) [3]

**Turnover number** (min[-1])

**Specific activity** (U/mg)
303 [3]; 1.06 [4]; 252 [5]; 910 [8]

**$K_m$-value** (mM)
32 (4-guanidinobutyrate, pH 10) [3]; 206 (delta-guanidinovalerate, pH 10)
[3]; 163 (epsilon-guanidinocaproate, pH 10) [3]; 33 (4-guanidinobutyrate)
[5]

**pH-optimum**
10 [3]; 9.5 [3, 5]; 7.5–8.5 (Penicillium roqueforti) [3]; 9.9 [3]; 10.2 [5]; 9.5 [1]

**pH-range**
7–11 [3]

**Temperature optimum** (°C)
50 (above) [3]; 40 (Penicillium roqueforti) [3]

**Temperature range** (°C)

---

## 3 ENZYME STRUCTURE

**Molecular weight**
178000 (gel permeation chromatography, Pseudomonas putida) [3]
190000 (sucrose density gradient centrifugation, Pseudomonas putida) [3]
252000 (lizard, sucrose density gradient centrifugation) [1]
180000–186000 (gel filtration, Pseudomonas sp. ATCC 14676) [5]
240000 (Flavobacterium) [5]

**Subunits**
Hexamer (6 × 33000–36000, SDS-PAGE, Pseudomonas sp. ATCC 14676) [5]

**Glycoprotein/Lipoprotein**
–

---

## 4 ISOLATION/PREPARATION

**Source organism**
Pseudomonas putida [3]; Chicken [1]; Lizard [1]; Ray [3]; Snail [3]; Streptomyces griseus [2]; Penicillium roqueforti [3]; Pseudomonas aeruginosa
[4, 8]; Pseudomonas sp. ATCC 14676 [5]; Flavobacterium [5]

**Source tissue**
Liver [1, 3]; Kidney [1]; Hepatopancreas [3]; Cell (bacteria) [3]

**Localisation in source**
Mitochondria [3]

**Purification**
Chicken [1]; Pseudomonas putida [3]; Pseudomonas aeruginosa [4, 8]; Pseudomonas sp. ATCC 14676 [5]

**Crystallization**
–

**Cloned**
–

**Renaturated**
–

---

# 5 STABILITY

**pH**
7.5–8.5 (Tris-HCl buffer) [5]; 6.5–9.0 (unstable above pH 9.0 and below 6.5) [5]; 7.5 (unstable, 4°C, 48 hours) [5]

**Temperature (°C)**
70 [5]; More (Helix pomatia: heat stable) [3]

**Oxidation**

**Organic solvent**

**General stability information**
$Mn^{2+}$ (stabilizes) [5]; $K_3PO_4$ and NaCl stabilize at pH 7.5–10 [5]

**Storage**
–20°C, 1 months [3]

---

# 6 CROSSREFERENCES TO STRUCTURE DATABANKS

**PIR/MIPS code**

**Brookhaven code**

# 7 LITERATURE REFERENCES

[1] Mora, J., Tarrab, R., Martuscelli, J., Soberóm, G.: Biochem. J., 96, 588–594 (1965)
[2] Van Thoai, N., Thome-Beau, F., Olomucki, A.: Biochim. Biophys. Acta, 115, 73–80 (1966)
[3] Chou, C.-S., Rodwell, V.W.: J. Biol. Chem., 247, 4486–4490 (1972)
[4] Yorifuji, T., Sugai, I.: Agric. Biol. Chem., 42, 1789–1790 (1978)
[5] Yorifuji, T., Kato, M., Kobayashi, T., Ozaki, S., Ueno, S.: Agric. Biol. Chem., 44, 1127–1134 (1980)
[6] Yorifuji, T., Shiritani, Y.: Agric. Biol. Chem., 46, 317–318 (1982)
[7] Yorifuji, T., Sugai, I., Matsumoto, H., Tabuchi, A.: Agric. Biol. Chem., 46, 1361–1367 (1982)
[8] Yorifuji, T., Kobayashi, T., Tabuchi, A., Shiritani, Y., Yonaha, K.: Agric. Biol. Chem., 47, 2825–2830 (1983)

# 1 NOMENCLATURE

**EC number**
3.5.3.8

**Systematic name**
N-Formimino-L-glutamate formiminohydrolase

**Recommended name**
Formiminoglutamase

**Synonymes**
N-Formiminoglutamate hydrolase

**CAS Reg. No.**
9054-92-6

---

# 2 REACTION AND SPECIFICITY

**Catalysed reaction**
N-Formimino-L-glutamate + $H_2O$ →
→ L-glutamate + formamide

**Reaction type**
Amidine hydrolysis

**Natural substrates**
N-Formimino-L-glutamate (intermediate in degradation of L-histidine) + $H_2O$ [1–3]

**Substrate spectrum**
1 N-Formimino-L-glutamate + $H_2O$ (ir)

**Product spectrum**
1 L-Glutamate + formamide

---

**Inhibitor(s)**
$Mn^{2+}$ (above 0.5 mM) [1]; EDTA [1, 2, 4]; $(NH_4)SO_4$ [2, 4]; Glucose [7]

**Cofactor(s)/prostethic group(s)**
Glutathione [1]

**Metal compounds/salts**
$Mn^{2+}$ [1, 2]

---

**Turnover number** (min$^{-1}$)

**Specific activity** (U/mg)
More [2, 4]

**K$_m$-value** (mM)
40 (N-formimino-L-glutamate) [1]; 530 (N-formiminoglutamate, pH 8.7) [2, 4]; 0.039 (N-formiminoglutamate, pH 7.4) [2, 4]

**pH-optimum**
8.5 [1]; 8.7 [2, 4]

**pH-range**
8.0–9.0 [1]

**Temperature optimum** (°C)
37 [1]

**Temperature range** (°C)

---

## 3 ENZYME STRUCTURE

**Molecular weight**
220000 (Bacillus subtilis, ultracentrifugation) [2]

**Subunits**

**Glycoprotein/Lipoprotein**
–

---

## 4 ISOLATION/PREPARATION

**Source organism**
Aerobacter aerogenes [1]; Pseudomonas fluorescens [1]; Bacillus subtilis [2, 4, 10]; Salmonella typhimurium [3]; Vibrio alginolyticus [5]; Alcaligenes eutrophus [6]; Pseudomonas putida [10]; Klebsiella aerogenes [10]

**Source tissue**
Cell [1–10]

**Localisation in source**
Cytoplasm [1–10]

**Purification**
Aerobacter aerogenes [1]; Bacillus subtilis [2, 4]; Salmonella typhimurium [3]; Pseudomonas putida [10]

## Crystallization
–

## Cloned
[3, 5, 9]

## Renaturated
[2]

---

## 5 STABILITY

**pH**

**Temperature (°C)**
4 [2, 4]; 22 [2, 4]; 37 (unstable) [2]

**Oxidation**

**Organic solvent**

**General stability information**

**Storage**
More [1]

---

## 6 CROSSREFERENCES TO STRUCTURE DATABANKS

**PIR/MIPS code**

**Brookhaven code**

---

## 7 LITERATURE REFERENCES

[1] Lund, P., Magasanik, B.: J. Biol. Chem., 240 (11) , 4316–4319 (1965)
[2] Kaminskas, E., Kimhi, Y., Magasanik, B.: J. Biol. Chem., 245 (14) , 3536–3544 (1970)
[3] Smith, G.R., Halpern, Y.S., Magasanik, B.: J. Biol. Chem., 246 (10) , 3320–3329 (1971)
[4] Magasanik, B., Kaminskas, E., Kimhi, Y.: Methods Enzymol., 17, Pt. B, 57–62 (Ed. Colowick) (1971)
[5] Bowden, G., Mothibeli, M.A., Robb, F.T., Woods, D.R.: J. Gen. Microbiol., 128, 2041–2045 (1982)
[6] Schlesier, M., Friedrich, B.: Arch. Microbiol., 132, 254–259 (1982)
[7] Bodasing, S.J., Robb, F.T., Wood, D.R.: FEMS Microbiol. Lett., 19, 175–177 (1983)
[8] Cooperman, J.M. in "Methods Enzym. Anal.", 3rd. Ed. (Bergmeyer H.U., ed.) 8, 514–521 (1985)
[9] Consevage, M.W., Porter, R.D., Phillips, A.T.: J. Bacteriol., 162 (1) , 138–146 (1985)
[10] Hu, L., Mulfinger, L.M., Phillips, A.T.: J. Bacteriol., 169 (10) , 4696–4702 (1987)

---

## 1 NOMENCLATURE

**EC number**
3.5.3.9

**Systematic name**
Allantoate amidinohydrolase (decarboxylating)

**Recommended name**
Allantoate deiminase

**Synonymes**
Deiminase, allantoate
Allantoate amidohydrolase [1, 2]

**CAS Reg. No.**
37289-13-7

---

## 2 REACTION AND SPECIFICITY

**Catalysed reaction**
Allantoate + $H_2O$ →
→ ureidoglycine + $NH_3$ + $CO_2$

**Reaction type**
Amidine hydrolysis

**Natural substrates**
Allantoate + $H_2O$ [1, 4]
Ureidoglycin + $H_2O$ [4]

**Substrate spectrum**
1 Allantoate + $H_2O$
2 Ureidoglycin + $H_2O$ [4]

**Product spectrum**
1 Ureidoglycine + $NH_3$ + $CO_2$
2 Ureidoglycolate + $NH_3$

---

**Inhibitor(s)**
$Zn^{2+}$ [1, 3]; $Cu^{2+}$ [1, 3]; EDTA [6]; $Cd^{2+}$ [3]; Borate [6]; Acetohydroxamate
[6]; $Mn^{2+}$ [3]; $Co^{2+}$ [3]; $Ni^{2+}$ [3]

**Cofactor(s)/prostethic group(s)**

---

**Metal compounds/salts**
$Mn^{2+}$ [1, 2, 6]; $Ca^{2+}$ [1]; $Co^{2+}$ [1]

---

**Turnover number** (min$^{-1}$)

**Specific activity** (U/mg)
210 [2]

**K$_m$-value** (mM)
1.0 [6]

**pH-optimum**
8.5 [1, 2]; 8–9 (crude extract) [6]

**pH-range**
7.5–9.5 [1]

**Temperature optimum** (°C)

**Temperature range** (°C)

---

## 3 ENZYME STRUCTURE

**Molecular weight**

**Subunits**

**Glycoprotein/Lipoprotein**
–

---

## 4 ISOLATION/PREPARATION

**Source organism**
Streptococcus allantoicus [1–3]; Arthrobacter allantoicus [1]; E. coli [1];
Bacillus fastidiosus [5]; Glycine max. [6]

**Source tissue**
Seed coat [6]

**Localisation in source**

**Purification**
Streptococcus allantoicus [2]

**Crystallization**
–

**Cloned**
–

2

Renaturated
–

## 5 STABILITY

**pH**
6 (complete inactivation, 0.2 mM $Mn^{2+}$) [2]; 4.3 (stable below) [2]; 8 (stable above) [2]; 8 (in absence of bivalent cations loss of activity above pH 8) [3]

**Temperature (°C)**
–20 (several months) [1]

**Oxidation**

**Organic solvent**

**General stability information**

**Storage**
Several months at –20°C [1]

## 6 CROSSREFERENCES TO STRUCTURE DATABANKS

**PIR/MIPS code**

**Brookhaven code**

## 7 LITERATURE REFERENCES

[1] Vogels, G.D.: Biochim. Biophys. Acta, 113, 277–291 (1966)
[2] Van Der Drift, C., Vogels, G.D.: Biochim. Biophys. Acta, 139, 162–168 (1967)
[3] Van Der Drift, C., Vogels, G.D.: Enzymologia, 36 (4–5), 278–286 (1969)
[4] Van Der Drift, C., De Windt, F.E., Vogels, G.D.: Arch. Biochem. Biophys., 136, 273–279 (1970)
[5] Bongaerts, G.P.A., Vogels, G.D.: J. Bacteriol., 125 (2), 689–697 (1976)
[6] Winkler, R.G., Polacco, J.C., Blevins, D.G., Randall, D.D.: Plant Physiol., 79, 787–793 (1985)

## 1 NOMENCLATURE

**EC number**
3.5.3.10

**Systematic name**
D-Arginase amidinohydrolase

**Recommended name**
D-Arginase

**Synonymes**

**CAS Reg. No.**
37289-14-8

## 2 REACTION AND SPECIFICITY

**Catalysed reaction**
D-Arginine + $H_2O$ →
→ D-ornithine + urea

**Reaction type**
Amidine hydrolysis

**Natural substrates**
D-Arginine + $H_2O$ [1]

**Substrate spectrum**
1 D-Arginine + $H_2O$ [1]
2 L-Arginine + $H_2O$ [1]
3 DL-Arginine (alpha-N-acetylated) + $H_2O$ (ir) [1]

**Product spectrum**
1 D-Ornithine + urea
2 L-Ornithine + urea
3 DL-Ornithine + urea (N-acetylated)

**Inhibitor(s)**
Heavy metal ions [1]; $F^-$ [1]; Lysine [1]; Ornithine [1]

**Cofactor(s)/prostethic group(s)**

**Metal compounds/salts**
$Mn^{2+}$ [1]; $Co^{2+}$ [1]; $Fe^{2+}$ [1]

**Turnover number** (min⁻¹)

**Specific activity** (U/mg)

**K$_m$-value** (mM)

**pH-optimum**
9.5 [1]

**pH-range**
7–10 [1]

**Temperature optimum** (°C)
50 [1]

**Temperature range** (°C)

---

## 3 ENZYME STRUCTURE

**Molecular weight**

**Subunits**

**Glycoprotein/Lipoprotein**
–

---

## 4 ISOLATION/PREPARATION

**Source organism**
Mouse [1]; Guinea-pig [1]; Rabbit [1]; Human [1]; Dog [1]; Hog [1]; Toad [1]; Rat [1]

**Source tissue**
Liver [1]; Kidney [1]; Mucosa [1]; Tissue [1]

**Localisation in source**

**Purification**

**Crystallization**
–

**Cloned**
–

**Renaturated**
–

## 5 STABILITY

**pH**

**Temperature (°C)**
 0 (20% loss in 24 hours) [1]; 37 (70% loss in 24 hours) [1]

**Oxidation**

**Organic solvent**

**General stability information**
 Unstable [1]

**Storage**

---

## 6 CROSSREFERENCES TO STRUCTURE DATABANKS

**PIR/MIPS code**

**Brookhaven code**

---

## 7 LITERATURE REFERENCES

[1] Nadai, Y.: J. Bacteriol., 45 (12) , 1011–1020 (1958)

# 1 NOMENCLATURE

**EC number**
3.5.3.11

**Systematic name**
Agmatine amidinohydrolase

**Recommended name**
Agmatinase

**Synonymes**
Agmatine ureohydrolase

**CAS Reg. No.**
37289-16-0

---

# 2 REACTION AND SPECIFICITY

**Catalysed reaction**
Agmatine + $H_2O$ →
→ putrescine + urea

**Reaction type**
Amidine hydrolysis

**Natural substrates**
Agmatine + $H_2O$ [4]

**Substrate spectrum**
1 Agmatine + $H_2O$

**Product spectrum**
1 Putrescine + urea

---

**Inhibitor(s)**
Urea [4]; L-Arginine (at concentrations above 14 mM) [4]; L-Ornithine (at agmatine concentrations above 14 mM) [4]; Putrescine (at agmatine concentrations above 14 mM) [4]; Chloroatranorin [5]; Evernic acid [5]; EDTA [8]; EGTA [8]

**Cofactor(s)/prostethic group(s)**

**Metal compounds/salts**

Turnover number (min⁻¹)

Specific activity (U/mg)
1.018 [4]; 34.75 [8]

K$_m$-value (mM)
6.4 [4]; 1.2 [8]

pH-optimum
6.9 [4]; 7.3 [8]

pH-range
4.5–8 [4]

Temperature optimum (°C)
35–40

Temperature range (°C)

---

## 3 ENZYME STRUCTURE

Molecular weight
320000 (gel filtration, Evernia prunastri) [4]
80000 (gel filtration, E. coli) [8]
38000 (SDS-PAGE, E. coli) [8]

Subunits
Dimer (E. coli, SDS-PAGE) [8]

Glycoprotein/Lipoprotein
–

---

## 4 ISOLATION/PREPARATION

Source organism
E. coli [1, 6, 7, 8]; Panus tigrinus [2]; Klebsiella aerogenes [2–3]; Evernia
prunastri [4, 5]

Source tissue

Localisation in source

Purification
Evernia prunastri [4]; E. coli [8]

Crystallization
–

## Cloned
[6]

## Renaturated
–

## 5 STABILITY

pH

Temperature (°C)

Oxidation

Organic solvent

General stability information

Storage

## 6 CROSSREFERENCES TO STRUCTURE DATABANKS

PIR/MIPS code
JV0067 (Escherichia coli)

Brookhaven code

## 7 LITERATURE REFERENCES

[1] Morris, D.R., Pardee, A.B.: J. Biol. Chem., 241, 3129–3135 (1966)
[2] Boldt, A., Miersch, J., Reinbothe, H.: Phytochemistry, 10, 731–738 (1971)
[3] Friedrich, B., Magasanik, B.: J. Bacteriol., 137 (3) , 1127–1133 (1979)
[4] Vicente, C., Legaz, M.E.: Physiol. Plant., 55, 335–339 (1982)
[5] Legaz, M.E., Vicente, C.: Plant Physiol., 71, 300–302 (1983)
[6] Tabor, C.W., Tabor, H., Hafner, E.W., Markham, G.D., Boyle, S.M.: Methods Enzymol., 94, 117–121 (1983)
[7] Shaibe, E., Metzer, E., Halpern, Y.S.: J. Bacteriol., 163 (3) , 933–937 (1985)
[8] Satishchandran, C., Boyle, S.M.: J. Bacteriol., 165 (3) , 843–848 (1986)

# 1 NOMENCLATURE

**EC number**
3.5.3.12

**Systematic name**
Agmatine iminohydrolase

**Recommended name**
Agmatine deiminase

**Synonymes**
Agmatine amidinohydrolase
Deiminase, agmatine

**CAS Reg. No.**
37289-17-1

---

# 2 REACTION AND SPECIFICITY

**Catalysed reaction**
Agmatine + $H_2O$ →
→ N-carbamoylputrescine + $NH_3$

**Reaction type**
Amidine hydrolysis

**Natural substrates**
Agmatine + $H_2O$ [3]

**Substrate spectrum**
1 Agmatine + $H_2O$

**Product spectrum**
1 N-Carbamoylputrescine + $NH_3$

---

**Inhibitor(s)**
N, N-Diguanidinobutane [3]; p-Chloromercuribenzoate [3]; Tryptamine [4];
Cadaverine [4]; $Mn^{2+}$ [9]; $Ba^{2+}$ [9]; $Cu^{2+}$ [5]; $Zn^{2+}$ [5]; $Co^{2+}$ [5]; $Fe^{2+}$ [5];
$Ni^{2+}$ [5]; $Ca^{2+}$ [5]; $Mg^{2+}$ [5]; Arcain [5]; p-Hydroxymercuribenzoate [5];
Putrescine (no effect [5, 9]) [4]; Spermidine (no effect [5]) [4, 9]; Spermine
(no effect [5]) [4, 9]; Iodoacetate [6]; N-Ethylmaleimide [6]; L-Arginine [8]

**Cofactor(s)/prostethic group(s)**

---

Enzyme Handbook © Springer-Verlag Berlin Heidelberg 1991
Duplication, reproduction and storage in data banks are only
allowed with the prior permission of the publishers

Metal compounds/salts

---

Turnover number (min⁻¹)

**Specific activity** (U/mg)
   More [2, 5, 6, 8, 9, 10]

$K_m$**-value** (mM)
   0.76 [4]; 0.19 [5]; 15 [9]

**pH-optimum**
   7.5 [2]; 6.5–7.5 [3]; 5.5–8.5 [4]; 6.5 [5]; More [6, 9]

**pH-range**
   4.5–10 [3]

**Temperature optimum** (°C)
   60 [5]; 28 [9]

**Temperature range** (°C)

---

## 3 ENZYME STRUCTURE

**Molecular weight**
   85000 (corn, gel filtration) [5]
   56000 (Latyrus sativus, gel filtration) [6]
   55000 (Lathyrus sativus, SDS-PAGE) [6]
   183000 (rice, gel filtration) [9]
   More [10]

**Subunits**
   Dimer (2 × 43000, corn) [5]
   Dimer (2 × 95000, rice) [9]

**Glycoprotein/Lipoprotein**
   –

---

## 4 ISOLATION/PREPARATION

**Source organism**
   Hordeum vulgare [1]; Glycine max [2]; Maize [3]; Sunflower [3]; Cabbage [3]; Groundnut [4]; Brussels sprout plants [3]; Corn [5]; Streptococcus faecalis [7]; Evernia prunastri [8]; Latyrus sativus (enzyme also has putrescine transcarbamylase, ornithine transcarbamylase, carbamate kinase and

in presence of inorganic – phosphate putrescine synthase activity) [6]; Rice
[9]; Cucumis sativus (multifunctional enzyme, enzyme also has putrescine
transcarbamylase, ornithine transcarbamylase, carbamate kinase and – in
presence of inorganic phosphate – putrescine synthase activity) [10]

## Source tissue
Leaf [3]; Seed [3]; Seedling [3, 6, 9, 10]; Cotyledons [4]; Shoots [5]; Thallus
[8]

## Localisation in source
Cytoplasm [5]

## Purification
Groundnut [4]; Corn [5]; Latyrus sativus [6]; Rice [9]; Evernia prunastri [8];
Cucumis sativus [10]

## Crystallization
–

## Cloned
–

## Renaturated
–

---

## 5 STABILITY

### pH

### Temperature (°C)
43 (50% loss of activity after 15 minutes, maize) [3]; 4 (48 hours, complete
loss of activity) [6]

### Oxidation

### Organic solvent

### General stability information
Purified enzyme, highly unstable even in the presence of glycerol,
dithiothreitol and $Mg^{2+}$ [6]; Loss of activity after dialysis and freeze-thawing
[6]

### Storage

# 6 CROSSREFERENCES TO STRUCTURE DATABANKS

PIR/MIPS code

Brookhaven code

# 7 LITERATURE REFERENCES

[1] Smith, T.A.: Phytochemistry, 4, 599–607 (1965)
[2] Le Rudulier, D., Goas, G.: Physiol. Veg., 18, 609–616 (1980)
[3] Smith, T.A.: Phytochemistry, 8, 2111–2117 (1969)
[4] Sindhu, R.K., Desai, H.V.: Phytochemistry, 18, 1937–1938 (1979)
[5] Yanagisawa, H., Suzuki, Y.: Plant Physiol., 67, 697–700 (1981)
[6] Srivenugopal, K.S., Adiga, P.R.: J. Biol. Chem., 256 (18), 9532–9541 (1981)
[7] Simon, J., Stalon, V.: J. Bacteriol., 152 (2), 676–681 (1982)
[8] Legaz, M.E., Iglesias, A., Vicente, C.: Z. Pflanzenphysiol., 110, 53–59 (1983)
[9] Chaudhuri, M.M., Ghosh, B.: Phytochemistry, 24 (10), 2433–2435 (1985)
[10] Prasad, G.L., Adiga, P.R.: J. Biosci., 10 (3) 373–391 (1986)

## 1 NOMENCLATURE

**EC number**
3.5.3.13

**Systematic name**
N-Formimino-L-glutamate iminohydrolase

**Recommended name**
Formiminoglutamate deiminase

**Synonymes**
Formiminoglutamate deiminase
Formiminoglutamic iminohydrolase

**CAS Reg. No.**
9054-85-7

## 2 REACTION AND SPECIFICITY

**Catalysed reaction**
N-Formimino-L-glutamate + $H_2O$ →
→ N-formyl-L-glutamate + $NH_3$

**Reaction type**
Amidine hydrolysis

**Natural substrates**
Formiminoglutamate + $H_2O$ (pathway of histidine dissimilation) [2 , 3]

**Substrate spectrum**
1 N-Formimino-L-glutamate + $H_2O$ (ir)

**Product spectrum**
1 N-Formyl-L-glutamate + $NH_3$

**Inhibitor(s)**
Tetranitromethane [3]

**Cofactor(s)/prostethic group(s)**

**Metal compounds/salts**

**Turnover number** (min$^{-1}$)

**Specific activity** (U/mg)
12.7 [3]; More [2]

**$K_m$-value** (mM)
0.1 (less than, formiminoglutamate) [3]

**pH-optimum**
7.2 [3, 4]

**pH-range**

**Temperature optimum** (°C)
37 [3, 4]

**Temperature range** (°C)

---

## 3 ENZYME STRUCTURE

**Molecular weight**
100000 (Pseudomonas sp., sedimentation equilibrium) [3]

**Subunits**
Dimer (2 × 53000, Pseudomonas sp., sedimentation equilibrium, guanidine hydrochloride, mercaptoethanol) [3]
Dimer (Pseudomonas sp., SDS-PAGE) [4]

**Glycoprotein/Lipoprotein**
–

---

## 4 ISOLATION/PREPARATION

**Source organism**
Pseudomonas sp. [1, 3, 4]; Streptomyces coelicolor [1, 2]

**Source tissue**
Cell [1]

**Localisation in source**
Cytoplasm [1, 2]

**Purification**
Streptomyces coelicolor [2]; Pseudomonas sp. [3, 4]

**Crystallization**
–

**Cloned**
[1, 2]

Renaturated

–

## 5 STABILITY

pH

Temperature (°C)

Oxidation

Organic solvent

General stability information

Storage
  More [1]

## 6 CROSSREFERENCES TO STRUCTURE DATABANKS

PIR/MIPS code

Brookhaven code

## 7 LITERATURE REFERENCES

[1] Hu, L., Mulfinger, L.M., Phillips, A.T.: J. Bacteriol. , 169 (10) , 4696–4702 (1987)
[2] Kendrick, K.E., Wheelis, M.: J. Gen. Microbiol., 128, 2029–2040 (1982)
[3] Wickner, R.B., Tabor, H.: J. Biol. Chem., 347 (5) , 1605–1609 (1972)
[4] Wickner, R.B., Tabor, H.: Methods Enzymol., 17, Pt. B, 80–84 (1971)

## 1 NOMENCLATURE

**EC number**
3.5.3.14

**Systematic name**
N-Amidino-L-aspartate amidinohydrolase

**Recommended name**
Amidinoaspartase

**Synonymes**
Amidinoaspartic amidinohydrolase

**CAS Reg. No.**
37325-60-3

## 2 REACTION AND SPECIFICITY

**Catalysed reaction**
N-Amidino-L-aspartate + $H_2O$ →
→ L-aspartate + urea (specific for the optical isomer) [2]

**Reaction type**
Amidine hydrolysis

**Natural substrates**
N-Amidino-L-aspartic acid [1]

**Substrate spectrum**
1 N-Amidino-L-aspartic acid (guanidinosuccinic acid) + $H_2O$ [1, 2]
2 N-Amidino-L-glutamate + $H_2O$ [1]
3 More (specific for the optical isomer [2], not: 3-guanidinopopionic acid [1], 4-guanidinobutyric acid [1, 2], 2-amino-4-guanidinopropionic acid [1], agmatine [1], arginine [1, 2], N-acetylarginine [1, 2], guanidinoacetic acid [1, 2], D-guanidinosuccinic acid [1, 2]) [1, 2]

**Product spectrum**
1 L-Aspartic acid + urea [1, 2]
2 L-Glutamate + urea
3 ?

## Inhibitor(s)

Urea (competitive) [1]; Aspartic acid (competitive) [1]; $Zn^{2+}$ (inhibits $Co^{2+}$ activation) [1]; $Cu^{2+}$ (inhibits $Co^{2+}$ activation) [1]; $Fe^{2+}$ (inhibits $Co^{2+}$ activation) [1]; EDTA (inhibits $Co^{2+}$ activation) [1]

## Cofactor(s)/prostethic group(s)

## Metal compounds/salts

$Co^{2+}$ (activation, temperature dependent reaction) [1]; $Mn^{2+}$ (activation, Pseudomonas putida) [2]

## Turnover number (min$^{-1}$)

## Specific activity (U/mg)

9.73 [1]

## $K_m$-value (mM)

8.3 [1]; 2.4 [1]; 20 [1]; 10.3 [2]

## pH-optimum

8 (broad) [1]; 9–10 [2]

## pH-range

## Temperature optimum (°C)

## Temperature range (°C)

## 3 ENZYME STRUCTURE

## Molecular weight

300000 (Pseudomonas chlororaphis, gel filtration) [1]
163000–210000 (Pseudomonas putida, gel filtration) [2]

## Subunits

Tetramer (4 × 39800, Pseudomonas putida) [2]

## Glycoprotein/Lipoprotein

–

## 4 ISOLATION/PREPARATION

## Source organism

Pseudomonas chlororaphis [1]; Pseudomonas putida [2]

## Source tissue

Cell [1, 2]

Localisation in source

Purification
Pseudomonas chlororaphis [1]; Pseudomonas putida [2]

Crystallization
–

Cloned
–

Renaturated
–

## 5 STABILITY

pH

Temperature (°C)

Oxidation

Organic solvent

General stability information

Storage
–15°C, several months

## 6 CROSSREFERENCES TO STRUCTURE DATABANKS

PIR/MIPS code

Brookhaven code

## 7 LITERATURE REFERENCES

[1] Milstien, S., Goldman, P.: J. Biol. Chem., 247, 6280–6283 (1972)
[2] Yorifuji, T., Furuyoshi, S.: Agric. Biol. Chem., 50 (5) , 1327–1328 (1986)

# 1 NOMENCLATURE

**EC number**
3.5.3.15

**Systematic name**
Protein-L-arginine iminohydrolase

**Recommended name**
Protein-arginine deiminase

**Synonymes**
Peptidylarginine deiminase [1]

**CAS Reg. No.**

---

# 2 REACTION AND SPECIFICITY

**Catalysed reaction**
Protein L-arginine + $H_2O$ →
→ protein L-citrulline + $NH_3$ (irreversible) [1]

**Reaction type**
Amidine hydrolysis

**Natural substrates**
Trichohyalin + $H_2O$ (hair follicle enzyme of guinea pig) [3]
Epidermal cell membrane + $H_2O$ (epidermis enzyme of rat or cow) [3]

**Substrate spectrum**
1 Protein L-arginine + $H_2O$ (ir) [1]
2 Alpha-N-benzoyl-L-arginine + $H_2O$ [1, 3]
3 N-Benzoylarginine amide + $H_2O$ [1, 3]
4 N-Benzoylglycylarginine + $H_2O$ [1, 3]
5 Alpha-N-benzoyl-L-arginine ethyl ester + $H_2O$ [1–3]
6 Egg albumin + $H_2O$ (after oxidative cleavage of disulfide bond) [1]
7 Bovine serum albumin + $H_2O$ [1]
8 Histone + $H_2O$ [1, 3]
9 Protamine + $H_2O$ [1]
10 Prekeratin + $H_2O$ (bovine, S-carboxymethylated) [2]
11 Myelin basic protein + $H_2O$ [2]
12 Polyarginine + $H_2O$ [3]
13 Keratin + $H_2O$ [3]
14 Trichohyalin + $H_2O$ [3]
15 More [1–3]

## Product spectrum
1 Protein L-citrulline + $NH_3$
2 N-Benzoyl citrulline + $NH_3$
3 N-Benzoylcitrullinamide + $NH_3$
4 N-Benzoyl glycylcitrulline + $NH_3$
5 Alpha-N-benzoylcitrulline ethyl ester + $NH_3$
6 ?
7 ?
8 ?
9 ?
10 ?
11 ?
12 Polycitrulline
13 ?
14 ?
15 ?

## Inhibitor(s)
Thiol inhibitors [1]; Monoiodoacetate [1]; p-Chloromercuribenzoate [1]; Carbonyl reagents (slight) [1]; Phenylhydrazine [1]; Hydroxyamine [1]; More (no product inhibition) [1]

## Cofactor(s)/prostethic group(s)

## Metal compounds/salts
$Ca^{2+}$ (required for activity, 10 mM) [1, 3]; $Sr^{2+}$ [1]; $Ba^{2+}$ (less effective than $Ca^{2+}$) [1]

## Turnover number (min$^{-1}$)

## Specific activity (U/mg)
46 [1]; 77 [2]

## $K_m$-value (mM)
0.33 (alpha-N-benzoyl-L-arginine ethyl ester) [2]

## pH-optimum
7.5 [1]

## pH-range
7.0–7.6 [3]

## Temperature optimum (°C)
50 [1]

## Temperature range (°C)

# 3 ENZYME STRUCTURE

**Molecular weight**
   48000 (rat, gel filtration) [1]
   85000 (bovine, SDS-gel electrophoresis) [2]
   125000 (gel filtration, bovine, brain) [2]
   69000 (gel filtration, bovine, epidermal) [2]
   115000 (rabbit muscle) [5]
   50000 (hair follicle) [3]

**Subunits**

**Glycoprotein/Lipoprotein**
   –

# 4 ISOLATION/PREPARATION

**Source organism**
   Rat [1]; Cow [2]; Guinea pig [4]; Bovine [2, 6]

**Source tissue**
   Skin (newborn rat, epidermis) [1]; Hair follicles [4]; Brain (mammalia) [2];
   Kidney (rat) [2]; Lung (rat) [2]

**Localisation in source**
   Membrane (horny cells, epidermis) [1]

**Purification**
   Rat [1]; Bovine (partially [6]) [2, 6]

**Crystallization**
   –

**Cloned**
   –

**Renaturated**
   –

# 5 STABILITY

**pH**

**Temperature (°C)**

**Oxidation**

**Organic solvent**

General stability information

Storage

---

## 6 CROSSREFERENCES TO STRUCTURE DATABANKS

**PIR/MIPS code**
   A34339 (rat)

**Brookhaven code**

---

## 7 LITERATURE REFERENCES

[1] Fujisaki, M., Sugawara, K.: J. Biochem., 89, 257–263 (1981)
[2] Kubilus, J., Baden, H.P.: Biochim. Biophys. Acta, 745, 285–291 (1983)
[3] Rothnagel, J.A., Rogers, G.E.: Methods Enzymol., 107, 624–631 (1984)
[4] Rogers, G.E., Harding, H.W.J., LLewellyn-Smith, I.J.: Biochim. Biophys. Acta, 495, 159–175 (1977)
[5] Sugawara, K., Oikawa, Y., Ochi, T.: J. Biochem., 91, 1065–1071 (1981)
[6] Kubilus, J., Waitkus, R.W., Baden, H.P.: Biochim. Biophys. Acta, 615, 246 (1980)

# 1 NOMENCLATURE

**EC number**
3.5.3.16

**Systematic name**
Methylguanidine amidinohydrolase

**Recommended name**
Methylguanidinase

**Synonymes**
Methylguanidine hydrolase

**CAS Reg. No.**
73200-93-8

---

# 2 REACTION AND SPECIFICITY

**Catalysed reaction**
Methylguanidine + $H_2O$ →
→ methylamine + urea

**Reaction type**
Amidine hydrolysis

**Natural substrates**
Methylguanidine + $H_2O$ [1]

**Substrate spectrum**
1 Methylguanidine + $H_2O$ [1]
2 More (poor substrates: ethylguanidine, n-propylguanidine,
  n-butylguanidine, agmatine, guanidine) [1]

**Product spectrum**
1 Methylamine + urea [1]
2 More (products of poor substrates: ethylamine + urea, n-propylamine +
  urea, n-butylamine + urea, 1,4-diaminobutane + urea, urea + $NH_3$) [1]

---

**Inhibitor(s)**
Trichloracetic acid [1]

**Cofactor(s)/prostethic group(s)**

**Metal compounds/salts**

---

Enzyme Handbook © Springer-Verlag Berlin Heidelberg 1991
Duplication, reproduction and storage in data banks are only
allowed with the prior permission of the publishers

Turnover number (min⁻¹)

Specific activity (U/mg)
102 [1]

K$_m$-value (mM)

pH-optimum

pH-range

Temperature optimum (°C)

Temperature range (°C)

---

## 3 ENZYME STRUCTURE

**Molecular weight**

**Subunits**

**Glycoprotein/Lipoprotein**
–

---

## 4 ISOLATION/PREPARATION

**Source organism**
Alcaligenes sp. N-42 [1]

**Source tissue**
Cell [1]

**Localisation in source**

**Purification**
Alcaligenes sp. N-42 [1]

**Crystallization**
–

**Cloned**
–

**Renaturated**
–

# 5 STABILITY

pH

Temperature (°C)

Oxidation

Organic solvent

General stability information

Storage

# 6 CROSSREFERENCES TO STRUCTURE DATABANKS

PIR/MIPS code

Brookhaven code

# 7 LITERATURE REFERENCES

[1] Nakajima, M., Shirokane, Y., Mizusawa, K.: FEBS Lett., 110, 43–46 (1980)

## 1 NOMENCLATURE

**EC number**
   3.5.4.1

**Systematic name**
   Cytosine aminohydrolase

**Recommended name**
   Cytosine deaminase

**Synonymes**
   Isocytosine deaminase
   Deaminase, cytosine

**CAS Reg. No.**
   9025-05-2

## 2 REACTION AND SPECIFICITY

**Catalysed reaction**
   Cytosine + $H_2O$ →
   → uracil + $NH_3$

**Reaction type**
   Amidine hydrolysis

**Natural substrates**
   Cytosine + $H_2O$ [1]

**Substrate spectrum**
   1  Cytosine + $H_2O$ [1]
   2  5-Methylcytosine + $H_2O$ (not found [4, 10]) [1, 5, 14]
   3  5-Fluoro-cytosine + $H_2O$ [3, 10, 14]
   4  5-Chloro-cytosine + $H_2O$ [3]
   5  5-Bromo-cytosine + $H_2O$ [3]
   6  5-Iodo-Cytosine + $H_2O$ [3]
   7  Creatinine + $H_2O$ (not found [16]) [15]

**Product spectrum**
   1  Uracil + $NH_3$
   2  5-Methyluracil + $NH_3$
   3  5-Fluoro-uracil + $NH_3$
   4  5-Chloro-uracil + $NH_3$

Enzyme Handbook © Springer-Verlag Berlin Heidelberg 1991
Duplication, reproduction and storage in data banks are only
allowed with the prior permission of the publishers

5 5-Bromo-uracil + $NH_3$
6 5-Iodo-uracil + $NH_3$
7 N-Methylhydantoin + $NH_3$

## Inhibitor(s)

CMP (inhibition depends on pH) [1, 2]; CDP (inhibition depends on pH) [1, 2]; CTP [1]; Thymidine [1]; TTP [1]; Guanosine [1]; GMP [1]; GDP [1]; GTP (activation [5]) [1]; Deoxyguanosine [6]; DeoxyGMP [6]; $Co^{2+}$ [6]; $Cd^{2+}$ [6]; $Hg^{2+}$ [6]; $Cu^{2+}$ [6]; $Ni^{2+}$ [6]; $Zn^{2+}$ [6]; $Mn^{2+}$ [7]; p-Chloromercuribenzoate [7]; $Fe^{3+}$ [10]; Orotidine-5-monophosphate [10, 11]; 5-Bromo-2-pyrimidone [12]; $Ag^+$ [14]; $Hg^+$ [14]; p-Choromercuriphenylsulfonate [14]; Mersaryl acid [14]

## Cofactor(s)/prostethic group(s)

## Metal compounds/salts

Phosphate [6, 7]; Pyrophosphate [6, 7, 10, 11]

## Turnover number ($min^{-1}$)

## Specific activity (U/mg)

3.1 [1]; 20 [4]; More [5]

## $K_m$-value (mM)

2.5 [1, 9]; 3.4 [6]; 4.5 (cytosine); 5.7 (5-methylcytosine) [7]; More [10, 12]

## pH-optimum

6.5 [1, 9]; 7–7.4 [3]; 9.5 [6]; 10.0 [6] 7.3–7.5 [10]

## pH-range

5–8.5 [1]; 7–11 [6]; 7 (no activity found) [7]

## Temperature optimum (°C)

45 [6]; 40–45 [7]; 45–50 [10]; 50 [14]

## Temperature range (°C)

30–70 (activity at 30°C and at 70°C: 2.5 times lower than at optimum) [6]; 75 (up to) [10]; More [14, 10]

# 3 ENZYME STRUCTURE

## Molecular weight

34000 (gel filtration, yeast) [1, 9]
580000 (Serratia marcescens, gel filtration) [4]
630000 (Pseudomonas aureofaciens, gel filtration) [5]
More [10, 14, 16]

## Subunits

Octamer (8 × 72000, Serratia marcescens) [4]
Dodecamer to hexadecamer (12 × 45000 or 16 × 45000, Pseudomonas aureofaciens) [5]
Tetramer (4 × 54000, Salmonella thyphimurium) [10]

## Glycoprotein/Lipoprotein

–

## 4 ISOLATION/PREPARATION

### Source organism

Yeast [1, 2, 9, 12]; Serratia marcescens [4, 6]; Pseudomonas aureofaciens [5, 7, 15]; Bacteria (distribution of enzyme in different species) [8]; Sallmonella typhimurium [10]; E. coli [13–15]; Proteus mirabilis [15], Pseudomonas chlororaphis [15]; Pseudomonas cruciviae [15]; Alcaligenes denitrificans [16]

### Source tissue

### Localisation in source

### Purification

E. coli (large scale purification) [13, 14]; Serratia marcescens [4]; Pseudomonas aureofaciens [5]; Yeast [9]; Salmonella typhimurium [10]

### Crystallization

–

### Cloned

–

### Renaturated

–

## 5 STABILITY

### pH

5–9 (4°C, for at least 48 hours) [1]; 6.5–9 (2 weeks, 4°C) [10]; More [5, 14]

### Temperature (°C)

4 (pH 5–9, for at least 48 hours) [1, 9]; 70 (10 minutes, 25% loss of activity) [4]; 60 (after 5 minutes, optimal pH, more than 70% loss of activity) [5]; More (heat stable at pH 7–9 [4]) [4, 5, 14]

## Oxidation
Unstable in absence of reducing reagents (Pseudomonas) [5]; High
stability against oxidation (Serratia) [7]

## Organic solvent

## General stability information

## Storage
4°C, pH 5–9, for at least 48 hours [1, 9]

---

## 6 CROSSREFERENCES TO STRUCTURE DATABANKS

### PIR/MIPS code

### Brookhaven code

---

## 7 LITERATURE REFERENCES

[1] Ipata, P.L., Marmocchi, F., Magni, G., Felicioli, R., Polidoro, G.: Biochemistry, 10 (23)
, 4270–4276 (1971)
[2] Balestreri, E., Felicioli, R.A., Ipata, P.L.: Biochim. Biophys. Acta, 293, 443–448 (1973)
[3] Holldorf, A.W. in "Methods Enzym. Anal." (Bergmeyer, H.U., Ed.) 3 (2) , 1964–1966
(1974)
[4] Sakai, T., Yu, T., Tabe, H., Omata, S.: Agric. Biol. Chem., 39 (8) , 1623–1629 (1975)
[5] Sakai, T., Yu, T., Taniguchi, K., Omata, S.: Agric. Biol. Chem., 39 (10) , 2015–2020
(1975)
[6] Yu, T., Sakai, T., Omata, S.: Agric. Biol. Chem., 40 (3) , 543–549 (1976)
[7] Yu, T., Sakai, T., Omata, S.: Agric. Biol. Chem., 40 (3) , 551–557 (1976)
[8] Sakai, T., Yu, T., Omata, S.: Agric. Biol. Chem., 40 (9) , 1893–1895 (1976)
[9] Ipata, P.L., Cercignani, G.: Methods Enzymol., 51, 394–401 (1978)
[10] West, T.P., Shanley, M.S., O'Donovan, G.A.: Biochim. Biophys. Acta, 791, 251–258
(1982)
[11] West, T.P.: Experientia, 41, 1563–1564 (1988)
[12] Kornblatt, M.J., Tee, D.S.: Eur. J. Biochem., 756, 297–300 (1986)
[13] Katsuragi, T., Sakai, T., Tonomura, K.: Agric. Biol. Chem., 50 (7) , 1713–1719 (1986)
[14] Katsuragi, T., Sakai, T., Matsumoto, K., Tonomura, K. : Agric. Biol. Chem., 50 (7) ,
1721–1730 (1986)
[15] Kim, J.M., Shimizu, S., Yamada, H.: Arch. Microbiol., 147, 58–63 (1987)
[16] Kim, J.M., Shimizu, S., Yamada, H.: FEBS Lett., 210 (1) , 77–80 (1987)

# 1 NOMENCLATURE

**EC number**
3.5.4.2

**Systematic name**
Adenine aminohydrolase

**Recommended name**
Adenine deaminase

**Synonymes**
Adenase
Adenine aminase
Deaminase, adenine
ADase [4]

**CAS Reg. No.**
9027-68-3

# 2 REACTION AND SPECIFICITY

**Catalysed reaction**
Adenine + $H_2O$ →
→ hypoxanthine + $NH_3$

**Reaction type**
Amidine hydrolysis

**Natural substrates**
Adenine + $H_2O$ (catabolism [5, 6], anaerobic conversion [7]) [5–7]
6-Chloropurine + $H_2O$ (in vivo utilization of 6-chloropurine) [9]

**Substrate spectrum**
1 Adenine + $H_2O$ (ir [7]) [1–9]
2 More (catalyzes replacement of the 6-substituted group of
6-chloropurine, 6-iodopurine, 6-bromopurine, 2, 6-diaminopurine,
2-amino-6-chloropurine, 6-amino-2-hydroxypurine and adenine with a
hydroxyl group [5], 5-methylcytosine) [4, 5, 7, 8]

**Product spectrum**
1 Hypoxanthine + $NH_3$ (ir [7]) [1–9]
2 ?

## Inhibitor(s)

4-Aminopyrazolo[3, 4-d]pyrimidine [5]; 6-Dimethylaminopurine [5]; $Cd^{2+}$ [5]; $Hg^{2+}$ [5]; $Mn^{2+}$ [5]; $Zn^{2+}$ [5]; Coformycin [3]; Deoxycoformycin [3]; Hypoxanthine [4]; $MnCl_2$ [8]; Adenosine [5]; AMP [5]; Purine [5]; 6-Methylaminopurine [5]; $Co^{2+}$ [5]; $Ni^{2+}$ (weak) [5]; Alpha-phenanthroline [5]; Monoiodoacetic acid [5]; p-Chloromercuribenzoate [5]; $NaN_3$ [5]; Mercurial compounds [5]

## Cofactor(s)/prostethic group(s)

More (no dialyzable cofactors required) [7]

## Metal compounds/salts

---

## Turnover number $(min^{-1})$

## Specific activity (U/mg)

More [6]; 3.7 [8]

## $K_m$-value (mM)

1.0 (adenine, Crithidia fasciculata) [3]; 1.1 (adenine, Leishmania tarentolae) [3]; 2.0 (adenine, Leishmania donovani) [3]; More [3–8]

## pH-optimum

6.0–7.5 [4]; 8.0 [3]; 9.0 (Tris-HCl buffer) [6]; 7.0 (glycine-NaOH buffer) [6]; 6.7–7.7 [8]; 7.0 (adenine, 6-bromopurine) [5]; 7.5 (6-chloropurine, 6-iodopurine, 2, 6-diaminopurine) [5]; 6.5–7.5 (2-amino-6-chloropurine) [5]

## pH-range

5–8.5 (5: about 30% of maximal activity, 8.5: about 20% of maximal activity) [4]; 6–9 [6]

## Temperature optimum (°C)

40 (about, not purified enzyme) [2]; 40–45 [6]

## Temperature range (°C)

25–55 [6]

---

# 3 ENZYME STRUCTURE

## Molecular weight

37000 (Pseudomonas synxantha, gel filtration) [6]

## Subunits

## Glycoprotein/Lipoprotein

–

## 4 ISOLATION/PREPARATION

**Source organism**
Pseudomonas synxantha [5, 6]; Streptomyces sp. (J-350P) [1]; Streptomyces viridiviolaceus [2]; Crithidia fasciculata [3, 4]; Leishmania
tarentolae [3]; Leishmania donovani [3]; Leishmania mexicana [3]; Leishmania braziliensis [3]; Candida utilis [5]; Azotobacter vinelandii [5, 7, 8];
Schizosaccharomyces pombe [9]; More (nonmammalian enzyme) [7]

**Source tissue**
Cell [8]

**Localisation in source**
Extracellular [1]; Intracellular [4]; Cytoplasm [4]; More [4]

**Purification**
Pseudomonas synxantha [6]; Azotobacter vinelandii [6]

**Crystallization**
–

**Cloned**
–

**Renaturated**
–

## 5 STABILITY

**pH**
7.0 (37°C, potassium phosphate buffer, unstable) [6]; 8.0 (most stable at
high pH) [3]; 6.5 (55°C, 80% loss of activity after 60 minutes) [8]

**Temperature (°C)**
50 (Leishmania, 95% loss of activity after 1 minute) [3]; 55 (pH 6.5, 80%
loss of activity after 60 minutes, Azotobacter [8], no appreciable loss of activity after 60 minutes, Crithidia [4]) [4, 8]; 65 (20% loss of activity after 5
minutes) [4]; 37 (pH 7.0, potassium phosphate buffer, unstable) [6]

**Oxidation**

**Organic solvent**

**General stability information**
Bovine serum albumin (protects against heat inactivation in dilute solutions) [4]; Dialysis (24 hours against 0.9% KCl, stable) [8]

## Storage

4°C, 1–2 weeks (Crithidia fasciculata and Leishmania tarentolae: stable for 1–2 weeks, enzymes from human pathogens: rapid loss of activity) [3]; –15°C (for at least 2 months) [8]; More [3]

## 6 CROSSREFERENCES TO STRUCTURE DATABANKS

PIR/MIPS code

Brookhaven code

## 7 LITERATURE REFERENCES

[1] Jun, H.K., Park, J.H.: Korean J. Appl. Microbiol. Bioeng., 12, 31–36 (1984)

[2] Allam, A.M., Elawamry, Z.A., Elzainy, T.A.: Egypt. J. Bot., 24, 149–152 (1981)

[3] Kidder, G.W., Nolan, L.L.: Proc. Natl. Acad. Sci. USA, 76, 3670–3672 (1979)

[4] Kidder, G.W., Dewey, V.C., Nolan, L.L.: Arch. Biochem. Biophys., 183, 7–12 (1977)

[5] Jun, H.-K., Sakai, T.: J. Ferment. Technol., 57, 294–299 (1979)

[6] Sakai, T., Jun, H.-K.: J. Ferment. Technol., 56, 257–265 (1978)

[7] Zielke, C.L., Suelter, C.H. in "The Enzymes", 3rd. Ed. (Boyer, P.D.) 4, 47–78 (1971) (Review)

[8] Heppel, L.A., Hurwitz, J., Horecker, B.L.: J. Am. Chem. Soc., 79, 630–633 (1957)

[9] Abbondandolo, A., Weyer, A., Heslot, H., Lambert, M.: J. Bacteriol., 108, 959–963 (1971)

# 1 NOMENCLATURE

**EC number**
   3.5.4.3

**Systematic name**
   Guanine aminohydrolase

**Recommended name**
   Guanine deaminase

**Synonymes**
   Guanase
   Guanine aminase
   Deaminase, guanine
   Guanine deaminase
   GAH [1]

**CAS Reg. No.**
   9033-16-3

# 2 REACTION AND SPECIFICITY

**Catalysed reaction**
   Guanine + $H_2O$ →
   → xanthine + $NH_3$

**Reaction type**
   Amidine hydrolysis

**Natural substrates**
   Guanine + $H_2O$ (purine metabolism [5], ammoniagenesis in the brain [6])
   [5, 6]

**Substrate spectrum**
   1 Guanine + $H_2O$ [1–14]
   2 8-Azaguanine + $H_2O$ [2, 6, 8]
   3 6-Thioguanine + $H_2O$ [8]
   4 More (low reaction rate with: 8-azaguanine, adenosine, guanosine,
     adenine, high substrate specificity) [2]

**Product spectrum**
   1 Xanthine + $NH_3$
   2 8-Azaxanthine + $NH_3$
   3 6-Thioxanthine + $NH_3$ [8 , 12]
   4 ?

### Inhibitor(s)

p-Chloromercuribenzoate [9, 12, 13]; 5-Aminoimidazole-4-carboxamide [9, 12]; Tetrahydrofolic acid [9, 12]; Iodoacetamide (slight) [9, 12]; $Zn^{2+}$ (rat: activation of isoenzyme B, inhibition of isoenzyme A [13]) [5, 13]; $Cu^{2+}$ (rat: activation of isoenzyme B, inhibition of isoenzyme A) [13]; $HgCl_2$ [13]; N-Ethylmaleimide [13]; Guanidine hydrochloride [13]; $Li^+$ [5]; $Ag^+$ [5]; $Pb^{2+}$ [5]; $Hg^{2+}$ [5]; $Fe^{2+}$ [5]; $F^-$ (inhibition decreases as pH increases) [1, 8]; Rose Bengal (photooxidation) [1]; Hypoxanthine [2]; Inosine [2]; 5-Aminoimidazole-4-carboxamide [2, 3, 6, 9, 12]; p-Hydroxymercuribenzoate [1, 3, 8]; Iodoacetic acid (weak) [3, 9, 12]; $Fe^{3+}$ [5, 13]; GTP (weak) [6]; 3-Deazaguanine [8]

### Cofactor(s)/prostethic group(s)

### Metal compounds/salts

$Mg^{2+}$ (rat: activates) [13]; $Zn^{2+}$ (rat: activates isoenzyme B, inhibits isoenzyme A) [13]; $Cu^{2+}$ (rat: activates isoenzyme B, inhibits isoenzyme A) [13]

### Turnover number (min$^{-1}$)

### Specific activity (U/mg)

More [4, 11, 13]; 35.5 [8]; 22.1 [1]; 21.5 [3]; 8.61 [2]; 51.2 [5]

### $K_m$-value (mM)

0.8 (6-thioguanine) [12]; 0.0074 (pH 8.0, guanine) [2]; 0.0153 (pH 6.0, guanine) [3]; 0.2 (pH 6.0, 8-azaguanine) [3]; 0.17 (guanine) [6]; 0.67 (8-azaguanine) [6]; More (effect of pH on $K_m$ [1, 3, 8]) [1, 3, 4, 7, 8, 9, 10, 11, 12, 13]

### pH-optimum

7.5–9.5 (broad) [1, 2]; 8.0 (sharp) [3]; 8.0–8.5 [4]; 7–8.5 (broad) [5]; 7.4–9.4 [6]; 7.4 (isoenzyme A) [7]; 9.0 (isoenzyme B) [7]; 8.0 [8]; 6.8 (guanine [9]) [9, 12]; 6.0 (8-azaguanine) [9]; 8 (near) [11]; 7–8 [13]

### pH-range

6.0–9.5 (6.0, 9.5: about 20% of maximal activity) [5]; 5.0–11.0 (5.0: about 40% of maximal activity at, 11.0: about 30% of maximal activity at) [2]

### Temperature optimum (°C)

37 (assay at) [5]; 30 (assay at) [3]; 25 (assay at) [4]

### Temperature range (°C)

## 3 ENZYME STRUCTURE

### Molecular weight
105000 (rat, brain, gel filtration) [6]
168000 (rat liver, gel filtration) [6]
92200 (non denaturing polyacrylamide gel electrophoresis, rabbit) [4]
55000 (human, SDS-PAGE [1], rabbit, gel filtration, SDS-PAGE [12]) [1, 12]
110000 (human [2], rabbit [8], pig [10], gel filtration) [2, 8, 10]
120000 (human, gel filtration) [3]
More [7, 9, 11]

### Subunits
Dimer (2 × 55000, human, SDS-PAGE [1], 2 × 59000, human SDS-PAGE [2],
2 × 49500, rabbit, SDS-PAGE [4], 2 × 52000, SDS-PAGE, rat [5], rabbit [8],
2 × 50000, SDS-PAGE, pig [10]) [1, 2, 4, 5, 8, 10]
Monomer (1 × 55000, gel filtration, SDS-PAGE, rabbit) [12]

### Glycoprotein/Lipoprotein
–

## 4 ISOLATION/PREPARATION

### Source organism
Human (2 isoenzymes, A, B [7]) [1, 2, 3, 7]; Rabbit [4, 8, 9, 11 , 12]; Rat (2
isoenzymes: A, B) [13]; Mouse [13]; Pig [10];
Clostridium cylindrosporum [14]

### Source tissue
Liver [1–9, 11–13]; Brain (low activity [5]) [4–7, 10, 13]; Intestine [4]; Lung
[5]; Kidney (low activity [5]) [5, 7]; Gut [7]

### Localisation in source
Soluble [5]; More (supernatant) [5, 13]

### Purification
Human [1, 2, 3, 7]; Rabbit [4, 8, 9, 11]; Rat [5, 6]; Pig [10]

### Crystallization
–

### Cloned
–

### Renaturated
–

# 5 STABILITY

## pH

## Temperature (°C)
50 (30 minutes) [13]; 60 (30 minutes, 60% loss of activity [13], 50% loss of activity after 30 minutes [2], 100% loss of activity after 30 minutes [9], 100% loss of activity after 25 minutes [12]) [2, 9, 12, 13]; 45 (17% loss of activity after 30 minutes) [3]; 56 (30 minutes, isoenzyme A: about 20% loss of activity, isoenzyme B: about 10% loss of activity) [7]

## Oxidation
Photooxidation (inactivated by photooxidation with 0.0025% Rose Bengal at acidic pH, protected partially by 5-aminoimidazole-4-carboxamide [1], inactivation, 0.001% Rose Bengal at pH 6.32 [8]) [1, 8]

## Organic solvent

## General stability information
Dilute solutions (irreversible denaturation by freezing) [10]; 2-Mercaptoethanol (stabilizes) [9]; Freezing (irreversible denaturation of dilute solutions) [10]

## Storage
More [13]; 0°C, pH 7.0, 0.1 mM dithiothreitol, 2 weeks [6]; 4°C, 0.1 M phosphate buffer, 2-mercaptoethanol, for at least 3 months; 0–4°C, 0.010–0.020 mg/ml protein concentration (rat, isoenzyme A: 100% loss of activity after 16 days, isoenzyme B: 30% loss of activity after 16 days) [13]

---

# 6 CROSSREFERENCES TO STRUCTURE DATABANKS

## PIR/MIPS code

## Brookhaven code

---

# 7 LITERATURE REFERENCES

[1] Kimm, S.-W., Park, J.-B., Kim, H.-L.: Korean J. Biochem., 19, 39–46 (1987)
[2] Kimm, S.-W., Park, J.-B., Lee, I.-S.: Korean J. Biochem., 17, 139–148 (1985)
[3] Gupta, N.K., Glantz, M.D.: Arch. Biochem. Biophys., 236, 266–276 (1985)
[4] Pugh, M. E., Bieber, A.L.: Comp. Biochem. Physiol., 77 B, 619–627 (1984)
[5] Kim, C.-J., Kimm, S.-W.: Korean J. Biochem., 14, 77–93 (1982)
[6] Miyamoto, S., Ogawa, H., Shiraki, H., Nakagawa, H.: J. Biochem., 91, 167–176 (1982)
[7] Kuzmits, R., Stemberger, H., Müller, M.M.: Adv. Exp. Med. Biol., 122B (Purine Metab. Man., Biochem. Immunol Cancer Res.) , 183–188 (1980)
[8] Bergstrom, J.D., Bieber, A.L.: Arch. Biochem. Biophys. , 194, 107–116 (1979)
[9] Glantz, M.D., Lewis, A.S.: Methods Enzymol., 51, 512–517 (1978) (Review)

[10] Rossi, C.A., Hakim, G., Solaini, G.: Biochim. Biophys. Acta, 526, 235–246 (1978)
[11] Fogle, P.J., Bieber, A.L.: Prep. Biochem., 5, 59–77 (1975)
[12] Lewis, A.S., Glantz, M.D.: J. Biol. Chem., 249, 3862–3866 (1974)
[13] Kumar, K.S., Sitaramayya, A., Krishnan, P.S.: Biochem. J., 128, 1079–1088 (1972)
[14] Rabinowitz, J.C., Barker, H.A.: J. Biol. Chem., 218, 161–173 (1956)

# 1 NOMENCLATURE

**EC number**
   3.5.4.4

**Systematic name**
   Adenosine aminohydrolase

**Recommended name**
   Adenosine deaminase

**Synonymes**
   Deoxyadenosine deaminase
   Deaminase, adenosine

**CAS Reg. No.**
   9026-93-1

# 2 REACTION AND SPECIFICITY

**Catalysed reaction**
   Adenosine + $H_2O$ →
   → Inosine + $NH_3$

**Reaction type**
   Amidine hydrolysis

**Natural substrates**
   Adenosine + $H_2O$ [7]
   Purin (compounds) + $H_2O$ [13]

**Substrate spectrum**
   1 Adenosine + $H_2O$
   2 Deoxyadenosine + $H_2O$ [2, 6]
   3 6-Chloropurine riboside + $H_2O$ [6]
   4 5-AMP + $H_2O$ [10]
   5 2-dAMP + $H_2O$ [10]
   6 ADP + $H_2O$ [10]
   7 ATP + $H_2O$ [10]
   8 More [11, 20]

## Product spectrum
1 Inosine + $NH_3$
2 Deoxyinosine + $NH_3$
3 ?
4 5-IMP + $NH_3$
5 2-dIMP + $NH_3$
6 IDP + $NH_3$
7 ITP + $NH_3$
8 ?

## Inhibitor(s)
p-Choromercuriphenylsulfonate [2, 3]; 2-Deoxyinosine [3]; Guanosine [3]; Purine riboside [3]; Adenine [3]; 2-Amino, 4-hydroxy-pteridine [3]; Folic acid [3], $Zn^{2+}$ [6]; $Hg^{2+}$ [6]; $Ag^+$ [6]; $Pb^{2+}$ [6]; EDTA [6]; Iodoacetic acid [6]; Iodoacetamide [6]; Aminoxyacetic acid [6]; $Cu^{2+}$ [10]; N-Ethylmaleimide [6]; $Cd^{2+}$ [10]; $Mn^{2+}$ [10]; 4-Amino-5-imidazole carboxamide ribonucleoside [11]; 4-Amino-5-imidazole carboxamide HCl [11]; 2, 6-Diaminopurinesulfate [11]; 6-Chloropurine [11]; Iodopurine [11]; Adenine [11]; Erythro-9-(–2-hydroxy-3-nonyl)adenine [12]; 1, 6-Dihydro-6-(hydroxymethyl)purine riboside [12]; Deoxycoformicin [12, 14]; Coformicin [21]

## Cofactor(s)/prostethic group(s)

## Metal compounds/salts
$NH_4Cl$ (stimulation) [3]

## Turnover number (min⁻¹)

## Specific activity (U/mg)
2.7 (larger enzyme) [2]; 47.4 (smaller enzyme) [2]; 180 [3]; More [4, 5, 6, 8, 10, 13, 18, 19, 20, 21]

## $K_m$-value (mM)
0.06 (adenosine) [2]; More [2–5, 7, 10, 13, 14, 15, 17, 18, 19]

## pH-optimum
7.0 [2, 5]; 5.5 (intermediate form of the enzyme) [7]; 7.0–7.4 (large and small form of the enzyme) [7]; 5.5–8.5 [10]; 7–10 [13]; More [11, 14, 15, 20, 21]

## pH-range
3.5–9 [10]

## Temperature optimum (°C)
37–40 [10]

## Temperature range (°C)

## 3 ENZYME STRUCTURE,

**Molecular weight**
35480 (bovine, gel filtration) [1]
230000 (human, gel filtration, large enzyme) [2]
47000 (human, gel filtration, small enzyme) [2]
130000 (Bay scallop, gel filtration, sucrose density centrifugation) [3]
More [4–7, 11, 12, 13, 16, 17, 18, 19, 20, 21]

**Subunits**
Monomer [16, 19]

**Glycoprotein/Lipoprotein**
–

## 4 ISOLATION/PREPARATION

**Source organism**
Bovine [1, 17]; Human (2 adenosine deaminases: large (2300000), small (47000) [2], 3 isoenzymes [4], 3 forms [7]) [2, 4, 7, 9, 11, 12, 18, 19]; Bay scallop [3]; Calf [5]; Streptomyces aureofaciens (3 forms) [10]; Micrococcus sodonensis [6]; Bacillus cereus [13]; Pseudomonas iodinum [8]; Plasmodium falciparum [14]; Pig [15]; Rat [16]; Azotobacter vinelandii [20]; E. coli [21]

**Source tissue**
Placenta [1]; Lung [2]; Stomach [2]; Digestive diverticulum [3]; Spleen [17]; Erythrocytes [4, 11]; Thymus [5, 12]; Seminal plasma [9]; Thyroid gland [15]; Hepatoma cells [16]; T and B-Lymphoblasts [17]; Leucemic granulocytes [19]; More [7]

**Localisation in source**
Membrane [5]; More [7]

**Purification**
Bovine [1, 17]; Bay scallop [3]; Calf [5]; Micrococcus sodonensis [6]; Pseudomonas iodinum [8]; Streptomyces aureofaciens [10]; Human [2, 4, 9, 11, 12, 18, 19]; Bacillus cereus [13]; Pig [15]; Rat [16]; Azotobacter vinelandii [20]; E. coli [21]

**Crystallization**
[8]

**Cloned**
–

**Renaturated**
–

## 5 STABILITY

pH
5–7 (at room temperature) [3]; 8–10 (highest stability, 4°C) [13]

Temperature (°C)
2–4 (12 months in 2-mercaptoethanol) [1]; 30 (1mM 2-mercaptoethanol, 1 week) [1]; 68 (1 hour, pH 5–7, little loss of activity) [3]; 23–37 (rapid loss of activity in dilute solutions) [6]; More (Mg$^{2+}$, Mn$^{2+}$: protection against denaturation at 37°C [6]) [2, 4, 10, 18]

Oxidation

Organic solvent
Ethyl alcohol (stabilizes) [8]; Glycerol (stabilizes) [13]; Ethylene glycol (stabilizes) [21]

General stability information
Large enzyme more stable than small one [2]; Dilute solutions (unstable) [6]; Freezing of dilute solutions (unstable) [6]; Mg$^{2+}$ (protection against denaturation at 37°C) [6]; Mn$^{2+}$ (protection against heat denaturation at 37°C [6], inhibition [10]) [6]; NH$_4^+$ (stabilizes) [13]; K$^+$ (stabilizes) [13]

Storage
2–4°C, 1mM 2-mercaptoethanol, 12 months 30°C, 1 week [1]; 10% loss of activity after 6 months, 50% glycerol, −18°C [10]

---

## 6 CROSSREFERENCES TO STRUCTURE DATABANKS

PIR/MIPS code
DUHUA (human); DUMSA (mouse); A21127 (human); A24638 (human)

Brookhaven code

---

## 7 LITERATURE REFERENCES

[1] Sim, M.K., Maguire, M.H.: Eur. J. Biochem., 23, 17–21 (1971)
[2] Akedo, H., Nishihara, H., Shinkai, K., Komatsu, K., Ishikawa, S.: Biochim. Biophys. Acta, 276, 257–271 (1972)
[3] Harbison, G.R., Fisher, J.R.: Arch. Biochem. Biophys., 154, 84–95 (1973)
[4] Osborne, W.R.A., Spencer, N.: Biochem. J., 133, 117–123 (1973)
[5] Rossi, C.A., Lucacchini, A.: Biochem. Soc. Trans., 2 (552nd Meeting, Galway), 1313–1315 (1974)
[6] Pickard, M.A.: Can. J. Biochem., 53, 344–353 (1974)
[7] Van Der Weyden, M.B., Kelley, W.N.: J. Biol. Chem., 251 (18), 5448–5456 (1976)

[8] Sakai, T., Jun, H.: FEBS Lett., 86 (2) , 174–178 (1978)
[9] James, M., Crabbe, C., Kavanagh, J.P.: Biochem. Soc. Trans., 5 (568th Meeting, Aberdeen) , 735–737 (1977)
[10] Rosinová, M., Zelinková, E., Zelinka, J.: Collect. Czech. Chem. Commun., 43, 2324–2329 (1978)
[11] Agarwal, R.P., Parks, R.E.: Methods Enzymol., 51, 502–507 (1978)
[12] Philips, A.V., Robbins, D.V., Coleman, M.S.: Biochemistry , 26, 2893–2903 (1987)
[13] Gabellieri, E., Bernini, S., Piras, L., Cioni, P., Balestreri, E., Cercignani, G., Felicioli, R.: Biochim. Biophys. Acta, 884, 490–496 (1986)
[14] Daddona, P.E., Wiesmann, W.P., Lambros, C., Kelley, W.N.: J. Biol. Chem., 259 (3) , 1472–1475 (1984)
[15] Jaroszewicz, L., Wyszynska, M.: Biochem. Biophys. Res. Commun., 109 (1) , 138–145 (1982)
[16] Hunt, S.W., Hoffee, P.A.: J. Biol. Chem., 257 (23) , 14239–14244 (1982)
[17] Lewis, A.S., McCalla, C., Murphy, L.: J. Biochem., 3, 378–386 (1981)
[18] Daddona, P.E.: J. Biol. Chem., 256 (23) , 12496–12501 (1981)
[19] Wiginton, D.A., Coleman, M.S., Hutton, J.J.: Biochem. J., 195, 389–397 (1981)
[20] Tsukada, T., Yoshino, M.: Arch. Microbiol., 128, 228–232 (1980)
[21] Nygaard, P.: Methods Enzymol., 51, 508–512 (1978)

# 1 NOMENCLATURE

**EC number**
3.5.4.5

**Systematic name**
Cytidine aminohydrolase

**Recommended name**
Cytidine deaminase

**Synonymes**
Deaminase, cytidine
Cytosine nucleoside deaminase

**CAS Reg. No.**
9025-06-3

---

# 2 REACTION AND SPECIFICITY

**Catalysed reaction**
Cytidine + $H_2O$ →
→ uridine + $NH_3$

**Reaction type**
Amidine hydrolysis

**Natural substrates**
Cytidine + $H_2O$ [5, 21]

**Substrate spectrum**
1 Cytidine + $H_2O$
2 Deoxycytidine + $H_2O$ [1, 9]
3 $N^4$-Methylcytidine + $H_2O$ (r) [3]
4 5, 6-Dihydrocytidine + $H_2O$ [8, 10]
5 More [11, 13, 18, 19, 21, 6]

**Product spectrum**
1 Uridine + $NH_3$
2 Deoxyuridine + $NH_3$
3 Uridine + methylamine (r)
4 5, 6-Dihydrouridine + $NH_3$
5 ?

### Inhibitor(s)

3, 4, 5, 6-Tetrahydrouridine (competitive) [3, 7]; EDTA (at dialysis [3], no effect [10]) [3]; Uridine (competitive) [3]; 5, 6-Dihydrouridine (competitive) [3]; Urea [6]; $Zn^{2+}$ [6]; $Cu^{2+}$ [6]; $Fe^{2+}$ [6]; $Fe^{3+}$ [6]; $Hg^+$ [6]; Nucleosides (competitive) [6, 9, 12, 17, 18, 20]; Cytidine (above 0.5 mM, substrate inhibition) [9]; $NH_4^+$ (no effect [3, 10]) [4]; N-Ethylmaleimide [11]; p-Hydroxymercuribenzoate [11, 14]; 5-(Chloromercuri)cytidine [15]; Phosphopyrimidine nucleoside I [16]

### Cofactor(s)/prostethic group(s)

### Metal compounds/salts

---

### Turnover number ($min^{-1}$)
3900 [15]

### Specific activity (U/mg)
8.9 [3, 10]; 344 [6]; 0.2 [9]; More [11, 15]

### $K_m$-value (mM)
0.17 (cytidine) [1]; 0.09 (cytosine desoxyriboside) [1]; 0.21 [3]; More [6, 8, 9, 10, 14, 17, 19]

### pH-optimum
6–11 [3]; 6–7.5 [6]; 7.2 [9]; 6.5–8.5 [14]

### pH-range
3–11 [3]

### Temperature optimum (°C)

### Temperature range (°C)

---

## 3 ENZYME STRUCTURE

### Molecular weight
73000 (E. coli, gel filtration, sucrose gradient centrifugation) [3, 10]
54000 (E. coli, gel filtration) [6]
57000 (wheat, gel filtration) [9]
74000 (mouse, gel filtration) [11]
More [14, 15, 17, 20]

### Subunits
Dimer (2 × 35000, E. coli [15], 2 × 33000, E. coli [17]) [15, 17]

**Glycoprotein/Lipoprotein**

–

---

## 4 ISOLATION/PREPARATION

**Source organism**

E. coli [1, 3, 6, 8, 10, 12, 15, 16, 17, 18, 20]; Wheat [1, 9]; Human [5, 7, 10, 19];
Mouse [11, 13]; Yeast [1, 9]; Crithidia fasciculata [14]; Trypanosoma cruzi
[14]; Fungi [4]; Dog [19]

**Source tissue**

Leaf [2]; Lymphocytes [5]; Liver [7, 10, 19]; Spleen [11]; Kidney [13]; More
(other tissues) [10]

**Localisation in source**

**Purification**

E. coli [3, 6, 10, 15, 17, 20]; Human [10]; Mouse [11]; Yeast [9]

**Crystallization**

–

**Cloned**

–

**Renaturated**

–

---

## 5 STABILITY

**pH**

5.5–9.5 (some hours, 4°C) [9]; 5 (rapid inactivation below) [9]; 4 (irreversible
loss of activity below) [10]

**Temperature (°C)**

40–62 (5 minutes) [9]; 21 (50% loss of activity after 1 day) [11]; More [11]

**Oxidation**

Unstable in air, even in frozen state [14]; Stabilization under anaerobic conditions [14]

**Organic solvent**

Acetone (inactivation) [1]

**General stability information**

Not inactivated by dialysis or freezing [1]

---

## Storage

45% loss of activity after 3 weeks at 4°C [3, 10]; 4 weeks at –20°C [11];
More [14]

## 6 CROSSREFERENCES TO STRUCTURE DATABANKS

### PIR/MIPS code
JE0022 (Bacillus subtilis)

### Brookhaven code

## 7 LITERATURE REFERENCES

[1] Wang, T.P., Sable, H.Z., Lampen, J.D.: J. Biol. Chem., 184, 17–28 (1950)
[2] Roberts, D.W.A.: J. Biol. Chem., 222, 259–270 (1956)
[3] Cohen, R.M., Wolfenden, R.: J. Biol. Chem., 246 (24) , 7561–7565 (1971)
[4] Wolfenden, R.: Biochemistry, 8, 2409–2412 (1969)
[5] Abell, C.W., Marchand, N.W.: Nature (New Biol.) , 244, 217–219 (1973)
[6] Hosono, H., Kuno, S.: J. Biochem., 74, 797–803 (1973)
[7] Wentworth, D.F., Wolfenden, R.: Biochemistry, 14 (23) , 5099–5104 (1975)
[8] Evans, B.E., Mitchell, G.N., Wolfenden, R.: Biochemistry, 14 (3) , 621–624 (1975)
[9] Ipata, P.L., Cercignani, G.: Methods Enzymol., LI, 394–401 (1978)
[10] Wentworth, D.F., Wolfenden, R.: Methods Enzymol., LI, 401–407 (1978)
[11] Rothman, I.K., Malathi, V.G., Silber, R.: Methods Enzymol., LI, 408–412 (1978)
[12] Ashley, G.W., Bartlett, P.A.: Biochem. Biophys. Res. Commun., 108 (4) , 1467–1474
(1982)
[13] Kára, J., Bártova, M., Ryba, M., Hrebabecky, H., Brokés, J., Novotny, L., Baránek, J.:
lect. Czech. Chem. Commun., 47, 2824–2830 (1982)
[14] Kidder, G.W.: J. Protozool., 31 (2) , 298–300 (1984)
[15] Ashley, G.W., Bartlett, P.A.: J. Biol. Chem., 259 (21) , 13615–13620 (1984)
[16] Ashley, G.W., Bartlett, P.A.: J. Biol. Chem., 259 (21) , 13621–13627 (1984)
[17] Vita, A., Amici, A., Cacciamani, T., Lanciotti, M., Magni, G.: Biochemistry, 24,
6020–6024 (1985)
[18] Holy, A., Ludzisa, A., Votruba, I., Sediva, K.: Collect. Czech. Chem. Commun., 50,
393–417 (1985)
[19] Fanucchi, M.P., Watanabe, K.A., Fox, J.J., Chou, T.: Biochem. Pharmacol., 35 (7) ,
1199–1201 (1986)
[20] Vita, A., Amici, A., Lanciotti, M., Cacciamani, T., Magni, G.: Ital. J. Biochem., 35 (3) ,
145A–147A (1986)
[21] Nygaard, P.: Adv. Exp. Med. Biol., 195B, 415–420 (1986)

# 1 NOMENCLATURE

**EC number**
3.5.4.6

**Systematic name**
AMP aminohydrolase

**Recommended name**
AMP deaminase

**Synonymes**
Deaminase, adenylate
Adenylic acid deaminase
Adenylic deaminase
AMP aminase
Adenylate deaminase
5-AMP deaminase
AMP aminohydrolase
Adenosine 5-monophosphate deaminase
5-Adenylate deaminase
Adenyl deaminase
5-Adenylic acid deaminase
5-AMP aminohydrolase
Adenosine 5-phosphate aminohydrolase
Adenylate desaminase
Adenosine 5-phosphate deaminase
Adenylate aminohydrolase
Adenosine monophosphate deaminase

**CAS Reg. No.**
9025-10-9

---

# 2 REACTION AND SPECIFICITY

**Catalysed reaction**
AMP + $H_2O$ →
→ IMP + $NH_3$

**Reaction type**
Amidine hydrolysis

**Natural substrates**
AMP + $H_2O$

## Substrate spectrum
1 AMP + $H_2O$
2 5-Deoxyadenylic acid + $H_2O$ (at 7% or less the rate for AMP) [3, 11]
3 Adenosine monosulfate + $H_2O$ [26]
4 Adenosine phosphoramidate + $H_2O$ [26]

## Product spectrum
1 IMP + $NH_3$
2 dIMP + $NH_3$
3 Inosine monosulfate + $NH_3$
4 ?

## Inhibitor(s)
$Mn^{2+}$ [3]; Protamine sulfate [3]; Orthophosphate [3]; Pyrophosphate [3];
$F^-$ [3]; $Zn^{2+}$ [3]; $Cu^{2+}$ [3]; $Fe^{3+}$ [3]; $Ag^+$ [3]; $Ca^{2+}$ [3]; $Cd^{2+}$ [3]; $Ni^{2+}$ [3];
$Hg^{2+}$ [5]; p-Hydroxymercuribenzoate [5]; Iodoacetate (not found [3]) [5];
EDTA [5, 17]; Halide anions [7]; 2, 3-Diphosphoglyceric acid [7]; 5-IMP (in
the absence of ATP) [7]; 2-AMP (in the absence of ATP) [7]; 3-AMP (in the
absence of ATP) [7]; GTP [8]; Polyphosphates [8]; Metaphosphates [8];
ADP [22]

## Cofactor(s)/prostethic group(s)

## Metal compounds/salts
$Zn^{2+}$ (human muscle AMP-deaminase probably a zinc metalloenzyme) [16]

## Turnover number ($min^{-1}$)
18300 (substrate concentration 0.045 mM) [3]; 598000 (calculated for
saturating concentration) [3]

## Specific activity (U/mg)
1.12 [4]; 140; 5.7 [7]; 976 [9]; 343.3 [11]; More [17, 21, 22, 25]

## $K_m$-value (mM)
1.4 [3]; 13 [5]; 1 [10]; More [12, 13, 17, 20, 21, 22, 24, 25]

## pH-optimum
6.4 (pH optimum about 6, depending on purification procedure and buffer)
[3]; 6.0–6.2 (succinate buffer) [5]; 5.6–6.8 (Tris-acetic acid buffer) [5]; 7.1 [6];
More [7, 8, 10, 11, 13, 14, 15, 17, 18, 19, 20, 24]

## pH-range
5.5–7.5 [3, 17]

## Temperature optimum (°C)

## Temperature range (°C)

## 3 ENZYME STRUCTURE

### Molecular weight

320000 (rabbit, calculation from sedimentation constant and diffusion coefficient) [1]
285000 (human, sucrose density centrifugation) [8]
276000 (chicken, sucrose density centrifugation) [9]
74000 (human, SDS-PAGE) [10]

### Subunits

Tetramer (4 × 80000 [15], 4 × 85000 [25], 4 × 69000, chicken, SDS-PAGE, [9]) [8, 9, 12, 15, 25]

### Glycoprotein/Lipoprotein

–

## 4 ISOLATION/PREPARATION

### Source organism

Rabbit [1–3]; Dog [4]; Pea [5]; Chicken [6, 9, 20]; Yeast [11, 12]; Spinach [13]; Pig [14, 21]; Orconectes limosus [17]; Human (4 isoenzymes [15]) [7, 8, 10, 15, 16, 25]; Chelon labrosus [18]; Scyliorhinus caniculus [19]; Trout [22]; Rat [23]; Rabbit (2 isoenzymes) [24]; Sheep [26]

### Source tissue

Skeletal muscle [1, 15, 16, 20, 24, 25]; Seeds [5]; Brain [4, 15 , 26]; Erythrocytes [6, 7, 8, 10, 15]; Leaf [13]; Thyroid gland [14]; Liver [15]; Tail muscle [17]; Heart [21]; Gill [22]; Small intestine [23]

### Localisation in source

### Purification

Rabbit [1]; Dog [4]; Chicken [6, 9]; Human [7, 8, 10]; Yeast [11, 12]; Spinach [13]; Pig [14]; Orconectes limosus [17]; Chelon labrosus [18]; Scyliorhinus caniculus [19]; Pig [21]; Trout [22]

### Crystallization

[1]

### Cloned

–

### Renaturated

–

## 5 STABILITY

pH
   6.8 (highest stability) [3]

Temperature (°C)
   0–4 (at high concentrations, 10 days) [6]; More [3]

Oxidation

Organic solvent

General stability information
   Repeated freezing and thawing (stable) [3]; ATP, ADP, GTP and alkali metal ions protect against heat inactivation [5]; Freezing (unstable) [10];
   Sulfhydryl reagents and monovalent cations in high concentrations stabilize [10]

Storage
   0–4°C or –20°C for up to 10 days, 30 mg/ml protein [6]; 4 months, –20°C,
   less than 10% loss of activity [7]

---

## 6 CROSSREFERENCES TO STRUCTURE DATABANKS

PIR/MIPS code
   A33365 (yeast, Saccharomyces cerevisiae); A27366 (skeletal muscle, rat)

Brookhaven code

---

## 7 LITERATURE REFERENCES

[1] Lee, Y.: J. Biol. Chem., 227, 987–992 (1957)
[2] Lee, Y.: J. Biol. Chem., 227, 993–998 (1957)
[3] Lee, Y.: J. Biol. Chem., 227, 999–1007 (1957)
[4] Mendicino, J., Muntz, J.A.: J. Biol. Chem., 223, 178–183 (1958)
[5] Turner, D.H., Turner, J.F.: Biochem. J., 79, 143–147 (1961)
[6] Kawamura, Y.: J. Biochem., 72, 21–28 (1972)
[7] Lian, C., Harkness, D.R.: Biochim. Biophys. Acta, 341, 27–40 (1978)
[8] Yun, S., Suelter, C.H.: J. Biol. Chem., 253 (2), 404–408 (1978)
[9] Kruckeberg, W.C., Lemley, S., Chilson, O.P.: Biochemistry, 17 (21), 4376–4383
    (1978)
[10] Nathans, G.R., Chang, D., Deuel, T.F.: Methods Enzymol., LI, 497–502 (1979)
[11] Yoshino, M., Murakami, K., Tsushima, K.: Biochim. Biophys. Acta, 570, 157–166
    (1979)
[12] Murakami, K.: J. Biochem., 86 (5), 1331–1336 (1979)
[13] Yoshino, M., Murakami, K.: Z. Pflanzenphysiol., 99, 331–338 (1980)
[14] Stelmach, H., Jarosewicz, L.: Biochem. Biophys. Res. Commun., 101 (1), 144–152
    (1981)

[15] Ogasawara, N., Goto, H., Yamada, Y., Watanabe, T., Asano, T.: Biochim. Biophys. Acta, 714, 298–306 (1982)
[16] Stankiewicz, A.: Int. J. Biochem., 13, 1177–1183 (1981)
[17] Stankiewicz, A.: Comp. Biochem. Physiol., 72B, 127–132 (1982)
[18] Raffin, J.P.: Comp. Biochem. Physiol., 75B, 465–469 (1983)
[19] Raffin, J.P.: Comp. Biochem. Physiol., 75B, 461–464 (1983)
[20] Kaletha, K.: Biochim. Biophys. Acta, 784, 90–92 (1984)
[21] Verwoerd, T., Harmsen, E., Achterberg, P., De Jong, J.W.: Adv. Exp. Med. Biol., 165 (Purine Metab., Man-4, Pt. B) , 501–504 (1984)
[22] Raffin, J.P.: J. Comp. Physiol., B, 154, 55–63 (1984)
[23] Spychala, I., Marszalek, J., Kucharczyk, E.: Biochim. Biophys. Acta, 880, 123–130 (1986)
[24] Raggi, A., Ranieri-Raggi, M.: Biochem. J., 242, 875–879 (1987)
[25] Kaletha, K., Nowak, G.: Biochem. J., 249, 255–261 (1988)
[26] Ito, K., Yamamoto, H., Mizugaki, M.: J. Biochem., 103 , 259–262 (1988)

# 1 NOMENCLATURE

**EC number**
3.5.4.7

**Systematic name**
ADP aminohydrolase

**Recommended name**
ADP deaminase

**Synonymes**
Deaminase, adenosine diphosphate
Adenosine diphosphate deaminase
Adenosinepyrophosphate deaminase
ADP-aminohydrolase

**CAS Reg. No.**
9027-79-6

# 2 REACTION AND SPECIFICITY

**Catalysed reaction**
ADP + $H_2O$ →
→ IDP + $NH_3$

**Reaction type**
Amidine hydrolysis

**Natural substrates**

**Substrate spectrum**
1 ADP + $H_2O$
2 ATP + $H_2O$ [2]
3 Adenosine sulfate + $H_2O$ [2]
4 3'-AMP + $H_2O$ [2]
5 5'-AMP + $H_2O$ [2]
6 dAMP + $H_2O$ [2]
7 3',5'-cAMP + $H_2O$ [2]
8 Adenosine + $H_2O$ [2]
9 6-Chloropurine ribotide + $H_2O$ [3]

**Product spectrum**
1 IDP + $NH_3$
2 ITP + $NH_3$
3 Inosine sulfate + $NH_3$
4 3'-IMP + $NH_3$
5 5'-IMP + $NH_3$
6 dIMP + $NH_3$
7 3',5'-IMP + $NH_3$
8 ?
9

**Inhibitor(s)**
p-Chloromercuribenzoate [2]; $Fe^{3+}$ [2]; $Ag^+$ [2]; $Zn^{2+}$ [2]; $Ca^{2+}$ [2]; $Mg^{2+}$
[2]; $Hg^{2+}$ [2]; 6-Chloropurine ribotide (inhibites deamination) [3];
Pyrophosphate (dechlorination) [3]

**Cofactor(s)/prostethic group(s)**

**Metal compounds/salts**

**Turnover number** (min⁻¹)

**Specific activity** (U/mg)

**$K_m$-value** (mM)
0.04 (Aspergillus ochraceus) [2]; 0.066 (Aspergillus melleus) [2]; More [3]

**pH-optimum**
3.4 [2, 3]

**pH-range**

**Temperature optimum** (°C)

**Temperature range** (°C)

**3 ENZYME STRUCTURE**

**Molecular weight**

**Subunits**

**Glycoprotein/Lipoprotein**
–

# 4 ISOLATION/PREPARATION

**Source organism**
  Rabbit [1]; Aspergillus ochraceus [2]; Aspergillus melleus [2, 3]

**Source tissue**
  Muscle [1]

**Localisation in source**
  Actomyosin [1]

**Purification**
  Aspergillus ochraceus [2]

**Crystallization**
  –

**Cloned**
  –

**Renaturated**
  –

---

# 5 STABILITY

**pH**
  5.0–7.0 [2]

**Temperature (°C)**

**Oxidation**

**Organic solvent**

**General stability information**

**Storage**

---

# 6 CROSSREFERENCES TO STRUCTURE DATABANKS

**PIR/MIPS code**

**Brookhaven code**

# 7 LITERATURE REFERENCES

[1] Deutsch, A., Nilsson, R.: Acta Chem. Scand., 8, 1898–1906 (1954)
[2] Chung, S.-T., Aida, K., Uemura, T.: J. Gen. Appl. Microbiol., 13, 335–347 (1967)
[3] Chung, S.-T., Ito, S., Aida, K., Uemura, T.: J. Gen. Appl. Microbiol., 14, 111–119 (1968)

## 1 NOMENCLATURE

**EC number**
3.5.4.8

**Systematic name**
4-Aminoimidazole aminohydrolase

**Recommended name**
Aminoimidazolase

**Synonymes**
4-Aminoimidazole hydrolase [1]

**CAS Reg. No.**
9025-17-6

## 2 REACTION AND SPECIFICITY

**Catalysed reaction**
4-Aminoimidazole + $H_2O$ →
→ unidentified product + $NH_3$

**Reaction type**
Amidine hydrolysis

**Natural substrates**
4-Aminoimidazole + $H_2O$

**Substrate spectrum**
1 4-Aminoimidazole + $H_2O$

**Product spectrum**
1 Unidentified product (containing glycine) + $NH_3$

**Inhibitor(s)**
EDTA [2]; 3-Amino-1, 2, 4-triazole [1]; 4-Hydroxy-1, 2, 3-triazole [1]; Histamine [1]

**Cofactor(s)/prostethic group(s)**

**Metal compounds/salts**
$Fe^{2+}$ [1]; More (divalent cations) [1]

**Turnover number** (min$^{-1}$)

Specific activity (U/mg)
   140 [1]

$K_m$-value (mM)
   1.8 (4-aminoimidazole) [1]

pH-optimum
   7.0 [1]

pH-range

Temperature optimum (°C)
   37 [1]

Temperature range (°C)

---

## 3 ENZYME STRUCTURE

Molecular weight

Subunits

Glycoprotein/Lipoprotein
   −

---

## 4 ISOLATION/PREPARATION

Source organism
   Clostridium cylindrosporum [1]

Source tissue
   Cell [1]

Localisation in source

Purification
   Clostridium cylindrosporum [1]

Crystallization
   −

Cloned
   −

Renaturated
   −

# 5 STABILITY

pH

Temperature (°C)

Oxidation

Organic solvent

General stability information

Storage

# 6 CROSSREFERENCES TO STRUCTURE DATABANKS

PIR/MIPS code

Brookhaven code

# 7 LITERATURE REFERENCES

[1] Rabinowitz, J.R., Pricer, W.E.: J. Biol. Chem., 222, 537–554 (1956)
[2] Rabinowitz, J.C.: J. Biol. Chem., 218, 175 (1956)

## 1 NOMENCLATURE

**EC number**
3.5.4.9

**Systematic name**
5, 10-Methenyltetrahydrofolate 5-hydrolase (decyclizing)

**Recommended name**
Methenyltetrahydrofolate cyclohydrolase

**Synonymes**
Citrovorum factor cyclodehydrase
Cyclohydrolase [12]
Formyl-methenyl-methylenetetrahydrofolate synthetase (combined)
More (trifunctional enzyme combines methylenetetrahydrofolate
dehydrogenase, methenyl tetrahydrofolate cyclohydrolase and
formyltetrahydrofolate synthetase activity, in some eucaryotic cells) [7–10]

**CAS Reg. No.**
9027-97-8

## 2 REACTION AND SPECIFICITY

**Catalysed reaction**
5, 10-Methenyltetrahydrofolate + $H_2O$ →
→ 10-formyltetrahydrofolate
More (in eucaryotes occurs as trifunctional enzyme also having
methylenetetrahydrofolate dehydrogenase ($NADP^+$) (EC 1.5.1.5) and
formate-tetrahydrofolate ligase (EC 6.3.4.3) activity, in some procaryotes
occurs as a bifunctional enzyme also having dehydrogenase (EC 1.5.1.5)
activity or formiminotetrahydrofolate cyclodeaminase (EC 4.3.1.4) activity)

**Reaction type**
Amidine hydrolysis

**Natural substrates**
5, 10-Methenyltetrahydrofolate + $H_2O$

## Substrate spectrum

1 5, 10-Methenyltetrahydrofolate + $H_2O$ (r)
2 More (in eucaryotes occurs as trifunctional enzyme also having methylenetetrahydrofolate dehydrogenase (NADP$^+$) (EC 1.5.1.5) and formate-tetrahydrofolate ligase (EC 6.3.4.3) activity, in some procaryotes occurs as a bifunctional enzyme also having dehydrogenase (EC 1.5.1.5) activity or formiminotetrahydrofolate cyclodeaminase (EC 4.3.1.4) activity)

## Product spectrum

1 10-Formyltetrahydrofolate
2 ?

## Inhibitor(s)

Methotrexate [11, 12, 15]; Aminopterin [11, 12]; Tubercidin [5]; Suramin [5]; Folate (derivatives, reduced form) [10, 11]; p-Chloromercuribenzoate [5, 11]; NADP$^+$ (trifunctional enzyme of some eucaryotic cells, bifunctional enzyme of some procaryotes) [10]; NAD$^+$ (trifunctional enzyme of some eucaryatic cells, bifunctional enzyme of some procaryotes) [7–10, 17–19]; ATP (trifunctional enzyme of some eucaryotic cells) [10]; Fe$^{2+}$ [5, 15]; Zn$^{2+}$ [5, 15]; Cu$^{2+}$ [5, 15]; Mg$^{2+}$ [5, 15]

## Cofactor(s)/prostethic group(s)

## Metal compounds/salts

## Turnover number (min$^{-1}$)

8700 [10]

## Specific activity (U/mg)

469 [2, 4]; 1.16 [1]; 0.42 [1]; 0.55 [1]

## K$_m$-value (mM)

0.074 (5, 10-methenyltetrahydrofolate) [12]; 0.04 (5, 10-methenyltetrahydrofolate) [12]; 0.25 (5, 10-methenyltetrahydrofolate) [14–16]; 0.22 (methylene tetrahydrofolate) [10]; More [7]

## pH-optimum

6.5 [5]; 6.6–7.6 [2, 4]; 7.7 [11]; 8.0 [15]

## pH-range

5.8–8 [2, 4]

Temperature optimum (°C)

Temperature range (°C)

---

## 3 ENZYME STRUCTURE

### Molecular weight
25500 (gel electrophoresis, Clostridium formicoaceticum) [2, 4]
41000 (sedimentation equilibrium, Clostridium formicoaceticum) [2, 4]
90000–100000 (gel electrophoresis, gel filtration, pig, sheep,
Saccharomyces cerevisiae) [6, 7, 10]
136000–150000 (gel filtration, pig, Saccharomyces cerevisiae) [7, 8, 10]
190000–226000 (gel filtration, mosquito, sheep) [5–7]

### Subunits
Monomer (1 × 100000, sheep, gel electrophoresis) [7, 10]
Dimer (2 × 25500, gel electrophoresis, Clostridium formicoaceticum) [2, 4]
Dimer (2 × 100000, gel electrophoresis, sheep) [7]
Dimer (2 × 90000–97000, gel electrophoresis) [6]

### Glycoprotein/Lipoprotein
–

---

## 4 ISOLATION/PREPARATION

### Source organism
Pig (trifunctional enzyme, monofunctional enzyme) [8–10, 12]; Rabbit
(monofunctional enzyme) [12]; Bovine [14]; Sheep (trifunctional enzyme)
[7, 16]; Mosquito (monofunctional enzyme) [5]; Pea [11]; Clostridium
acidi-urici [13]; Clostridium cyclindrosporum [13]; Clostridium
formicoaceticum (monofunctional enzyme) [2, 4]; Clostridium
thermoautotrophicum [3]; Clostridium thermoaceticum (bifunctional enzyme) [17]; E. coli (bifunctional enzyme) [18]; Saccharomyces cerevisiae
[7]; Butyribacterium methylotrophicum [1]; More [11, 12]

### Source tissue
Liver [7–10, 12, 14]; Seedling [11]

### Localisation in source
Soluble [11]; Mitochondria [11]

### Purification
Pig (trifunctional enzyme) [10]; Clostridium formicoaceticum (monofunctional enzyme) [2, 4]; Clostridium thermoaceticum (bifunctional enzyme)
[17]

---

**Crystallization**

–

**Cloned**

–

**Renaturated**

–

## 5 STABILITY

**pH**

**Temperature (°C)**
45 (stable up to) [4]; 60 (5 minutes, 91% loss of activity) [11]

**Oxidation**

**Organic solvent**
Glycerol (stabilizes) [11]

**General stability information**
Glycerol stabilizes [11]

**Storage**
Purified enzyme, pH 7.3, 20% glycerol, 1 week [10]; Purified enzyme, pH 7,
5°C, 1 week [2, 4]; Lyophilized [4]

## 6 CROSSREFERENCES TO STRUCTURE DATABANKS

**PIR/MIPS code**

**Brookhaven code**

## 7 LITERATURE REFERENCES

[1] Kerby, R., Zeikus, J.G.: J. Bacteriol., 169 (12) , 5605–5609 (1987)
[2] Ljungdahl, L.G., Clark, J.E.: Methods Enzymol., 122, 385–391 (1986)
[3] Clark, J.E., Ragsdale, S.W., Ljungdahl, L.G., Wiegel, J.: J. Bacteriol., 151 (1) ,
    507–509 (1982)
[4] Clark, J.E., Ljungdahl, L.G.: J. Biol. Chem., 257 (7) , 3833–3836 (1982)
[5] Jaffe, J.J., Curin, L.R., Smith, R.B.: Comp. Biochem. Physiol., 66 (B) , 597–600 (1980)
[6] Smith, G.K., Mueller, T., Wasserman, G.F., Taylor, W. D., Benkovic, S.J.:
    Biochemistry, 19 (18) , 4313–4321 (1980)
[7] Paukert, J.L., Rabinowitz, J.C.: Methods Enzymol., 66, 616–626 (1980)
[8] Tan, L.U.L., Mackenzie, R.E.: Can. J. Biochem., 57, 806–812 (1979)
[9] Cohen, L., Mackenzie, R.E.: Biochim. Biophys. Acta, 522, 311–317 (1978)

[10] Tan, L.U.L., Drury, E.J., Mackenzie, R.E.: J. Biol. Chem., 252 (3) , 1117–1122 (1977)

[11] Suzuki, N., Iwai, K.: Plant Cell Physiol., 14, 319–327 (1973)

[12] Tabor, H., Wyngarden, L.: J. Biol. Chem., 234 (7) , 1830–1846 (1959)

[13] Schnell, E., Rochow, E.G.: J. Am. Chem. Soc., 78, 4176–4178 (1956)

[14] Lambrozo, L., Greenberg, D.M.: J. Biol. Chem., 234, 1830–1846 (1959)

[15] Lombrozo, L., Greenberg, D.M.: Arch. Biochem. Biophys., 118, 297–304 (1967)

[16] Paukert, J.L., Straus, L.D., Rabinowitz, J.C.: J. Biol. Chem., 251, 5104–5111 (1976)

[17] O'Brien, W.E., Brewer, J.M., Ljungdahl, L.G.: J. Biol. Chem., 248, 403–408 (1973)

[18] Dev, I.K., Harvey, R.J.: J. Biol. Chem., 253, 4245–4253 (1978)

[19] Ljungdahl, L.G., O'Brien, W.E., Moore, M.R., Liu, M. T.: Methods Enzymol., 66, 599–609 (1980)

## 1 NOMENCLATURE

**EC number**
3.5.4.10

**Systematic name**
IMP 1, 2-hydrolase (decyclizing)

**Recommended name**
IMP cyclohydrolase

**Synonymes**
Inosinicase [1]
Inosinate cyclohydrolase

**CAS Reg. No.**
9013-81-4

---

## 2 REACTION AND SPECIFICITY

**Catalysed reaction**
5-Formamido-1-(5-phosphoribosyl)imidazole-4-carboxamide →
→ IMP + $H_2O$

**Reaction type**
Internal C-N condensation, $H_2O$-Elimination

**Natural substrates**
5-Formamido-1-(5-phosphoribosyl)imidazole-4-carboxamide

**Substrate spectrum**
1 5-Formamido-1-(5-phosphoribosyl)imidazole-4-carboxamide [1 , 4]
2 Trans-alpha,
   beta-diformamido-beta-(5'-phosphoribosylamino)acrylamide [4]

**Product spectrum**
1 IMP + $H_2O$ [1, 4]
2 IMP + $H_2O$ [4]

---

**Inhibitor(s)**

**Cofactor(s)/prostethic group(s)**

**Metal compounds/salts**

---

Turnover number (min$^{-1}$)

Specific activity (U/mg)
  More [1, 2, 3]

K$_m$-value (mM)
  1.28 (5'-phosphoribosyl-5-formamido-4-imidazole carboxamide) [3]

pH-optimum
  7.4 [1]; 7.25 [2]; 8.5 [3]

pH-range
  7.0–7.8 [1]

Temperature optimum (°C)
  38 [1]; 37 [2]

Temperature range (°C)

---

## 3 ENZYME STRUCTURE

Molecular weight
  350000 (mouse, gel filtration) [3]

Subunits
  Octamer (8 × 46000, mouse, gel filtration) [3]

Glycoprotein/Lipoprotein
  –

---

## 4 ISOLATION/PREPARATION

Source organism
  Pig [1]; Salmonella typhimurium [2]; Mouse [3]; Chicken [4]; Rat [4];
  Bacillus subtilus [5]

Source tissue
  Liver [1, 3, 4]; Ehrlich ascites tumor cells [3]

Localisation in source

Purification
  Pig [1]; Mouse [3]

Crystallization
  –

Cloned
  [5]

Renaturated
–

---

## 5 STABILITY

pH

Temperature (°C)
50 (up to) [3]

Oxidation

Organic solvent

General stability information

Storage
Several months at −15°C [1]; 2 years at −25°C [3]

---

## 6 CROSSREFERENCES TO STRUCTURE DATABANKS

PIR/MIPS code

Brookhaven code

---

## 7 LITERATURE REFERENCES

[1] Flaks, J.G., Erwin, M.J., Buchanan, J.M.: J. Biol. Chem., 229, 603–612 (1957)
[2] Gots, S.J., Dalal, F.R., Shumas, S.R.: J. Bacteriol., 99 (2) , 441–449 (1969)
[3] Geiger, R., Guglielmi, H.: Hoppe-Seyler's Z. Physiol. Chem., 356, 819–825 (1975)
[4] Baggott, J.E., Krumdieck, C.L.: Biochemistry, 18 (16) , 3501–3506 (1979)
[5] Ebbole, J.D., Zalkin, H.: J. Biol. Chem., 262 (17) , 8274–8287 (1987)

# 1 NOMENCLATURE

**EC number**
3.5.4.11

**Systematic name**
2-Amino-4-hydroxypteridine aminohydrolase

**Recommended name**
Pteridin deaminase

**Synonymes**
Acrasinase [1]

**CAS Reg. No.**
9025-04-1

# 2 REACTION AND SPECIFICITY

**Catalysed reaction**
2-Amino-4-hydroxypteridine + $H_2O$ →
→ 2, 4-dihydroxypteridine + $NH_3$

**Reaction type**
Amidine hydrolysis

**Natural substrates**
Pterins + $H_2O$

**Substrate spectrum**
1 2-Amino-4-hydroxypteridine + $H_2O$
2 Isoxanthopterin + $H_2O$ [7]
3 Tetrahydropterin + $H_2O$ [7]
4 Folic acid + $H_2O$ [3, 4]
5 Acrasin + $H_2O$ [1]
6 Pterins + $H_2O$ (variety of pterins) [5, 6]
7 More (the animal enzyme is specific for pterin, isoxanthopterin and
tetrahydropterin) [3, 5, 6]

**Product spectrum**
1 2, 4-Dihydroxypteridine + $NH_3$
2 7-Hydroxy-2, 4-dihydroxypteridine + $NH_3$
3 2, 4-Dihydroxypteridine + $NH_3$
4 ?
5 ?
6 ?
7 ?

**Inhibitor(s)**
  Azaguanine [5–7]; 2, 4-Dihydroxypteridine [7]; KCN [7];
  p-Chloromercuribenzoate [5, 6]; ZnCl$_2$ [4]

**Cofactor(s)/prostethic group(s)**

**Metal compounds/salts**

---

**Turnover number** (min$^{-1}$)

**Specific activity** (U/mg)
  0.825 [4]

**K$_m$-value** (mM)
  0.03 (2-amino-4-hydroxypteridine) [2, 7]; 1.6 (6-carboxypteridine) [6]; 1.3
  (6-carboxypteridin) [5]; 0.057 (6-carboxypteridin) [3]

**pH-optimum**
  6.3–6.7 [8]; 6.6 [11]; 6.5 [7]; 7.0 [4]; 7.3 [5, 6]; 8.0 [10]; 8.3 [3]

**pH-range**

**Temperature optimum** (°C)

**Temperature range** (°C)

---

## 3 ENZYME STRUCTURE

**Molecular weight**
  110000 (gel filtration, Bacillus megaterium) [6]

**Subunits**

**Glycoprotein/Lipoprotein**
  –

---

## 4 ISOLATION/PREPARATION

**Source organism**
  Rat [7]; Honeybee [7]; Bacillus megaterium [5, 6]; Bacillus subtilis [5]; Alcaligenes metalcaligenes [8]; Alcaligenes faecalis [9]; Bombyx mori [10, 11];
  Pseudomonas sp. [3]; Aspergillus sp. [4, 9]; Dictyostelium lacteum [1]; Dictyostelium discoideum [2]

**Source tissue**
  Liver [7]; Larvae [7]; Cell [5, 6, 9, 10, 11]; Supernatent of washed cells [2]

## Localisation in source
Cytoplasm [7]; Extracellular [2]

## Purification
Bacillus megaterium [5, 6]

## Crystallization
–

## Cloned
–

## Renaturated
–

---

## 5 STABILITY

### pH

### Temperature (°C)

### Oxidation

### Organic solvent

### General stability information

### Storage
Partially purified enzyme, 90% loss of activity after 16 days, 4°C [7]; Partially purified enzyme, –20°C or –80°C, several months [6]

---

## 6 CROSSREFERENCES TO STRUCTURE DATABANKS

### PIR/MIPS code

### Brookhaven code

---

## 7 LITERATURE REFERENCES

[1] Van Haastert, P.J.M., DeWit, R.J.W., Grijpama, Y., Konijn, T.M.: Proc. Natl. Acad. Sci. USA, 79, 6270–6274 (1982)
[2] Wurster, B., Bek, F., Butz, U.: J. Bacteriol., 148 (1) , 183–192 (1981)
[3] Bacher, A., Rappold, H.: Methods Enzymol., 66, 652–656 (1980)
[4] Kusakabe, H., Kodama, K., Machida, H., Kuninaka, A.: Agric. Biol. Chem., 43 (9) , 1983–1984 (1979)

[5] Tsusue, M., Takikawa, S., Yokokawa, C. K.: Dev. Biochem., 4, 153–158 (1978)
[6] Takikawa, S., Kitayama-Yokokawa, C., Tsusue, M.: J. Biochem., 85, 785–790 (1979)
[7] Rembold, H., Simmersbach, F.: Biochim. Biophys. Acta, 184, 589–596 (1969)
[8] Levenberg, B., Hayaishi, O.: J. Biol. Chem., 234, 955–961 (1959)
[9] McNutt, W.S.: J. Biol. Chem., 238, 116–121 (1963)
[10] Tsusue, M.: Experientia, 23, 116–117 (1967)
[11] Gyure, W.L.: Insect Biochem., 4, 113–121 (1974)

## 1 NOMENCLATURE

**EC number**
3.5.4.12

**Systematic name**
dCMP aminohydrolase

**Recommended name**
dCMP deaminase

**Synonymes**
Deaminase, deoxycytidylate
Deoxycytidylate deaminase
Deoxy-CMP-deaminase
Deoxycytidylate aminohydrolase
Deoxycytidine monophosphate deaminase
Deoxycytidine-5'-phosphate deaminase
Deoxycytidine-5'-monophosphate aminohydrolase

**CAS Reg. No.**
9026-92-0

---

## 2 REACTION AND SPECIFICITY

**Catalysed reaction**
dCMP + $H_2O$ →
→ dUMP + $NH_3$

**Reaction type**
Amidine hydrolysis

**Natural substrates**
dCMP + $H_2O$ (pyrimidine deoxynucleotide pathway [1, 8], DNA
biosynthesis, regulatory enzyme [10]) [1, 8, 10]

**Substrate spectrum**
1 dCMP + $H_2O$
2 More (halogenated deoxycytidylates: very poor substrates [4], CMP: poor
substrate [3], 1-beta-D-arabinose analogue is an effective substrate in
presence of dCTP [3], 5-mercury derivative [7], halogenated derivatives
are hydrolyzed [8]) [3, 4, 7, 8]

## Product spectrum
1 dUMP + NH$_3$
2 ?

## Inhibitor(s)
5-Bromo-UTP; Glutaraldehyde [2]; Guanidine hydrochloride [14]; dUTP [15]; dAMP [2, 6]; dTTP (enzyme from infected cells is more resistant than enzyme from non-infected cells [11], inhibition requires Mg$^{2+}$ or Mn$^{2+}$ [12]) [2, 11, 12, 13, 15]; More [7]

## Cofactor(s)/prostethic group(s)
More (dCTP and Mg$^{2+}$ activate [3, 4, 12], Mg$^{2+}$ in absence of dCTP: no effect [4, 12], dCTP, Zn$^{2+}$ and 2-mercaptoethanol required, Mg$^{2+}$ cannot substitute for Zn$^{2+}$ [9], dCTP activates [11, 12, 15]) [3, 4, 9, 11, 12]

## Metal compounds/salts
Mg$^{2+}$ (completely dependent on Mg$^{2+}$ [8], dCTP and Mg$^{2+}$ activate, Mg$^{2+}$ in absence of dCTP: no effect [4, 12], dCTP, Zn$^{2+}$ and 2-mercaptoethanol required, Mg$^{2+}$ cannot substitute for Zn$^{2+}$ [9]) [4, 12]; Zn$^{2+}$ (dCTP, Zn$^{2+}$ and mercaptoethanol required, Mg$^{2+}$ cannot substitute for Zn$^{2+}$) [9]; Mn$^{2+}$ (activates, less effective than Mg$^{2+}$) [8]; Ca$^{2+}$ (activates, less effective than Mg$^{2+}$) [8]

## Turnover number (min$^{-1}$)

## Specific activity (U/mg)
430 [13]; 9.81 [3]; More [8, 12, 15]

## K$_m$-value (mM)
More [3, 8, 15]; 50 (allosteric activators: CTP, dCDP) [3]; 1.0 (dCMP) [4]; 0.25 (dCTP, allosteric effector) [3]; 0.1 (dCMP) [8]; 10 (CMP) [8]; 0.14 (dCMP) [9]

## pH-optimum
5.3 (absence of dCTP) [8, 13]; 8.3 (presence of dCTP) [8, 13]; 7.5–9.5 [4]; 6.5–7.3 [3]; 7.8 [9]; 8.4 [12]

## pH-range
More [3]; 7.0–8.5 (50% of maximal activity at pH 7.0 and 8.5) [9]; 7.0–9.0 (7.0: 60% of maximal activity, 9.0: 96% of maximal activity) [12]; More [3]

## Temperature optimum (°C)
37 (assay at) [3, 13]

## Temperature range (°C)

## 3 ENZYME STRUCTURE

### Molecular weight
170000 (Bacillus subtilis, gel filtration) [9]
109500 (human, gel filtration) [3]
124000 (E. coli, T2r$^+$ bacteriophage-induced enzyme, sedimentation equilibrium) [14]
130000 (gel filtration, hamster, BHK-21) [12]

### Subunits
Hexamer (E. coli [8], 6 × 20200, sedimentation in guanidine hydrochloride [14] /[1, 8, 14], chicken [1], donkey [2]) [1, 2, 8, 14]
Dimer (human, 2 × 15300, SDS-PAGE) [3]

### Glycoprotein/Lipoprotein
—

## 4 ISOLATION/PREPARATION

### Source organism
Bacillus subtilis [9]; Lactobacillus acidophilus [15]; E. coli (T2r$^+$ bacteriophage-induced enzyme [13, 14], T2-phage infected [1, 8]) [1, 8, 13, 14]; Saccharomyces cerevisiae [4]; Donkey [2, 6]; Human [3]; Mycoplasma mycoides (subsp. mycoides) [5]; Hamster (BHK-21 cells, enzyme from non-infected cells and cells infected by the virus herpes simplex [11]) [11, 12]

### Source tissue
Spleen [3, 6]; Cell [8, 15]; Cultured cells (BHK-21) [11, 12]; Kidney (BHK-21 cells, baby hamster kidney cells) [11, 12]

### Localisation in source

### Purification
Human [3]; Hamster (BHK-21 cells, enzyme from non-infected cells and cells infected by Herpes simplex) [11]; E. coli (T2r$^+$ bacteriophage-induced enzyme [13]) [8, 13]; Bacillus subtilis [9]; Lactobacillus acidophilus (partial) [15]

### Crystallization
[1]

### Cloned
—

### Renaturated
[14]

## 5 STABILITY

pH

**Temperature (°C)**
37 (enzyme from infected cells more stable than enzyme from non-infected cells) [11]

**Oxidation**

**Organic solvent**

**General stability information**
2-Mercaptoethanol (stabilizes) [10]; Glycerol (stabilizes) [10]; Ethyleneglycol (stabilizes against thermal and UV inactivation) [10, 12]; $Zn^{2+}$ (protects against heat inactivation at 50°C); Thiols (stabilizes [8], enzyme highly unstable in absence of thiols [9]) [8, 9]; dCTP (stabilizes against heat inactivation, protein denaturants and proteolytic enzymes [8], protects against inactivation) [9, 10]

**Storage**
0–4°C, pH 7, 0.1 M 2-mercaptoethanol, several months [8]; Frozen in absence of 2-mercaptoethanol (indefinitely stable) [8]; 0°C (half-life: 30 minutes) [9]

---

## 6 CROSSREFERENCES TO STRUCTURE DATABANKS

**PIR/MIPS code**
DUBPC2 (bacteriophage T2); A25230 (yeast, Saccharomyces cerevisiae)

**Brookhaven code**

---

## 7 LITERATURE REFERENCES

[1] Maley, F., Belfort, M., Maley, G.: Adv. Enzyme Regul., 22, 413–430 (1984) (Review)
[2] Nucci, R., Raia, C.A., Vaccaro, C., Sepe, S., Scarano, E., Rossi, M.: J. Mol. Biol., 124, 133–145 (1978)
[3] Ellims, P.H., Kao, A.Y., Chabner, B.A.: J. Biol. Chem. , 256, 6335–6340 (1981)
[4] McIntosh, E.M., Haynes, R.H.: J. Bacteriol., 158, 644–649 (1984)
[5] Neale, G.A.M., Mitchell, A., Finch, L.R.: J. Bacteriol., 156, 1001–1005 (1983)
[6] Mastrantonio, S., Nucci, R., Vaccaro, C., Rossi, M., Whitehead, E. P.: Eur. J. Biochem., 137, 421–427 (1983)
[7] Raia, C.A., Nucci, R., Vaccaro, C., Sepe, S., Rella, R.: J. Mol. Biol., 157, 557–570 (1982)
[8] Maley, G.F.: Methods Enzymol., 51, 412–418 (1978) (Review)
[9] Mollgaard, H., Neuhard, J.: J. Biol. Chem., 253, 3536–3542 (1978)

[10] Dosseva. I.M., Tomov, T.H.: Dokl. Bolg. Akad. Nauk, 28, 214–244 (1975)
[11] Rolton, H.A., Keir, H.: Biochem. J., 143, 403–409 (1974)
[12] Rolton, H.A., Keir, H.M.: Biochem. J., 141, 211–217 (1974)
[13] Maley, G.F., Guarino, D.U., Maley, F.: J. Biol. Chem. , 247, 931–939 (1972)
[14] Maley, G.F., MacColl, R., Maley, F.: J. Biol. Chem., 247, 940–945 (1972)
[15] Sergott, R.C., Debeer, L.J., Bessman, M.J.: J. Biol. Chem., 246, 7755–7758 (1971)

# 1 NOMENCLATURE

**EC number**
3.5.4.13

**Systematic name**
dCTP aminohydrolase

**Recommended name**
dCTP deaminase

**Synonymes**
Deoxycytidine triphosphate deaminase
5-Methyl-dCTP deaminase (XP-12 infected Xanthomonas oryzae cells) [1]

**CAS Reg. No.**
37289-18-2

---

# 2 REACTION AND SPECIFICITY

**Catalysed reaction**
dCTP + $H_2O \rightarrow$
$\rightarrow$ dUTP + $NH_3$; 5-Methyl-dCTP + $H_2O \rightarrow$
$\rightarrow$ dTTP + $NH_3$ (in XP-12 infected Xanthomonas oryzae cells) [1]

**Reaction type**
Amidine hydrolysis

**Natural substrates**
dCTP + $H_2O$
5-Methyl-dCTP + $H_2O$ [1]

**Substrate spectrum**
1 dCTP + $H_2O$
2 5-Methyl-dCTP + $H_2O$ (5-methyl-dCTP deaminase) [1]

**Product spectrum**
1 dUTP + $NH_3$
2 dTTP + $NH_3$ [1]

---

**Inhibitor(s)**
dTTP [3–5]; dUTP [3]; EDTA [4]; $Ni^{2+}$ (at high concentrations) [3];
p-Chloromercuribenzoate [2, 3]; More [3, 4]

**Cofactor(s)/prostethic group(s)**

---

**Metal compounds/salts**
$Mn^{2+}$ (activation) [2, 3, 5]; $Mg^{2+}$ (activation) [2–5]; $Ca^{2+}$ (activation) [3, 5];
$Co^{2+}$ (activation) [2, 3]

**Turnover number** (min$^{-1}$)

**Specific activity** (U/mg)
0.041 [4]; 0.98 [2, 3]

**$K_m$-value** (mM)
0.05–0.1 (dCTP) [4]

**pH-optimum**
6.0 [3]; 6.65–6.85 [5]; 6.8 [2, 3]; 6–8 [4]

**pH-range**

**Temperature optimum** (°C)

**Temperature range** (°C)

## 3 ENZYME STRUCTURE

**Molecular weight**
125000 (gel filtration, Bacillus subtilis) [4]
82000 (gel filtration, Salmonella typhimurium) [3]

**Subunits**
Monomer (gel filtration, Bacillus subtilis [4], Salmonella typhimurium [3]) [3, 4]

**Glycoprotein/Lipoprotein**
–

## 4 ISOLATION/PREPARATION

**Source organism**
Bacillus subtilis (infected with phage PBS1 [5], with phage PBS2 [4]) [4, 5];
Salmonella typhimurium [2, 3]; E. coli [6]; Xanthomonas oryzae (infected
with Bacteriophage XP-12) [1]

**Source tissue**
Cell [1–5]

**Localisation in source**

**Purification**
Salmonella typhimurium [2, 3]

Crystallization
–

Cloned
–

Renaturated
–

---

## 5 STABILITY

pH

**Temperature** (°C)
75 (inactivated after a few minutes) [3]

**Oxidation**

**Organic solvent**

**General stability information**
Ethylene glycol stabilizes [3]

**Storage**
Partially purified enzyme, pH 6.8, –50°C, 2 months [2, 3]

---

## 6 CROSSREFERENCES TO STRUCTURE DATABANKS

**PIR/MIPS code**

**Brookhaven code**

---

## 7 LITERATURE REFERENCES

[1] Wang, R.Y.-H., Ehrlich, M.: J. Virol., 42 (1) , 42–48 (1982)
[2] Neuhard, J.: Methods Enzymol., 51, 418–423 (1978)
[3] Beck, C.F., Eisenhardt, A.R., Neuhard, J.: J. Biol. Chem., 250 (2) , 609–616 (1975)
[4] Price, A.R.: J. Virol., 14 (5) , 1314–1317 (1974)
[5] Tomita, F., Takahashi, J.: Biochim. Biophys. Acta, 179 , 18–27 (1969)
[6] O'Donovan, G.A., Edlin, G., Fuchs, J.A., Neuhard, J., Thomassen, E.: J. Bacteriol., 105, 666–672 (1971)

## 1 NOMENCLATURE

**EC number**
3.5.4.14

**Systematic name**
Deoxycytidine aminohydrolase

**Recommended name**
Deoxycytidine deaminase

**Synonymes**
Deaminase, deoxycytidine

**CAS Reg. No.**
37259-56-6

## 2 REACTION AND SPECIFICITY

**Catalysed reaction**
Deoxycytidine + $H_2O \rightarrow$
$\rightarrow$ deoxyuridine + $NH_3$

**Reaction type**
Amidine hydrolysis

**Natural substrates**

**Substrate spectrum**
1 Deoxycytidine + $H_2O$
2 Cytidine + $H_2O$ [1]
3 5-Methyldeoxycytidine + $H_2O$ [1]
4 5-Bromodeoxycytidine + $H_2O$ [1]
5 Arabinosylcytosine + $H_2O$ [1]

**Product spectrum**
1 Deoxyuridine + $NH_3$
2 Uridine + $NH_3$
3 5-Methyldeoxyuridine + $NH_3$
4 5-Bromodeoxyuridine + $NH_3$
5 Arabinosyluridine + $NH_3$

**Inhibitor(s)**
p-Chloromercuribenzoate [1]; Mercaptoethanol [1]; Dithiothreitol [1];
Deoxyuridine [1]; Uridine [1]; 5-Bromodeoxyuridine [1]

**Cofactor(s)/prostethic group(s)**

**Metal compounds/salts**

---

**Turnover number (min$^{-1}$)**

**Specific activity (U/mg)**
  0.7 [1]

**$K_m$-value (mM)**
  More [1]

**pH-optimum**

**pH-range**

**Temperature optimum (°C)**

**Temperature range (°C)**

---

## 3 ENZYME STRUCTURE

**Molecular weight**
  77000 (gel filtration, Zea mays) [1]

**Subunits**

**Glycoprotein/Lipoprotein**
  –

---

## 4 ISOLATION/PREPARATION

**Source organism**
  Zea mays [1]; Syrian hamster [2]; Mouse [3, 4]

**Source tissue**
  Leaf [1]; Small intestine [4]; Kidney [3]; Liver [3]; Lung [3]

**Localisation in source**
  Membrane (cytoplasmic) [5]

**Purification**
  Zea mays [1]

**Crystallization**
  –

2

Cloned

–

Renaturated

–

## 5 STABILITY

**pH**
4.5 (50% loss of activity after 30 minutes, 0°C) [1]; 9.0 (50% loss of activity after 30 minutes, 0°C) [1]

**Temperature** (°C)
70 (up to, 5 minutes) [4]; 4 (50% loss of activity after 30 days) [1]

**Oxidation**

**Organic solvent**

General stability information

**Storage**
50% loss of activity after 30 days at 4°C [1]

## 6 CROSSREFERENCES TO STRUCTURE DATABANKS

**PIR/MIPS code**

**Brookhaven code**

## 7 LITERATURE REFERENCES

[1] Le Floc'h, F., Guillot, A.: Phytochemistry, 13, 2503–2509 (1974)
[2] Cullen, B.R., Bick, M.D.: Biochim. Biophys. Acta, 517, 158–168 (1978)
[3] Chan, T., Lakhchaura, B.D., Hsu, T.: Biochem. J., 210, 367–371 (1983)
[4] Kang, M.S., Rhee, J.G., Cho, J.M.: Korean J. Zool., 17 (3) , 107–116 (1974)
[5] Taketo, A., Kuno, S.: J. Biochem., 72, 1557–1563 (1972)

## 1 NOMENCLATURE

**EC number**
3.5.4.15

**Systematic name**
Guanosine aminohydrolase

**Recommended name**
Guanosine deaminase

**Synonymes**
Guanosine aminase

**CAS Reg. No.**
9067-85-0

## 2 REACTION AND SPECIFICITY

**Catalysed reaction**
Guanosine + $H_2O$ →
→ xanthosine + $NH_3$

**Reaction type**
Amidine hydrolysis

**Natural substrates**
Guanosine + $H_2O$

**Substrate spectrum**
1 Guanosine + $H_2O$
2 Deoxyguanosine + $H_2O$ [2, 3]
3 8-Azaguanosine + $H_2O$ [2, 3]

**Product spectrum**
1 Xanthosine + $NH_3$
2 8-Azadeoxyxanthosine + $NH_3$
3 8-Azaxanthosine + $NH_3$

**Inhibitor(s)**

**Cofactor(s)/prostethic group(s)**

**Metal compounds/salts**
    $Hg^{2+}$ (activates) [3]; $Cu^{2+}$ (activates) [3]; $Co^{2+}$ (activates); $Mn^{2+}$ (activates)
    [3]; $Cd^{2+}$ (activates) [3]; $Al^{3+}$ (activates) [3]

**Turnover number** (min$^{-1}$)

**Specific activity** (U/mg)

**$K_m$-value** (mM)
    0.036 (guanosine) [2, 3]; 0.062 (deoxyguanosine) [2, 3]; 0.122
    (8-azaguanosine) [2, 3]

**pH-optimum**
    6.0–6.5 [2, 3]

**pH-range**
    4.5–9.0 [3]

**Temperature optimum** (°C)
    45 [3]

**Temperature range** (°C)
    55 (inactivated after 5 minutes) [3]

## 3 ENZYME STRUCTURE

**Molecular weight**
    100000–200000 (gel filtration, Pseudomonas convexa) [2, 3]

**Subunits**

**Glycoprotein/Lipoprotein**
    –

## 4 ISOLATION/PREPARATION

**Source organism**
    Pseudomonas convexa [2, 3]; Rat [1]

**Source tissue**
    Cell [3]; Brain [1]

**Localisation in source**

**Purification**
    Pseudomonas convexa [2]

**Renaturated**

–

---

**Crystallization**

–

**Cloned**

–

---

## 5 STABILITY

pH

Temperature (°C)

Oxidation

Organic solvent

General stability information

Storage

---

## 6 CROSSREFERENCES TO STRUCTURE DATABANKS

PIR/MIPS code

Brookhaven code

---

## 7 LITERATURE REFERENCES

[1] Davies, L.P., Taylor, K.M.: J. Neurochem., 33, 951–952 (1979)
[2] Zielke, C.L., Suelter, C.H. in "The Enzymes", 3rd Ed. (Boyer, P.D., Ed.) , Vol.4, 47–78 (1971) (Review)
[3] Ishida, Y., Shirafuji, H., Kida, M., Yoneda, M.: Agric. Biol. Chem., 33 (3) , 384–390 (1969)

## 1 NOMENCLATURE

**EC number**
    3.5.4.16

**Systematic name**
    GTP 7, 8–8, 9-dihydrolase

**Recommended name**
    GTP cyclohydrolase I

**Synonymes**
    Hydrolase, guanosine triphosphate cyclo-
    GTP cyclohydrolase
    Guanosine triphosphate cyclohydrolase
    Guanosine triphosphate 8-deformylase
    Dihydroneopterin triphosphate synthase
    GTP 8-formylhydrolase

**CAS Reg. No.**
    37289-19-3

---

## 2 REACTION AND SPECIFICITY

**Catalysed reaction**
    GTP + 2 $H_2O$ →
    → formate + 2-amino-4-hydroxy-6-(erythro-
    1, 2, 3-trihydroxypropyl)dihydropteridine triphosphate

**Reaction type**
    Amidine hydrolysis
    More (the reaction involves hydrolysis of two C-N bonds and isomerization
    of the pentose unit, the recyclization may be non-enzymatic)

**Natural substrates**
    GTP + $H_2O$ (first step in biosynthesis of tetrahydrobiopterin (BH4) [1], first
    step in pathway of pterins [2, 6, 7, 10, 11]) [1, 2, 6, 7, 10, 11]

**Substrate spectrum**
    1 GTP + $H_2O$ [1–14]
    2 Beta-gamma-methyleneguanosine 5'-triphosphate + $H_2O$ [8]

Enzyme Handbook © Springer-Verlag Berlin Heidelberg 1991
Duplication, reproduction and storage in data banks are only
allowed with the prior permission of the publishers

**Product spectrum**
  1 Formate + 2-amino-4-hydroxy-6-(erythro-
    1, 2, 3-trihydroxypropyl)dihydropteridine triphosphate [1–14]
  2 Beta-gamma-methylene-7, 8-dihydroneopterin 3'-triphosphate + formate
    [8]

**Inhibitor(s)**
  $PO_4^{3-}$ [6, 10]; GDP [12]; $Hg^{2+}$ [6, 9, 11, 12, 14]; $Sr^{2+}$ [9]; $Pb^{2+}$ [9]; $Sn^{2+}$ [9];
  $Cd^{2+}$ [9, 11, 12]; dGTP [10]; Guanosine 5'-tetraphosphate [10]; $Al^{3+}$ [12];
  $SO_4^{2-}$ [12]; p-Chloromercuribenzoate [12]; $Fe^{2+}$ [14]; Ascorbate [14]; $Ca^{2+}$
  [5, 6, 9]; UTP [2]; ITP [5]; NaCl [5]; KCl [5]; $Mg^{2+}$ [5, 6]; $Mn^{2+}$ [5, 14]; ATP [2,
  5, 10, 12]; 8-Aminoguanosine triphosphate [4]; Divalent cations [5]; GTP
  (substrate inhibition above 0.2 mM) [5, 6]; L-Erythro-
  5, 6, 7, 8-tetrahydrobiopterin [4]; L-Erythro-7, 8-dihydrobiopterin [4];
  L-Sepiapterin [4]; $Zn^{2+}$ [6, 12]; $Cu^{2+}$ [6, 9, 11, 14]; $Fe^{3+}$ [6]

**Cofactor(s)/prostethic group(s)**
  No cofactors required [4, 5, 6, 13]

**Metal compounds/salts**
  $K^+$ (activates) [11]; $Na^+$ (activates) [11]; $Li^+$ (activates) [11]; More (no metals required) [13, 14]

---

**Turnover number** (min⁻¹)

**Specific activity** (U/mg)
  67 [11]; More [2, 4, 5, 9, 10, 12, 13, 14]

**$K_m$-value** (mM)
  0.06 (GTP) [3]; 0.031 (GTP) [4]; 980 (GTP) [9]; More [6, 7, 10, 11, 12, 14]

**pH-optimum**
  8.6 [12]; 7.6 [13]; 7.8 [3]; 7.3 (phosphate buffer: 2 optima, 7.3 and 8.0) [5];
  8.0 (phosphate buffer: 2 optima, 7.3 and 8.0 [5]) [5, 14]; 8.0–8.4 [9];
  More [6, 7]

**pH-range**

**Temperature optimum** (°C)
  42 [14]; 60 [5]; 60–65 [9]

**Temperature range** (°C)
  38–78 (half-maxima at) [9]

# 3 ENZYME STRUCTURE

## Molecular weight

105000 (Bacillus stearothermophilus, gel filtration) [6]
More [2]
575000 (gel filtration, Drosophila melanogaster) [5]
650000 (Comamonas sp.) [6]
210000 (E. coli, gel filtration) [10]
200000 (Serratia indica, gel filtration, D-I, D-II) [11]
170000 (Serratia indica, gel filtration) [12]
300000 (E. coli, larger than 300000, gel filtration) [14]

## Subunits

Polymer (x × 39000, Drosophila melanogaster, SDS-PAGE) [5]
Decamer (10 × 20000, Lactobacillus plantarum) [6]
Tetramer (E. coli, SDS-PAGE, 4 × 51000) [10]
Octamer (Serratia indica, 8 × 25000, SDS-PAGE) [11]

## Glycoprotein/Lipoprotein

–

# 4 ISOLATION/PREPARATION

## Source organism

Human [1, 3, 4]; E. coli [2, 6, 8, 10, 14]; Rat [3, 6]; Drosophila melanogaster
[5]; Mammalia [6]; Comamonas sp. [6, 13]; Lactobacillus plantarum [6];
Serratia indica (multiple forms, D-I, D-II [11]) [6, 11, 12]; Bacillus
stearothermophilus [6, 9]; More [6, 7]

## Source tissue

Blood [1]; Lymphocytes [1]; Monocytes [1]; Granulocytes [1]; Kidney [6];
Liver [3, 4, 6]; Brain [3, 6]

## Localisation in source

## Purification

Drosophila melanogaster [5]; Bacillus stearothermophilus [9]; E. coli [2, 10,
14]; Serratia indica [11]; Comamonas sp. [13]

## Crystallization

–

## Cloned

–

## Renaturated

–

# 5 STABILITY

**pH**
> 8.4 (stability maximum) [13]; More [13]

**Temperature** (°C)
> 22 (no significant loss of activity after 5 hours) [3]; 80 (human liver enzyme, half-life: 2 minutes) [4]; 82 (E. coli, half-life: 7 minutes) [10]

**Oxidation**

**Organic solvent**

**General stability information**
> Bovine serum albumin (stabilizes) [5]

**Storage**
> −80°C, 6 months (stable) [3]; −70°C [4]; 4°C (extreme instability of purified enzyme) [4]

---

# 6 CROSSREFERENCES TO STRUCTURE DATABANKS

**PIR/MIPS code**

**Brookhaven code**

---

# 7 LITERATURE REFERENCES

[1] Schoedon, G., Curtis, H.-C., Niederwieser, A.: Biochem. Biophys. Res. Commun., 148, 1232–1236 (1987)
[2] Ferre, J., Yim, J.J., Jacobson, K.B.: J. Chromatogr., 357, 283–292 (1986)
[3] Sawada, M., Horikoshi, T., Masada, M., Akino, M., Sugimoto, T., Matsuura, S., Nagatsu, T.: Anal. Biochem., 154, 361–366 (1986)
[4] Blau, N., Niederwieser, A.: Biochim. Biophys. Acta, 880, 26–31 (1986)
[5] Weisberg, E.P., O'Donnell, J.M.: J. Biol. Chem., 261, 1453–1458 (1986)
[6] Blau, N., Niederwieser, A.: J. Clin. Chem. Clin. Biochem., 23, 169–176 (1985) (Review)
[7] Blau, N., Niederwieser, A.: Biochem. Clin. Aspects Pteridines, 3, 77–92 (1984) (Review)
[8] Ferre, J., Jacobson, K.B.: Arch. Biochem. Biophys., 233, 475–480 (1984)
[9] Suzuki, Y., Yasui, T., Abe, S.: J. Biochem., 86, 1679–1685 (1979)
[10] Yim, J.J., Brown, G.M.: J. Biol. Chem., 251, 5087–5094 (1976)
[11] Kohashi, M., Itadani, T., Iwai, K.: Agric. Biol. Chem., 44, 271–278 (1980)
[12] Kobashi, M., Hariu, H., Iwai, K.: Agric. Biol. Chem., 40, 1597–1603 (1976)
[13] Cone, J.E., Plowman, J., Guroff, G.: J. Biol. Chem., 249, 5551–5558 (1974)
[14] Burg, A.W., Brown, G.M.: J. Biol. Chem., 243, 2349–2358 (1968)

## 1 NOMENCLATURE

**EC number**
3.5.4.17

**Systematic name**
Adenosine-phosphate aminohydrolase

**Recommended name**
Adenosine-phosphate deaminase

**Synonymes**
Adenylate deaminase [1]
Adenine nucleotide deaminase [2]
Adenosine (phosphate) deaminase [3]

**CAS Reg. No.**
37289-20-6

## 2 REACTION AND SPECIFICITY

**Catalysed reaction**
5'-AMP + $H_2O$ →
→ 5'-IMP + $NH_3$

**Reaction type**
Amidine hydrolysis

**Natural substrates**
5'-AMP + $H_2O$

**Substrate spectrum**
1 Adenosine + $H_2O$ [1, 2]
2 Adenosine phosphates + $H_2O$ [1, 2]
3 $NAD^+$ + $H_2O$ [1, 2]
4 dATP + $H_2O$ [2]
5 dADP + $H_2O$ [2]
6 dAMP + $H_2O$ [2]
7 Deoxyadenosine + $H_2O$ [2]

Enzyme Handbook © Springer-Verlag Berlin Heidelberg 1991
Duplication, reproduction and storage in data banks are only
allowed with the prior permission of the publishers

**Product spectrum**
1 Inosine + $NH_3$ [1, 2]
2 Inosine phosphates + $NH_3$ [1, 2]
3 Nicotinamide-hypoxanthine-dinucleotide + $NH_3$ [1, 2]
4 dITP + $NH_3$ [2]
5 dIDP + $NH_3$ [2]
6 dIMP + $NH_3$ [2]
7 Deoxyinosine + $NH_3$ [2]

**Inhibitor(s)**
$Mn^{2+}$ (at neutral or alkaline pH) [1, 2]; $F^-$ [1, 2]; $Fe^{3+}$ [1, 2]; $CN^-$ [2]; $Cl^-$ [2]; $Co^{2+}$ [2]; $Zn^{2+}$ [2]; $Hg^{2+}$ [2];
p-Substituted mercuribenzoate [1, 2]; More [2]

**Cofactor(s)/prostethic group(s)**

**Metal compounds/salts**
$Ca^{2+}$ [1]; $Mg^{2+}$ [1]; $Ba^{2+}$ [1]; $Mn^{2+}$ (activates at acidic pH) [1]

**Turnover number** (min$^{-1}$)
690 (ATP) [2]; 630 (ADP) [2]; 710 (AMP) [2]; More [2]

**Specific activity** (U/mg)
More [1]

**$K_m$-value** (mM)
0.047 (5'-AMP) [1]; 0.047 (ADP) [1]; 0.066 (ATP) [1]; 0.56 (adenosine) [1]; 0.072 (NAD) [1]; 0.285 (ATP) [2]; 0.3 (ADP) [2]; 0.25 (AMP) [2]; More [2]

**pH-optimum**
6.0–6.8 (5'-AMP) [1, 2]; 6.0 (ADP) [1, 2]; 6.0 (ATP) [1, 2]; 7.1 (adenosine) [1]; 5.6 (NAD$^+$) [1]; 6.0 [3]

**pH-range**
4.0–8.0 [2]

**Temperature optimum** (°C)
55 [3]

**Temperature range** (°C)

## 3 ENZYME STRUCTURE

**Molecular weight**
30000–60000 (Desulfovibrio desulfuricans, gel chromatography) [2]

**Subunits**

Glycoprotein/Lipoprotein
–

## 4 ISOLATION/PREPARATION

**Source organism**
Porphyra crispata [1]; Desulfovibrio desulfuricans [2]; Aspergillus sp. [3]

**Source tissue**

**Localisation in source**

**Purification**
Porphyra crispata [1]; Desulfovibrio desulfuricans [2]

**Crystallization**
–

**Cloned**
–

**Renaturated**
–

## 5 STABILITY

**pH**
5.0–6.0 [3]

**Temperature (°C)**
65 (up to) [1]; 60 (up to) [2]

**Oxidation**

**Organic solvent**

**General stability information**

**Storage**
1 month at –10°C [1]; Indefinitely at –20°C [2]; 3 days at 25°C [2]

## 6 CROSSREFERENCES TO STRUCTURE DATABANKS

**PIR/MIPS code**

**Brookhaven code**

# 7 LITERATURE REFERENCES

[1] Su, J.C., Li, C.C., Ting, C.C.: Biochemistry, 5 (2) , 536–543 (1966)
[2] Yates, M.G.: Biochim. Biophys. Acta, 171, 299–310 (1969)
[3] Rokugawa, K., Fujishima, A., Kuninaka, A., Yoshino, H.: J. Ferment. Technol., 58 (6) , 583–585 (1980)

## 1 NOMENCLATURE

**EC number**
3.5.4.18

**Systematic name**
ATP aminohydrolase

**Recommended name**
ATP deaminase

**Synonymes**
Adenosine triphosphate deaminase

**CAS Reg. No.**
37289-21-7

## 2 REACTION AND SPECIFICITY

**Catalysed reaction**
$ATP + H_2O \rightarrow$
$\rightarrow ITP + NH_3$

**Reaction type**
Amidine hydrolysis
Dechlorination [5]

**Natural substrates**
$ATP + H_2O$

**Substrate spectrum**
1  $ATP + H_2O$
2  $AMP + H_2O$
3  6-Chloropurine riboside $+ H_2O$
4  6-Chloropurine ribotide $+ H_2O$
5  Adenosine tetraphosphate $+ H_2O$ [5, 6]
6  $ADP + H_2O$
7  ADP-ribose $+ H_2O$
8  ADP-glucose $+ H_2O$
9  $dATP + H_2O$
10  $dADP + H_2O$
11  $dAMP + H_2O$
12  3'5'-Cyclic AMP $+ H_2O$

13  $NAD^+ + H_2O$
14  $FAD + H_2O$
15  Coenzyme A $+ H_2O$
16  Adenosine $+ H_2O$
17  3-Iso-AMP $+ H_2O$
18  5'-Adenosine monosulfate $+ H_2O$
19  More (not: 2'-AMP, NADP, adenine) [5]

## Product spectrum

1  $ITP + NH_3$
2  $IMP + NH_3$
3  Inosine $+ Cl^-$
4  $IMP + Cl^-$
5  $ITPP + NH_3$
6  $IDP + NH_3$
7  IDP-ribose $+ NH_3$
8  IDP-glucose $+ NH_3$
9  $dITP + NH_3$
10  $dIDP + NH_3$
11  $dIMP + NH_3$
12  3'5'-Cyclic IMP $+ NH_3$
13  Nicotinamide-hypoxanthine-dinucleotide $+ NH_3$
14  Flavine-inosine dinucleotide $+ NH_3$
15  Inosine-3',5'-bisphosphate pantetheine-4'-phosphate $+ NH_3$
16  Inosine $+ NH_3$
17  ?
18  5'-Inosine monosulfate $+ NH_3$
19  ?

## Inhibitor(s)

$Fe^{3+}$ [5, 6]; $Ag^+$ [4, 6]; $Cu^{2+}$ [6]; $Zn^{2+}$ [6]; $Ca^{2+}$ [6]; $Mn^{2+}$ [6];
p-Chloromercuribenzoate [6]; Phosphates [5]; Pyrophosphate (inhibites
dechlorination) [3]; 6-Chloropurine ribotide (inhibites deamination of ATP)
[3]; Cacodylate [1]; Maleate [1]; Succinate [1]; Citrate [1]; Beta,
beta-dimethylglutarate [1]

## Cofactor(s)/prostethic group(s)

## Metal compounds/salts

$Mg^{2+}$ (activates) [1]; $Ca^{2+}$ (activates) [1]; $Ba^{2+}$ (activates) [1]

## Turnover number (min⁻¹)

## Specific activity (U/mg)

## $K_m$-value (mM)

0.04 (ATP) [6]; 0.017 (ATP) [4]; 0.022 (ATP) [4]; 0.083 (ADP) [3]; 0.52 (adenosine) [3]; 1.2 (6-chloropurine ribotide) [3]; 4.2 (6-chloropurine riboside) [3]; More [1]

## pH-optimum

5.0 [4, 6]; 4.8 [4]; 5.3 [4]; 3.4 (dechlorinating activity) [3]; 3.0 (immobilized enzyme) [2]; More [1]

## pH-range

## Temperature optimum (°C)

40 [1]

## Temperature range (°C)

---

## 3 ENZYME STRUCTURE

## Molecular weight

## Subunits

## Glycoprotein/Lipoprotein

—

---

## 4 ISOLATION/PREPARATION

## Source organism

Microsporum audouini [5, 6, 2]; Aspergillus glaucus [4]; Aspergillus repens [4]; Aspergillus sp. [5]; Porphyra crispata [7]; Desulfovibrio desulfuricans [1, 8]; More [4]

## Source tissue

Culture medium [6]

## Localisation in source

Cytoplasm [6]; Soluble [6]

## Purification

Microsporum audouini [2, 6]; Aspergillus glaucus [4]; Aspergillus repens [4]

## Crystallization

—

## Cloned

—

---

Renaturated

–

## 5 STABILITY

pH
  5.5–6.5 [4]; 5.5–7.0 [4]; 5.0–7.5 [4]

Temperature (°C)

Oxidation

Organic solvent

General stability information

Storage
  Purified enzyme, 4°C, 25 days [1]

## 6 CROSSREFERENCES TO STRUCTURE DATABANKS

PIR/MIPS code

Brookhaven code

## 7 LITERATURE REFERENCES

[1] Zielke, C.L., Suelter, C.H. in "The Enzymes", 3rd. Ed. (Boyer, P.D., Ed.) Vol.4 , 47–78
    (1971) (Review)
[2] Chung, S.-T., Hamano, M., Aida, K., Uemura, T.: Agric. Biol. Chem., 32 (10) ,
    1287–1291 (1968)
[3] Ching, S.-T., Ito, S., Aida, K., Uemura, T.: J. Gen. Appl. Microbiol., 14, 111–119 (1968)
[4] Chung, S.-T., Aida, K., Uemura, T.: J. Gen. Appl. Microbiol., 13, 335–347 (1967)
[5] Chung, S.-T., Aida, K., Uemura, T.: J. Gen. Appl. Microbiol., 13, 237–245 (1967)
[6] Chung, S.-T., Aida, K.: J. Biochem., 61 (1) , 1–9 (1967)
[7] Su, J.C., Li, C.C., Ting, C.C.: Biochemistry, 5, 536 (1966)
[8] Yates, M.G.: Biochim. Biophys. Acta, 171, 299 (1969)

## 1 NOMENCLATURE

**EC number**
3.5.4.19

**Systematic name**
$N^1$-(5-Phospho-D-ribosyl)-AMP 1, 6-hydrolase

**Recommended name**
Phosphoribosyl-AMP cyclohydrolase

**Synonymes**
PRAMP-cyclohydrolase [1]
Phosphoribosyladenosine monophosphate cyclohydrolase

**CAS Reg. No.**
37289-22-8

---

## 2 REACTION AND SPECIFICITY

**Catalysed reaction**
$N^1$-(5-Phospho-D-ribosyl)-AMP + $H_2O$ →
→
5-(5-phospho-D-ribosylaminoformimino)-1-(5-phosphoribosyl)imidazole-4-carboxamide (the Neurospora crassa enzyme also catalyzes the reactions of E.C. 1.1.1.23 and E.C. 3.6.1.31)

**Reaction type**
Amidine hydrolysis

**Natural substrates**
$N^1$-(5-Phospho-D-ribosyl)-AMP + $H_2O$

**Substrate spectrum**
1 $N^1$-(5-Phospho-D-ribosyl)-AMP + $H_2O$

**Product spectrum**
1 5-(5-Phospho-D-ribosylaminoformimino)-1-(5-phosphoribosyl) imidazole-4-carboxamide

---

**Inhibitor(s)**

**Cofactor(s)/prostethic group(s)**

**Metal compounds/salts**

---

**Turnover number** (min$^{-1}$)

**Specific activity** (U/mg)
    More [1, 2]

**K$_m$-value** (mM)

**pH-optimum**

**pH-range**

**Temperature optimum** (°C)
    37 [1]; 30 [2]

**Temperature range** (°C)

---

## 3 ENZYME STRUCTURE

**Molecular weight**
    126000 (Neurospora crassa, amino acid analysis) [1]
    190000 (Saccharomyces cerevisiae, gel electrophoresis) [2]
    48000 (Salmonella thypimurium, sedimentation) [3]

**Subunits**
    Dimer (2 × 95000, genetic complementation, Saccharomyces cerevisiae [2])
    [2, 4]

**Glycoprotein/Lipoprotein**
    –

---

## 4 ISOLATION/PREPARATION

**Source organism**
    Saccharomyces cerevisiae [2]; Salmonella typhimurium [3]; Neurospora
    crassa (enzyme also catalyzes reactions of E.C. 1.1.1.23 and E.C. 3.6.1.31)
    [1]

**Source tissue**

**Localisation in source**

**Purification**
    Neurospora crassa [1]; Saccharomyces cerevisiae [2]

**Crystallization**
    –

**Cloned**
    –

Renaturated

–

## 5 STABILITY

pH

Temperature (°C)

Oxidation

Organic solvent

General stability information

Storage

## 6 CROSSREFERENCES TO STRUCTURE DATABANKS

### PIR/MIPS code

SHBY (phosphoribosyl-ATP pyrophosphohydrolase/histidinol dehydrogenase EC 1.1.1.23, yeast, Saccharomyces cerevisiae); SHNC (phosphoribosyl-ATP pyrophosphohydrolase/histidinol dehydrogenase EC 1.1.1.23, Neurospora crassa); YNECHI (phosphoribosyl-ATP pyrophosphatase EC 3.6.1.31, Escherichia coli); C26022 (phosphoribosyl-ATP pyrophosphatase EC 3.6.1.31, Escherichia coli); B26022 (phosphoribosyl-ATP pyrophosphatase EC 3.6.1.31, Salmonella typhimurium); C26022 (phosphoribosyl-ATP pyrophosphatase EC 3.6.1.31, Escherichia coli); B26022 (phosphoribosyl-ATP pyrophosphatase EC 3.6.1.31, Salmonella typhimurium)

### Brookhaven code

## 7 LITERATURE REFERENCES

[1] Minson, A.C., Creaser, E.H.: Biochem. J., 114 (49) , 49–56 (1969)
[2] Keesey, J.K., Bigelis, R., Fink, G.R.: J. Biol. Chem., 254 (15) , 7427–7433 (1979)
[3] Whitefield, H.J., Smith, D.W.E., Martin, R.G.: J. Biol. Chem., 239 (10) , 3288–3291 (1964)
[4] Fink, G.R.: Genetics, 53, 445–459 (1966)

## 1 NOMENCLATURE

**EC number**
3.5.4.20

**Systematic name**
1-(4-Amino-2-methylpyrimid-5-yl-methyl)-3-(beta-hydroxy-ethyl)-2-methyl-pyrimidinium-bromide aminohydrolase

**Recommended name**
Pyrithiamin deaminase

**Synonymes**
Pyrithiamine deaminase

**CAS Reg. No.**
37289-23-9

## 2 REACTION AND SPECIFICITY

**Catalysed reaction**
1-(4-Amino-2-methylpyrimid-5-ylmethyl)-3-(beta-hydroxy-ethyl)-2-methylpyrimidinium bromide + $H_2O$ →
→
1-(4-hydroxy-2-methylpyrimid-5-ylmethyl)-3-(beta-hydroxyethyl)-2-methylpyrimidinium bromide + $NH_3$

**Reaction type**
Amidine hydrolysis

**Natural substrates**
1-(4-Amino-2-methylpyrimid-5-ylmethyl)-3-(beta-hydroxyethyl)-2-methylpyrimidinium bromide + $H_2O$

**Substrate spectrum**
1   1-(4-Amino-2-methylpyrimid-5-ylmethyl)-3-(beta-hydroxyethyl)-2-methylpyrimidinium bromide + $H_2O$

**Product spectrum**
1   1-(4-Hydroxy-2-methylpyrimid-5-ylmethyl)-3-(beta-hydroxyethyl)-2-methyl pyrimidinium bromide + $NH_3$

**Inhibitor(s)**

**Cofactor(s)/prostethic group(s)**
No cofactor [2]

**Metal compounds/salts**

**Turnover number** (min$^{-1}$)

**Specific activity** (U/mg)
More [2]

**K$_m$-value** (mM)

**pH-optimum**
7.5–7.8 [2]

**pH-range**
5.0–8.5 [2]

**Temperature optimum** (°C)
37 [2]

**Temperature range** (°C)

## 3 ENZYME STRUCTURE

**Molecular weight**

**Subunits**

**Glycoprotein/Lipoprotein**
–

## 4 ISOLATION/PREPARATION

**Source organism**
Staphylococcus aureus (mutant) [1, 2]

**Source tissue**

**Localisation in source**
Cell wall (exoenzyme) [2]

**Purification**

**Crystallization**
–

Cloned

–

Renaturated

–

## 5 STABILITY

pH

Temperature (°C)

Oxidation

Organic solvent

General stability information

Storage

## 6 CROSSREFERENCES TO STRUCTURE DATABANKS

PIR/MIPS code

Brookhaven code

## 7 LITERATURE REFERENCES

[1] Sinha, A.K., Chatterjee, G.C.: Enzymologia, 35 (5) , 298–302 (1968)
[2] Sinha, A.K., Chatterjee, G.C.: Biochem. J., 107, 165–169 (1968)

# 1 NOMENCLATURE

**EC number**
3.5.4.21

**Systematic name**
Creatinine iminohydrolase

**Recommended name**
Creatinine deiminase

**Synonymes**
Creatinine hydrolase
Creatinine desiminase
Cytosine deaminase (in some microorganisms creatinine deiminase activity associated with cytosine deaminase) [1]

**CAS Reg. No.**
37289-15-9

---

# 2 REACTION AND SPECIFICITY

**Catalysed reaction**
Creatinine + $H_2O \rightarrow$
$\rightarrow$ N-methylhydantoin + $NH_3$

**Reaction type**
Amidine hydrolysis

**Natural substrates**
Creatinine + $H_2O$
Cytosine + $H_2O$

**Substrate spectrum**
1 Creatinine + $H_2O$
2 Cytosine + $H_2O$ [4]
3 5-Fluorocytosine + $H_2O$ [4]

**Product spectrum**
1 N-Methylhydantoin + $NH_3$
2 Uracil + $NH_3$
3 S-Fluorouracil + $NH_3$

## Inhibitor(s)

Sulfhydryl reagents [6, 9, 10]; p-Chloromercuribenzoate [6, 9, 10]; $Hg^+$ [6, 9, 10]; $Ag^+$ [6, 9, 10]; $Cu^{2+}$ [6, 9, 10]; Metal ions (divalent) [10]

## Cofactor(s)/prostethic group(s)

## Metal compounds/salts

$Zn^{2+}$ [4]; $Fe^{2+}$ [4]; $Ni^{2+}$ [4]; $Co^{2+}$ [4]

---

## Turnover number ($min^{-1}$)

## Specific activity (U/mg)

36.1 [10]; 34 [9]; 39.8 [4]; More [1]

## $K_m$-value (mM)

18 (creatinine) [10]; 1.27 (creatinine) [8, 9]; 14.3 (creatinine) [7]; 0.62 (cytosine) [4]; 0.42 (5-fluorocytosine) [4]; More [5, 4]

## pH-optimum

7.0–8.5 [4]; 7.5–8.0 (immobilized enzyme) [5]; 7.5–10.0 [6 , 9]; 8.5 [10]

## pH-range

## Temperature optimum (°C)

37 [7]; 60 [6]

## Temperature range (°C)

30–40 [7]

---

## 3 ENZYME STRUCTURE

## Molecular weight

195000–200000 (sedimentation equilibrium, gel filtration, Corynebacterium lilium) [6, 9]
472480 (gel electrophoresis, Flavobacterium filamentosus) [4]

## Subunits

Hexamer (6 × 44300, SDS-PAGE) [4]

## Glycoprotein/Lipoprotein

–

---

## 4 ISOLATION/PREPARATION

## Source organism

Clostridium paraputrificum [10]; Corynebacterium lilium [6 , 8, 9]; Cryptococcus neoformans [7]; Cryptococcus bacillisporus [7]; Flavobacterium filamentosus [4]; Pseudomonas putida [1–3]; E. coli [1]; More [1, 2, 8]

**Source tissue**
  Cell [6–10]

**Localisation in source**
  Cytoplasm [7]; Soluble [7]

**Purification**
  Corynebacterium lilium [6, 8, 9]; Flavobacterium filamentosum [4]

**Crystallization**
  [6, 8, 9]

**Cloned**
  –

**Renaturated**
  –

---

**5 STABILITY**

**pH**
  5–11.0 [6, 9]

**Temperature** (°C)
  60 (stable up to) [6, 9]

**Oxidation**

**Organic solvent**

**General stability information**

**Storage**
  Crude extract, frozen, 8 months [10]; Purified enzyme, 5°C, 6 months [6]

---

**6 CROSSREFERENCES TO STRUCTURE DATABANKS**

**PIR/MIPS code**

**Brookhaven code**

# 7 LITERATURE REFERENCES

[1] Kim, J.M., Shimizu, S., Yamada, H.: Arch. Microbiol., 147, 58–63 (1987)

[2] Shimizu, S., Kim, J.M., Shinmen, Y., Yamada, H.: Arch. Microbiol., 145, 322–328 (1986)

[3] Yamada, H., Shimizu, S., Kim, J.M., Shinmen, Y., Sakai, T.: FEMS Microbiol. Lett., 30, 337–340 (1985)

[4] Esders, T.W., Lynn, S.Y.: J. Biol. Chem., 260 (7) , 3915–3922 (1985)

[5] Tabata, M., Kido, T., Totani, M., Murachi, T.: Anal. Biochem., 134, 44–49 (1983)

[6] Uwajima, T., Terada, O.: Agric. Biol. Chem., 44 (8) , 1787–1792 (1980)

[7] Polacheck, J., Kwon-Chung, K.J.: J. Bacteriol., 142 (1) , 15–20 (1980)

[8] Uwajima, T., Terada, O.: Agric. Biol. Chem., 41 (2) , 339–344 (1977)

[9] Uwajima, T., Terada, O.: Agric. Biol. Chem., 40 (5) , 1055–1056 (1976)

[10] Szulmajster, J.: Biochim. Biophys. Acta, 30, 154–163 (1958)

## 1 NOMENCLATURE

**EC number**
3.5.4.22

**Systematic name**
1-Pyrroline-4-hydroxy-2-carboxylate aminohydrolase (decyclizing)

**Recommended name**
1-Pyrroline-4-hydroxy-2-carboxylate deaminase

**Synonymes**
HPC deaminase [1]

**CAS Reg. No.**
9054-77-7

## 2 REACTION AND SPECIFICITY

**Catalysed reaction**
1-Pyrroline-4-hydroxy-2-carboxylate + $H_2O$ →
→ 2,5-deoxypentanoate + $NH_3$

**Reaction type**
Amidine hydrolysis

**Natural substrates**
1-Pyrroline-4-hydroxy-2-carboxylate (L-form) + $H_2O$

**Substrate spectrum**
1  1-Pyrroline-4-hydroxy-2-carboxylate (L-form) + $H_2O$ [1–4]
2  1-Pyrroline-4-hydroxy-2-carboxylate (D-form) + $H_2O$ [1, 3]

**Product spectrum**
1  2,5-Dioxopentanoate + $NH_3$
2  ?

**Inhibitor(s)**
$HgCl_2$ [1]; $CuSO_4$ [1]; $AgNO_3$ [1]; p-Substituted-mercuribenzoate [1]

**Cofactor(s)/prostethic group(s)**

**Metal compounds/salts**

**Turnover number** (min$^{-1}$)

Enzyme Handbook © Springer-Verlag Berlin Heidelberg 1991
Duplication, reproduction and storage in data banks are only
allowed with the prior permission of the publishers

**Specific activity** (U/mg)
   More [1, 3]

**K$_m$-value** (mM)
   1.4 (1-pyrroline-4-hydroxy-2-carboxylate, L-form) [1, 3]

**pH-optimum**
   6.5–7.5 [1, 3]

**pH-range**
   6.0–9.0 [1]

**Temperature optimum** (°C)
   37 [1]

**Temperature range** (°C)
   0–40 [1]

---

## 3 ENZYME STRUCTURE

**Molecular weight**
   62000 (Pseudomonas striata, sedimentation) [1, 3]

**Subunits**

**Glycoprotein/Lipoprotein**
   –

---

## 4 ISOLATION/PREPARATION

**Source organism**
   Pseudomonas striata [1, 3]; Pseudomonas putida [4]

**Source tissue**

**Localisation in source**

**Purification**
   Pseudomonas striata [1, 3]

**Crystallization**
   –

**Cloned**
   –

**Renaturated**
   –

## 5 STABILITY

pH

**Temperature** (°C)
70 (stable up to) [1, 3]

Oxidation

Organic solvent

**General stability information**
Several hours at room temperature [1]

**Storage**
1 month at −15°C [1, 3]

---

## 6 CROSSREFERENCES TO STRUCTURE DATABANKS

PIR/MIPS code

**Brookhaven code**

---

## 7 LITERATURE REFERENCES

[1] Singh, R.M.M., Adams, E.: J. Biol. Chem., 240 (11), 4344–4351 (1965) (Review)
[2] Singh, R.M.M., Adams, E.: J. Biol. Chem., 240 (11), 4352–4356 (1965)
[3] Adams, E.: Methods Enzymol., 17, Pt. B, 266–306 (1971)
[4] Koo, P.H., Adams, E.: J. Biol. Chem., 249 (6), 1704–1716 (1974)

## 1 NOMENCLATURE

**EC number**
3.5.4.23

**Systematic name**
Blasticidin-S aminohydrolase

**Recommended name**
Blasticidin-S deaminase

**Synonymes**
Deaminase, blasticidin S

**CAS Reg. No.**
54576-55-5

## 2 REACTION AND SPECIFICITY

**Catalysed reaction**
Blasticidin S + $H_2O$ →
→ deaminohydroxyblasticidin S + $NH_3$

**Reaction type**
Amidine hydrolysis
Deamination (deamination of the cytosine moiety of antibiotics)

**Natural substrates**
Blasticidin S + $H_2O$ [5]

**Substrate spectrum**
1 Blasticidin S + $H_2O$ [1]
2 Cytomycin + $H_2O$ [1, 2]
3 Acetylblasticidin S + $H_2O$ [1]

**Product spectrum**
1 Deaminohydroxyblasticidin S + $NH_3$
2 Deaminohydroxycytomycin + $NH_3$
3 Acetylblasticidin + $NH_3$

**Inhibitor(s)**
p-Chloromercuribenzoate [1]; $Cu^{2+}$ [1]; $Hg^{2+}$ [1]; $Zn^{2+}$ [2]; $Fe^{2+}$ [2]; Rose Bengal (photooxidation in presence of Rose Bengal) [3]; Methylene Blue (photooxidation in presence of methylene blue) [3]; More (competitive inhibitors) [4]

Cofactor(s)/prostethic group(s)

Metal compounds/salts

---

Turnover number (min$^{-1}$)

Specific activity (U/mg)

$K_m$-value (mM)
    More [3]; 0.021 [4]

pH-optimum
    10 [1]; 10–11.5 [2]; 8–9 [5]

pH-range
    5–12 [1]

Temperature optimum (°C)
    60–70 [1]; 35–70 [2]

Temperature range (°C)

---

## 3 ENZYME STRUCTURE

Molecular weight
    30000 (Aspergillus terreus, SDS-PAGE, sedimentation analysis) [1]
    46000 (Aspergillus terreus, gel filtration, with NaCl) [1]
    55000 (Aspergillus terreus, gel filtration, without NaCl) [1]
    55000 (Aspergillus terreus) [2]

Subunits

Glycoprotein/Lipoprotein
    –

---

## 4 ISOLATION/PREPARATION

Source organism
    Aspergillus terreus [1–4]; Bacillus cereus [5]

Source tissue

Localisation in source

Purification
    Aspergillus terreus [1–4]

## Crystallization
—

## Cloned
[6]

## Renaturated
—

---

## 5 STABILITY

### pH
4.5 (rapid inactivation) [1]; 7 [2]; 4.0 (inactivated after 1 hour) [2]

### Temperature (°C)
50 (70% loss of activity after 30 minutes, bovine serum albumin and mercaptoethanol protect) [1]

### Oxidation

### Organic solvent

### General stability information
Photooxidation in presence of Rose Bengal or Methylene Blue [3]

### Storage
−20°C, 50% glycerol + 5 mM mercaptoethanol, 3 months (no loss of activity) [1]

---

## 6 CROSSREFERENCES TO STRUCTURE DATABANKS

### PIR/MIPS code

### Brookhaven code

---

## 7 LITERATURE REFERENCES

[1] Yamaguchi, I., Shibata, H., Seto, H., Misato, T.: J. Antibiot., 28 (1) , 7–14 (1975)
[2] Misato, T., Yamaguchi, I.: Jpn. Kokai, JP48099383 JP72–33359, 19 Pp. (1973)
[3] Yamaguchi, I., Misato, T.: Agric. Biol. Chem., 49 (12) , 3355–3361 (1985)
[4] Yamaguchi, I., Sheto, H., Misato, T.: Pestic. Biochem. Physiol., 25, 54–62 (1986)
[5] Endo, T., Furuta, K., Kaneko, A., Katsiki, T., Kobayashi, K.: J. Antibiot., XL (12) ,
    1791–1793 (1987)
[6] Kamakura, T., Kobayashi, I., Tanaka, T., Yamaguchi, I. , Endo, T.: Agric. Biol. Chem.,
    51 (11) , 3165–3168 (1987)

# 1 NOMENCLATURE

**EC number**
3.5.4.24

**Systematic name**
Sepiapterin aminohydrolase

**Recommended name**
Sepiapterin deaminase

**Synonymes**

**CAS Reg. No.**
62213-22-3

# 2 REACTION AND SPECIFICITY

**Catalysed reaction**
Sepiapterin + $H_2O$ →
→ xanthopterin-$B_2$ + $NH_3$

**Reaction type**
Amidine hydrolysis

**Natural substrates**
Sepiapterin + $H_2O$

**Substrate spectrum**
1 Sepiapterin + $H_2O$
2 Isosepiapterin + $H_2O$ (more slowly) [3]

**Product spectrum**
1 Xanthopterin-$B_2$ + $NH_3$
2 ?

**Inhibitor(s)**
Xanthopterin [1–3]; Pterin [1]; 2-Amino-4-hydroxypteridine [2]; Biopterin [2];
Amethopterin [1]; Aminopterin [1]; p-Chloromercuribenzoate [1]; KF [1];
8-Azaguanine [1]; Phenylmethylsulfonylfluoride [1]

**Cofactor(s)/prostethic group(s)**

**Metal compounds/salts**

Turnover number (min$^{-1}$)

Specific activity (U/mg)

K$_m$-value (mM)
    0.59 (sepiapterin) [2]

pH-optimum
    8.0 [2]

pH-range
    6–10 [3]

Temperature optimum (°C)

Temperature range (°C)

---

3 ENZYME STRUCTURE

Molecular weight

Subunits

Glycoprotein/Lipoprotein
    –

---

4 ISOLATION/PREPARATION

Source organism
    Silkworm (Bombyx mori, mutant lemon) [1–3]

Source tissue
    Larva [1–3]

Localisation in source
    Integument [1, 2]; Fat body [2]; Midgut [2]; Malphigian tube [2]; Silk gland
    (posterior) [2]; Gonads [2]

Purification
    Silkworm (mutant lemon) [1–3]

Crystallization
    –

Cloned
    –

Renaturated
    –

## 5 STABILITY

pH

Temperature (°C)

Oxidation

Organic solvent

General stability information

Storage
   Purified enzyme, 3°C, 1 month [2]

---

## 6 CROSSREFERENCES TO STRUCTURE DATABANKS

PIR/MIPS code

Brookhaven code

---

## 7 LITERATURE REFERENCES

[1] Tsusue, M., Mazda, T.: Experientia, 33 (7) , 854–855 (1977)
[2] Tsusue, M.: J. Biochem., 69, 781–788 (1971)
[3] Tsusue, M.: Experientia, 23 (2) , 116–117 (1967)

## 1 NOMENCLATURE

**EC number**
3.5.4.25

**Systematic name**
GTP 7, 8–8, 9-dihydrolase (pyrophosphate-forming)

**Recommended name**
GTP cyclohydrolase II

**Synonymes**
Guanosine triphosphate cyclohydrolase II
GTP-8-formylhydrolase [1]

**CAS Reg. No.**
56214-35-8

---

## 2 REACTION AND SPECIFICITY

**Catalysed reaction**
GTP + 3 $H_2O$ →
→ formate + 2, 5-diamino-6-hydroxy-4-(5-phosphoribosylamino)pyrimidine
+ pyrophosphate (two C-N bonds are hydrolyzed, releasing formate, with
simultaneous hydrolysis of the terminal pyrophosphate)

**Reaction type**
Amidine hydrolysis

**Natural substrates**
GTP + $H_2O$

**Substrate spectrum**
1 GTP + $H_2O$

**Product spectrum**
1 Formate + 2, 5-diamino-6-hydroxy-4-(5-phosphoribosylamino)
pyrimidine + pyrophosphate

---

**Inhibitor(s)**
Pyrophosphate (inorganic) [1–3]; $Cu^{2+}$ (Pichia guilliermondii) [1]; $Zn^{2+}$
(Pichia guilliermondii) [1]

**Cofactor(s)/prostethic group(s)**

---

**Metal compounds/salts**
$Mg^{2+}$ [1–4]; $Mn^{2+}$ [3]

**Turnover number** ($min^{-1}$)

**Specific activity** (U/mg)
0.223 [2, 3]

**$K_m$-value** (mM)
0.041 (GTP) [1–3]; 0.022 (GTP) [1]

**pH-optimum**
8.4 [2, 3]; 8.2 [1]

**pH-range**
7.2–10.5 [2, 3]

**Temperature optimum** (°C)
42 [2]

**Temperature range** (°C)

## 3 ENZYME STRUCTURE

**Molecular weight**
44000 (gel filtration, E. coli) [1–3]
170000 (gel filtration, Pichia guilliermondii) [1, 4]

**Subunits**
Monomer (1 × 44000, E. coli, gel filtration) [3]

**Glycoprotein/Lipoprotein**
–

## 4 ISOLATION/PREPARATION

**Source organism**
E. coli [1–3]; Pichia guilliermondi [1, 4]

**Source tissue**
Cell [2, 3]

**Localisation in source**

**Purification**
E. coli [2, 3]

Crystallization
–

Cloned
–

Renaturated
–

---

## 5 STABILITY

pH

Temperature (°C)
60 (inactivated) [3]

Oxidation

Organic solvent

General stability information

Storage
Purified enzyme, pH 8, EDTA, –20°C [2, 3]

---

## 6 CROSSREFERENCES TO STRUCTURE DATABANKS

PIR/MIPS code

Brookhaven code

---

## 7 LITERATURE REFERENCES

[1] Blau, N., Niederwieser, A.: J. Clin. Chem. Clin. Biochem., 23, 169–176 (1985) (Review)
[2] Foor, F., Brown, G.M.: Methods Enzymol., 66, 303–307 (1980)
[3] Foor, F., Brown, G.M.: J. Biol. Chem., 250 (9) , 3545–3551 (1975)
[4] Shavlovsky, G.M., Logvinenko, E.M., Zakalsky, A.E.: Biokhimiya, 48, 837–843 (1983)

## 1 NOMENCLATURE

EC number
3.5.4.26

Systematic name
2, 5-Diamino-6-hydroxy-4-(5-phosphoribosylamino)-pyrimidine
2-aminohydrolase

Recommended name
Diaminohydroxyphosphoribosylaminopyrimidine deaminase

Synonymes

CAS Reg. No.

## 2 REACTION AND SPECIFICITY

Catalysed reaction
2, 5-Diamino-6-hydroxy-4-(5-phosphoribosylamino)-pyrimidine + $H_2O$ →
→ 5-amino-6-(5-phosphoribosylamino)uracil + $NH_3$

Reaction type
Amidine hydrolysis

Natural substrates
2, 5-Diamino-6-hydroxy-4-(5-phosphoribosylamino)-pyrimidine + $H_2O$

Substrate spectrum
1 2, 5-Diamino-6-hydroxy-4-(5-phosphoribosylamino)-pyrimidine + $H_2O$
2 2, 5-Diamino-6-oxy-4-(5-phosphoribosylamino)-pyrimidine + $H_2O$ [1]
3 More (not: 2, 5-diamino-6-oxy-4-(ribosylamino)-pyrimidine) [1]

Product spectrum
1 5-Amino-6-(5-phosphoribosylamino)uracil + $NH_3$ [1]
2 ?
3 ?

Inhibitor(s)
More (not: GTP, GMP, AMP, CTP, FAD, FMN, riboflavin, 6-hydroxy-2, 4, 5-tri-
aminopyrimidine) [1]

Cofactor(s)/prostethic group(s)

Metal compounds/salts

Turnover number (min$^{-1}$)

Specific activity (U/mg)

K$_m$-value (mM)

pH-optimum
  9.1 [1]

pH-range

Temperature optimum (°C)
  39 [1]

Temperature range (°C)

---

## 3 ENZYME STRUCTURE

Molecular weight
  80000 (E. coli, gel filtration) [1]

Subunits

Glycoprotein/Lipoprotein
  –

---

## 4 ISOLATION/PREPARATION

Source organism
  E. coli B [1]

Source tissue

Localisation in source

Purification
  E. coli B [1]

Crystallization
  –

Cloned
  –

Renaturated
  –

## 5 STABILITY

pH

Temperature (°C)

Oxidation

Organic solvent

General stability information

Storage
1 month, −6°C, pH 8.0 [1]

---

## 6 CROSSREFERENCES TO STRUCTURE DATABANKS

PIR/MIPS code

Brookhaven code

---

## 7 LITERATURE REFERENCES

[1] Burrows, R.B., Brown, G.M.: J. Bacteriol., 136 (2) , 657–667 (1978)

## 1 NOMENCLATURE

**EC number**
3.5.5.1

**Systematic name**
Nitrile aminohydrolase

**Recommended name**
Nitrilase

**Synonymes**
Acetonitrilase
Benzonitrilase [4]

**CAS Reg. No.**
9024-90-2

## 2 REACTION AND SPECIFICITY

**Catalysed reaction**
A nitrile + $H_2O$ →
→ a carboxylate + $NH_3$; A nitrile + $H_2O$ →
→ an amide [1, 5, 7–10]

**Reaction type**
Nitrile hydrolysis (N-bond hydrolysis)

**Natural substrates**
Indoleacetonitrile + $H_2O$

**Substrate spectrum**
1 Indoleacetonitrile + $H_2O$
2 Benzonitrile + $H_2O$
3 Acetonitrile + $H_2O$
4 3, 5-Dibromo-4-hydroxy-benzonitrile (Bromoxynil) + $H_2O$
5 Acrylnitrile + $H_2O$
6 Nitriles (aliphatic) + $H_2O$
7 Nitriles (aliphatic) + $H_2O$
8 Nitriles (aromatic) + $H_2O$
9 Nitriles (aromatic) + $H_2O$
10 More [3, 4, 8, 13, 14]

**Product spectrum**

1 Indoleacetic acid + $NH_3$
2 Benzoic acid + $NH_3$
3 Acetic acid + $NH_3$
4 3, 5-Dibromo-4-hydroxy-benzoic acid + $NH_3$
5 Acrylamide
6 Carboxylate + $NH_3$
7 Amide (aliphatic)
8 Carboxylate (aromatic) + $NH_3$
9 Amide (aromatic)
10 ?

**Inhibitor(s)**

SH-reagents (glutathione reverses inhibition) [4, 13–17]; Heavy metal ions [4, 13, 15, 17]; $Ag^{2+}$ [4, 13, 15, 17]; Urea [13, 15]; $Hg^{2+}$ [4, 13, 15, 17]; $Cu^{2+}$ [4, 13, 15, 17]; $Zn^{2+}$ [4, 13, 15, 17]; Diisopropylfluorophosphate [14]; $CaCl_2$ [4]

**Cofactor(s)/prostethic group(s)**

**Metal compounds/salts**

**Turnover number** ($min^{-1}$)

**Specific activity** (U/mg)

1.66 [13]; 1.31 [4]; 0.122 [4]; More [8]

**$K_m$-value** (mM)

0.051 (indoleacetonitrile) [17]; 0.52 (indoleacetonitrile) [10]; 4 (benzonitrile) [15, 16]; 1.1 (benzonitrile) [15]; 0.039 (benzonitrile) [13]; 0.25 (acetonitrile) [14]; 2.6 (3-cyanopyridine) [10]

**pH-optimum**

7.0 [14]; 7.0–7.4 [6]; 7.1 [17]; 7.5 [4, 8, 10]; 7.7 [1]; 7.8–9.1 [13]; 7.95–8.05 [16]; 8.0 [15]; 8.5 [1, 4]

**pH-range**

5.5–8 [17]; 5–10 [13, 14]

**Temperature optimum** (°C)

20–40 [15]; 30–35 [14]; 40 [4]; 30 [4]; 35 [1]

**Temperature range** (°C)

5–35 [17]

## 3 ENZYME STRUCTURE

**Molecular weight**
  23000 (gel filtration, Arthrobacter sp.) [4]
  30000 (gel filtration, Arthrobacter sp.) [4]
  44500–47000 (gel filtration, gel electrophoresis, Nocardia sp., in absence of substrate) [15, 16]
  560000 (gel filtration, Nocardia sp., in presence of substrate) [15, 16]
  620000 (gel filtration, Fusarium sp.) [13]
  76000 (gel electrophoresis, Fusarium sp.) [13]

**Subunits**
  Monomer (1 × 30000, gel eletrophoresis) [4]
  Octamer (8 × 76000, gel electrophoresis) [13]
  Polymer (12 × 44500–47000, gel electrophoresis) [15, 16]

**Glycoprotein/Lipoprotein**
  –

---

## 4 ISOLATION/PREPARATION

**Source organism**
  Barley [17]; Chinese cabbage [10]; Fusarium sp. [8, 13, 17]; Gibberella fujikuroi [17]; Aspergillus niger [17]; Penicillium chrysogenum [17]; Nocardia sp. [6, 15, 16]; Pseudomonas sp. [18]; Corynebacterium sp. [19]; Brevibacterium [5, 7, 11, 12, 14]; Arthrobacter sp. [4, 9]; Klebsiella pneumonia [2, 3]; Rhodococcus sp. [1]; More [17]

**Source tissue**
  Leaves [17]; Seedlings [10]; Cell [16]

**Localisation in source**
  Cytoplasm [5, 6]; Soluble [5, 6]

**Purification**
  Nocardia sp. [15, 16]; Fusarium solani [13]; Arthrobacter sp. [4]

**Crystallization**
  –

**Cloned**
  –

**Renaturated**
  –

## 5 STABILITY

**pH**

**Temperature (°C)**
35 (unstable above) [14, 17]; 0 (unstable at) [16]; 30 (unstable above) [14]; 45 (up to, 20 minutes) [4]; 45 (unstable above) [4]

**Oxidation**

**Organic solvent**

**General stability information**
Highly unstable [4]; Freezing (unstable) [13]

**Storage**
More [13]

---

## 6 CROSSREFERENCES TO STRUCTURE DATABANKS

**PIR/MIPS code**
A28658 (Klebsiella pneumoniae)

**Brookhaven code**

---

## 7 LITERATURE REFERENCES

[1] Watanabe, I.: Methods Enzymol., 136, 523–530 (1987)
[2] Stalker, D.M., McBride, K.E.: J. Bacteriol., 169 (3) , 955–960 (1987)
[3] McBride, K.E., Kenny, J.W., Stalker, D.M.: Appl. Environ. Microbiol., 52 (2) , 325–330 (1986)
[4] Bandyopadhyay, A.K., Nagasawa, T., Asano, Y., Fujishiro, K., Tani, Y., Yamada, H.: Appl. Environ. Microbiol., 51 (2) , 302–306 (1986)
[5] Miller, J.M., Knowles, C.J.: FEMS Microbiol. Lett., 21, 147–151 (1984)
[6] Collins, P.A., Knowles, C.J.: J. Gen. Microbiol., 129, 711–718 (1983)
[7] Bui, K., Arnaud, A., Galzy, P.: Enzyme Microb. Technol., 4, 195–197 (1982)
[8] Kuwahara, M., Yanase, H., Ishida, Y., Kikuchi, Y.: J. Ferment. Technol., 58 (6) , 573–577 (1980)
[9] Asano, Y., Tani, Y., Yamada, H.: Agric. Biol. Chem., 44 (9) , 2251–2252 (1980)
[10] Rausch, T., Hilgenberg, W.: Phytochemistry, 19, 747–750 (1980)
[11] Jallageas, J.C., Arnaud, A., Galzy, P.: Anal. Biochem., 95, 436–443 (1979)
[12] Jallageas, J.C., Arnaud, A., Galzy, P.: J. Chromatogr., 166, 181–187 (1978)
[13] Harper, D.B.: Biochem. J., 167, 685–692 (1977)
[14] Arnaud, A., Galzy, P., Jallageas, J.C.: Agric. Biol. Chem., 41 (11) , 2183–2191 (1977)
[15] Harper, D.B.: Biochem. J., 165, 309–319 (1977)
[16] Harper, D.B.: Biochem. Soc. Trans., 4, 502–505 (1976)
[17] Thimann, K.V., Mahadevan, S.: Arch. Biochem. Biophys. , 105, 133–141 (1964)
[18] Robinson, W.G., Hook, R.H.: J. Biol. Chem., 239, 4263–4267 (1964)
[19] Mimura, A., Kawano, T., Yamaga, K.: J. Ferment. Technol., 47, 631–638 (1964)

# 1 NOMENCLATURE

**EC number**
3.5.5.2

**Systematic name**
Ricinine aminohydrolase

**Recommended name**
Ricinine nitrilase

**Synonymes**

**CAS Reg. No.**
9075-40-5

---

# 2 REACTION AND SPECIFICITY

**Catalysed reaction**
N-Methyl-3-cyano-4-methoxy-2-pyridone + $H_2O$ →
→ 3-carboxy-4-methoxy-2-pyridone + $NH_3$

**Reaction type**
Nitrile hydrolysis

**Natural substrates**
Ricinine + $H_2O$

**Substrate spectrum**
1 N-Methyl-3-cyano-4-methoxy-2-pyridone + $H_2O$
2 3-Cyano-2-pyridones + $H_2O$ [1]
3 More [1]

**Product spectrum**
1 3-Carboxy-4-methoxy-N-methyl-2-pyridone + $NH_3$
2 3-Carboxy-2-pyridones (corresponding) [1]
3 ?

---

**Inhibitor(s)**
p-Hydroxymercuribenzoate [2, 5]; Diisopropylfluorophosphate [2, 5]; NaCN
[2, 5]

**Cofactor(s)/prostethic group(s)**

**Metal compounds/salts**

---

**Turnover number** (min$^{-1}$)

**Specific activity** (U/mg)
More [2]

**K$_m$-value** (mM)
0.019 (ricinine) [2, 5]; 0.05 (N-ethyl-3-cyano-4-methoxy-2-pyridone) [2]

**pH-optimum**
7.4–8.8 [2, 5]

**pH-range**
6.0–11.0 [2]

**Temperature optimum** (°C)
37 [2]

**Temperature range** (°C)

---

## 3 ENZYME STRUCTURE

**Molecular weight**

**Subunits**

**Glycoprotein/Lipoprotein**
–

---

## 4 ISOLATION/PREPARATION

**Source organism**
Pseudomonas sp. [1, 2, 5]; Ricinus communis [3, 4, 6]

**Source tissue**

**Localisation in source**

**Purification**
Pseudomonas spec. [2]

**Crystallization**
–

**Cloned**
–

**Renaturated**
–

# 5 STABILITY

**pH**
 9.5–10.5 (at 0°C) [2, 5]

**Temperature (°C)**

**Oxidation**

**Organic solvent**

**General stability information**

**Storage**

---

# 6 CROSSREFERENCES TO STRUCTURE DATABANKS

**PIR/MIPS code**

**Brookhaven code**

---

# 7 LITERATURE REFERENCES

[1] Robinson, W.G., Hook, R.H.: J. Biol. Chem., 239 (12) , 4257–4262 (1964)
[2] Hook, R.H., Robinson, W.G.: J. Biol. Chem., 239 (12) , 4263–4267 (1964)
[3] Waller, G.R., Lee, J.L.: Plant Physiol., 44, 522–526 (1969)
[4] Skursky, L., Burleson, D., Waller, G.R.: J. Biol. Chem., 244 (12) , 3238–3242 (1969)
[5] Robinson, W.G., Hook, R.H.: Methods Enzymol., Vol.17, Pt. B, 244–248 (1971)
[6] Waller, G.R., Skursky, L.: Plant Physiol., 50, 622–626 (1972)

# 1 NOMENCLATURE

**EC number**
3.5.5.3

**Systematic name**
Cyanate aminohydrolase

**Recommended name**
Cyanate hydrolase

**Synonymes**
Cyanase

**CAS Reg. No.**
37289-24-0

# 2 REACTION AND SPECIFICITY

**Catalysed reaction**
Cyanate + $H_2O$ →
→ $CO_2$ + $NH_3$

**Reaction type**
Nitrile hydrolysis

**Natural substrates**
Cyanate + $H_2O$ [2]

**Substrate spectrum**
1  Cyanate + $H_2O$ [1–2, 4–6, 8–10, 12, 13]

**Product spectrum**
1  $CO_2$ + $NH_3$ [1–2, 4–6, 8–10, 12, 13]

**Inhibitor(s)**
Chloride [3, 8, 10]; Bromide [3, 8]; Nitrite [3, 8]; Nitrate [3, 8]; Azide [3, 8, 10];
Cyanide [3]; Thiocyanate [3]; Cyanate [8]; Bicarbonate [8]; Acetate [8];
Malonate [8, 10]; Formate [8]; Oxalate [8, 10]; Oxalacetate [8, 10]; Sulfite
[10]

**Cofactor(s)/prostethic group(s)**

**Metal compounds/salts**

**Turnover number** (min$^{-1}$)

**Specific activity** (U/mg)
  More [6, 9]

**$K_m$-value** (mM)
  29 (cyanate) [2]; 2 (cyanate) [3]; 0.6 (cyanate, presence of bicarbonate) [6]

**pH-optimum**
  6–8 [2]; 5–6 [5]; 7.4 [6]; 7.2 [8]; 8 [9]

**pH-range**
  6–8 [2]

**Temperature optimum** (°C)
  37 [4, 6, 9, 15]; 26 [8, 10]

**Temperature range** (°C)

---

## 3 ENZYME STRUCTURE

**Molecular weight**
  141000 (E. coli, gradient centrifugation) [3]
  150000 (E. coli, gradient centrifugation, gel filtration) [6]
  136000 (E. coli, amino acid analysis) [14]

**Subunits**
  Oligomer (8–12 × 15000, E. coli, gel electrophoresis [6], 8–10 × 17000,
  E. coli [13, 14]) [6, 13, 14]

**Glycoprotein/Lipoprotein**
  –

---

## 4 ISOLATION/PREPARATION

**Source organism**
  E. coli [2–4, 6, 7, 9, 10–15]; Guinea pig [5]; Rat [5]; Flavobacterium sp. [9]

**Source tissue**
  Liver [5]; Kidney [5]

**Localisation in source**

**Purification**
  E. coli [2–4, 6, 7, 9, 10–15]; Guinea pig [5]; Rat [5]

**Crystallization**
  [14]

## Cloned
[11, 15]

## Renaturated
(removal of denaturants: urea, guanidine hydrochloride) [13]

---

# 5 STABILITY

## pH

## Temperature (°C)
66 (1 hour: 60% activity, 2 hours: 3% activity) [3]; 55 [13]; 100 (destroyed by boiling) [2]

## Oxidation

## Organic solvent
Acetone (stabilization) [3]; Ethanol (stabilization) [3]

## General stability information
Azide stabilizes against denaturation by urea [13]

## Storage
−20°C, some weeks [2]; −20°C, some months [6]

---

# 6 CROSSREFERENCES TO STRUCTURE DATABANKS

## PIR/MIPS code

## Brookhaven code

---

# 7 LITERATURE REFERENCES

[1] Rotini, O.F.: Ric. Sci., 26, 2786 (1956)
[2] Taussig, A.: Biochim. Biophys. Acta, 44, 510 (1960)
[3] Taussig, A.: Can. J. Biochem., 43, 1063 (1965)
[4] Taussig, A., Ronnen, E.: Can. J. Biochem., 48, 790 (1970)
[5] Collins, C.A., Anderson, P.M.: Biochem. Biophys. Res. Commun., 79, 1255 (1977)
[6] Anderson, P.M.: J. Biol. Chem. , 258, 276 (1983)
[7] Chin, C.C.Q., Anderson, P.M., Wold, F.: J. Biol. Chem. , 258, 276 (1983)
[8] Anderson, P.M., Little, R.M.: Biochemistry, 25, 1621 (1986)
[9] Guilloton, M., Karst, F.: J. Gen. Microbiol., 133, 645 (1987)

---

Enzyme Handbook © Springer-Verlag Berlin Heidelberg 1991
Duplication, reproduction and storage in data banks are only
allowed with the prior permission of the publishers

[10] Anderson, P.M., Johnson, W.V., Endrizzi, J.A., Little, R.H., Korte, J.J.: Biochemistry, 26, 3938 (1987)
[11] Sung, Y.-C., Parsell, D., Anderson, P.M., Fuchs, J.A. : J. Bacteriol., 169, 2639 (1987)
[12] Johnson, W.V., Anderson, P.M.: J. Biol. Chem., 262, 9021 (1987)
[13] Little, R.M., Anderson, P.M.: J. Biol. Chem., 262, 10120 (1987)
[14] Kim, K.H.: J. Mol. Biol., 198, 137 (1987)
[15] Sung, Y.-C., Anderson, P.M., Fuchs, J.A.: J. Bacteriol., 169, 5524 (1987)

# 1 NOMENCLATURE

**EC number**
3.5.99.1

**Systematic name**
Riboflavin hydrolase

**Recommended name**
Riboflavinase

**Synonymes**
Riboflavine hydrolase

**CAS Reg. No.**
9024-79-7

---

# 2 REACTION AND SPECIFICITY

**Catalysed reaction**
Riboflavin + $H_2O \rightarrow$
$\rightarrow$ ribitol + lumichrome

**Reaction type**
Hydrolytic C-N bond cleavage

**Natural substrates**
D-Riboflavin + $H_2O$

**Substrate spectrum**
1 Riboflavin + $H_2O$
2 Riboflavin (derivatives) + $H_2O$ [1]
3 More [2, 3]

**Product spectrum**
1 Ribitol + lumichrome
2 Alcohol (corresponding) + lumichrome
3 ?

---

**Inhibitor(s)**
Riboflavin 5'-phosphate [4]; More (high substrate levels [2, 3], anaerobic conditions [1–3]) [1–3]

**Cofactor(s)/prostethic group(s)**
Glutathione (reduced) [4]

---

**Metal compounds/salts**
$Mg^{2+}$ [4]; $Li^+$ [4]; $Mn^{2+}$ [4]; $Cd^{2+}$ [4]

**Turnover number** ($min^{-1}$)

**Specific activity** (U/mg)
More [4]

**$K_m$-value** (mM)
More [2, 3]

**pH-optimum**
5.8–7.0 [1]; 7.5 [4]

**pH-range**
5.2–8.0 [1]; 6.0–9.0 [4]

**Temperature optimum** (°C)
30 [2, 3]; 37 [4]

**Temperature range** (°C)

## 3 ENZYME STRUCTURE

**Molecular weight**

**Subunits**

**Glycoprotein/Lipoprotein**
–

## 4 ISOLATION/PREPARATION

**Source organism**
Pseudomonas riboflavina [1–3]; Crinum longifolium [4]

**Source tissue**
Bulbs [4]

**Localisation in source**

**Purification**
Crinum longifolium [4]

**Crystallization**
–

Cloned

–

Renaturated

–

## 5 STABILITY

pH

Temperature (°C)
  55 (stable up to) [4]

Oxidation

Organic solvent

General stability information

Storage

## 6 CROSSREFERENCES TO STRUCTURE DATABANKS

PIR/MIPS code

Brookhaven code

## 7 LITERATURE REFERENCES

[1] Yanagita, T., Foster, J.W.: J. Biol. Chem., 221, 593–607 (1956)
[2] Yang, C.S., McCormick, D.B.: Biochim. Biophys. Acta, 132, 511–513 (1967)
[3] Yang, C.S., McCormick, D.B.: Methods Enzymol., 18, Pt. B, 571–573 (1971)
[4] Kumar, S.A., Vaidyanathan, C.S.: Biochim. Biophys. Acta, 89, 127- (1964)

# 1 NOMENCLATURE

**EC number**
3.5.99.2

**Systematic name**
Thiamin hydrolase

**Recommended name**
Thiaminase

**Synonymes**
Thiaminase II

**CAS Reg. No.**
9024-80-0

# 2 REACTION AND SPECIFICITY

**Catalysed reaction**
Thiamin + $H_2O$ →
→ 4-amino-5-hydroxymethyl-2-methylpyrimidine +
5-(2-hydroxymethyl)-4-methylthiazole

**Reaction type**
Hydrolytic C-N bond cleavage

**Natural substrates**
Thiamin + $H_2O$

**Substrate spectrum**
1 Thiamin + $H_2O$ (r)
2 Pyrimidinemethylaniline + $H_2O$ (r) [8]
3 Thiothiamine + $H_2O$ [3, 6]
4 2-Northiamine + $H_2O$ [1]
5 Dimethialium + $H_2O$ [1]
6 Pyrithiamine + $H_2O$ [1]
7 Oxythiamine + $H_2O$ [1]
8 More (not: thiaminpyrophosphate) [3]

## Product spectrum

1 4-Amino-5-hydroxymethyl-2-methylpyrimidine +
   5-(2-hydroxyethyl)-4-methylthiazole
2 Pyrimidinemethanol + aniline
3 2-Thiothiazole + pyrimidine [3]
4 2-Norhydroxymethylpyrimidine + hydroxyethylthiazole
5 ?
6 ?
7 ?
8 ?

## Inhibitor(s)

SH-reagents [6, 7]; Heavy metal ions [6, 7]; $Hg^{2+}$ [6, 7]; $Ag^{2+}$ [6, 7]; $Cu^{2+}$
[6, 7]; $Zn^{2+}$ [6, 7]; $Fe^{2+}$ [6, 7]

## Cofactor(s)/prostethic group(s)

## Metal compounds/salts

## Turnover number (min$^{-1}$)

## Specific activity (U/mg)

77 [6, 7]

## $K_m$-value (mM)

0.003 (thiamin) [4, 6, 7]

## pH-optimum

8 [8]; 8.6 [4, 6, 7]; 7.0 [6]

## pH-range

## Temperature optimum (°C)

60 [8]; 40–45 [6]; 30–40 [6]; 37 [4]

## Temperature range (°C)

## 3 ENZYME STRUCTURE

## Molecular weight

100000 (sedimentation equilibrium, Bacillus aneurinolyticus) [6, 7]

## Subunits

## Glycoprotein/Lipoprotein

–

## 4 ISOLATION/PREPARATION

**Source organism**
Bacillus aneurinolyticus [2, 4–8]; Trichosporon aneurinolyticum [6, 7];
Candida aneurinolytica [6, 7]; Staphylococcus aureus [7]; Micrococcus
pyrogenes [6]; E. coli [6]; Oospora [3]; Saccharomyces cerevisiae [1]

**Source tissue**
Cell [1]

**Localisation in source**
Cytoplasm [1]; Soluble [1]

**Purification**
Bacillus aneurinolyticus [6, 7]

**Crystallization**
[7]

**Cloned**
–

**Renaturated**
–

## 5 STABILITY

**pH**

**Temperature (°C)**

**Oxidation**

**Organic solvent**

**General stability information**

**Storage**

## 6 CROSSREFERENCES TO STRUCTURE DATABANKS

**PIR/MIPS code**

**Brookhaven code**

# 7 LITERATURE REFERENCES

[1] Kimura, Y., Iwashima, A.: Experientia, 43, 888–890 (1987)
[2] Abe, M., Nishimune, T., Ito, S., Kimoto, M., Hayashi, R.: FEMS Microbiol. Lett., 34, 129–133 (1986)
[3] Murata, K.: Am. N.Y. Acad. Sci. (Thiamin Twenty Years Prog.) , 378, 146–156 (1982) (Review)
[4] Edwin, E.E.: Methods Enzymol., 62, 113–117 (1979)
[5] Fujita, A.: J. Vitaminol., 18, 67–72 (1972) (Review)
[6] Wittliff, J.L., Airth, R.L.: Methods Enzymol., 18, Pt. A., 234–238 (1970)
[7] Ikehata, H.: J. Gen. Appl. Microbiol., 6 (1) , 30–39 (1960)
[8] Fujita, H.: J. Gen. Appl. Microbiol., 6 (1) , 30–39 (1954)

# 1 NOMENCLATURE

**EC number**
3.7.1.1

**Systematic name**
Oxaloacetate acetylhydrolase

**Recommended name**
Oxalacetase

**Synonymes**
Oxaloacetase
Oxalacetic hydrolase

**CAS Reg. No.**
9024-89-9

# 2 REACTION AND SPECIFICITY

**Catalysed reaction**
Oxaloacetate + $H_2O$ →
→ oxalate + acetate

**Reaction type**
Hydrolytic C-C bond cleavage

**Natural substrates**
Oxalacetate + $H_2O$

**Substrate spectrum**
1 Oxalacetate + $H_2O$

**Product spectrum**
1 Oxalate + acetate

**Inhibitor(s)**

**Cofactor(s)/prostethic group(s)**

**Metal compounds/salts**
$Mn^{2+}$ [1, 3, 6]

**Turnover number (min$^{-1}$)**

**Specific activity** (U/mg)
  0.2 [1]

**K$_m$-value** (mM)
  0.22 (oxalacetate) [3]; 1–2 (oxalacetate) [1]; 0.021 (Mn$^{2+}$) [3]

**pH-optimum**
  7.0–9.0 [1, 3]

**pH-range**

**Temperature optimum** (°C)

**Temperature range** (°C)

---

## 3 ENZYME STRUCTURE

**Molecular weight**
  420000 (Aspergillus niger, zonal centrifugation) [3]

**Subunits**

**Glycoprotein/Lipoprotein**
  —

---

## 4 ISOLATION/PREPARATION

**Source organism**
  Aspergillus niger (inducible enzyme) [2, 3, 4, 6]; Whetzelinia (Sclerotinia)
  sclerotiorum [5]; Streptomyces cattleya [1]; Endothia parasitica [1]

**Source tissue**
  Cell

**Localisation in source**

**Purification**
  Aspergillus niger [3]

**Crystallization**
  —

**Cloned**
  —

**Renaturated**
  —

## 5 STABILITY

pH
  6.75 (unstable below) [1]

Temperature (°C)

Oxidation

Organic solvent

General stability information

Storage
  Lyophilized, −18°C, 3 months [3]; More [3]

## 6 CROSSREFERENCES TO STRUCTURE DATABANKS

PIR/MIPS code

Brookhaven code

## 7 LITERATURE REFERENCES

[1] Houck, D.R., Inamine, E.: Arch. Biochem. Biophys., 259 (1) , 58–65 (1987)
[2] Müller, H.-M.: Zentralbl. Mikrobiol., 141, 461–469 (1986)
[3] Lenz, H., Wunderwald, P., Eggerer, H.: Eur. J. Biochem., 65, 225–236 (1976)
[4] Müller, H.-M.: Arch. Mikrobiol., 103, 185–189 (1975)
[5] Maxwell, D.P.: Physiol. Plant Pathol., 3, 279–288 (1973)
[6] Hayaishi, O., Shimazono, H., Katagiri, M., Saito, Y.: J. Am. Chem. Soc., 78, 5126–5127 (1956)

# 1 NOMENCLATURE

**EC number**
3.7.1.2

**Systematic name**
4-Fumarylacetoacetate fumarylhydrolase

**Recommended name**
Fumarylacetoacetase

**Synonymes**
Beta-diketonase
Fumarylacetoacetate hydrolase

**CAS Reg. No.**
9032-59-1

---

# 2 REACTION AND SPECIFICITY

**Catalysed reaction**
4-Fumarylacetoacetate + $H_2O$ →
→ acetoacetate + fumarate

**Reaction type**
Hydrolytic C-C bond cleavage

**Natural substrates**
Fumarylacetoacetate + $H_2O$ [5]

**Substrate spectrum**
1 4-Fumarylacetoacetate + $H_2O$
2 3,5-Dioxy acids + $H_2O$
3 2,4-Dioxo acids + $H_2O$
4 More [4, 9]

**Product spectrum**
1 Acetoacetate + fumarate
2 Acetoacetate
3 Pyruvate
4 ?

## Inhibitor(s)
Anions (monovalent) [5]; p-Hydroxymercuribenzoate [1, 4];
5, 5 '-Dithiobis(2-nitrobenzoic acid) [1]; Malonate [8]; Oxalate [8]; Citrate
[8]; Benzoylacetone [8]

## Cofactor(s)/prostethic group(s)

## Metal compounds/salts

## Turnover number (min$^{-1}$)

## Specific activity (U/mg)

## K$_m$-value (mM)
1.38 (triacetic acid) [8]; 12–89 (several diketo acids) [4]

## pH-optimum
7.2–7.9 [4, 9]; 6.5–8.2 (depending on buffer) [8]

## pH-range
5.5–9 [9]

## Temperature optimum (°C)

## Temperature range (°C)

## 3 ENZYME STRUCTURE

### Molecular weight
76000–96000 (bovine, gel filtration) [4]

### Subunits
Dimer (bovine, 2 × 38000–43000) [4]

### Glycoprotein/Lipoprotein
–

## 4 ISOLATION/PREPARATION

### Source organism
Rat [2, 9]; Rabbit [9]; Cat [9]; Mouse [9]; Bovine [4, 6, 8]; Human [3]

### Source tissue
Liver [2, 4, 7, 8, 9]; Kidney [9]

### Localisation in source
Soluble

## Purification
Bovine [4, 8]; Rat [2]

## Crystallization
–

## Cloned
–

## Renaturated
–

---

## 5 STABILITY

**pH**
6.0–8.5 [8]

**Temperature (°C)**
50 (5 minutes, stable up to) [8]

**Oxidation**

**Organic solvent**

**General stability information**

**Storage**
Frozen for at least 3 months [8]

---

## 6 CROSSREFERENCES TO STRUCTURE DATABANKS

**PIR/MIPS code**

**Brookhaven code**

---

## 7 LITERATURE REFERENCES

[1] Palcic Nagainis, M., Pu, W., Cheng, B., Taylor, K.E., Schmidt, D.E.Jr.: Biochim. Biophys. Acta, 657, 203–211 (1981)
[2] Doshi, K.S., Schmidt, D.E.Jr.: Can. J. Biochem., 56, 866–868 (1977)
[3] Lindblad, B., Lindstedt, S., Streeu, G.: Proc. Natl. Acad. Sci. USA, 74 (10) , 4641–4645 (1977)
[4] Mahuran, D.J., Angus, R.H., Braun, C.V., Sim, S.S., Schmidt, D.E.Jr.: Can. J. Biochem., 55 (1) , 1–8 (1977)

---

[5] Braun, C.V., Schmidt, D.E.Jr.: Biochemistry, 12 (24) , 4878–4881 (1973)
[6] Hsiang, H.H., Sim, S.S., Mahuran, D.J., Schmidt, D.E. Jr.: Biochemistry, 11 (11) , 2098–2102 (1972)
[7] Edwards, S.W., Knox, W.E.: J. Biol. Chem., 220, 79–91 (1956)
[8] Connors, W.M., Stotz, E.: J. Biol. Chem., 178, 881–890 (1949)
[9] Meister, A., Greenstein, J.P.: J. Biol. Chem., 175, 573–588 (1948)

# 1 NOMENCLATURE

**EC number**
3.7.1.3

**Systematic name**
L-Kynurenine hydrolase

**Recommended name**
Kynureninase

**Synonymes**

**CAS Reg. No.**
9024-78-6

# 2 REACTION AND SPECIFICITY

**Catalysed reaction**
L-Kynurenine + $H_2O$ →
→ anthranilate + L-alanine

**Reaction type**
Hydrolytic C-C bond cleavage
Elimination (beta-elimination, transamination, C-C bond hydrolysis) [1, 14]

**Natural substrates**
L-Kynurenine + $H_2O$
L-3-Hydroxy-kynurenine + $H_2O$

**Substrate spectrum**
1 L-Kynurenine + $H_2O$
2 L-3'-Hydroxy-kynurenine + $H_2O$
3 (3-Acrylcarbonyl)-alanines + $H_2O$ [7]
4 L-Alanine + $H_2O$
5 S-Cysteine (conjugates) + $H_2O$ [1]

**Product spectrum**
1 Anthranilate + L-alanine
2 3-Hydroxy-anthranilate + L-alanine
3 3-Acylcarbonylate + L-alanine
4 Pyruvate + $NH_3$
5 Pyruvate + $NH_3$ + thiol

## Inhibitor(s)

Carbonyl reagents [11, 15, 17]; Amines [16]; Diamines [16]; Thiol reagents [11, 15]; L-Alanine; L-Ornithine (inducible type) [14]; 3-Hydroxy-anthranilate (constitutive type) [8]

## Cofactor(s)/prostethic group(s)

Pyridoxal-phosphate [18, 19]

## Metal compounds/salts

$Mg^{2+}$ (activates) [17]

## Turnover number (min$^{-1}$)

## Specific activity (U/mg)

2.7 (L-kynurenine) [9]; 9.7 (L-kynurenine) [7]; 0.23–0.3 (L-kynurenine) [5]; More [5, 7, 9]

## $K_m$-value (mM)

0.4 (L-kynurenine) [19]; 0.006 (L-kynurenine) [17]; 0.035 (L-kynurenine) [9]; 0.003 (L-3-hydroxy-kynurenine) [17]; 0.00667 (L-3-hydroxy-kynurenine) [13]; 0.059 (L-3-hydroxy-kynurenine) [13]; 0.0006 (pyridoxal phosphate) [17]; 0.00023 (pyridoxal phosphate) [15]; 0.00014 (pyridoxal phosphate) [12]; More [5, 11]

## pH-optimum

8 (kynurenine) [15, 19]; 8.1 (kynurenine) [17]; 8.5 (kynurenine) [10, 11, 12]; 8.4–8.8 (kynurenine) [5]; 7.8–8.5 (3'-hydroxy-kynurenine) [5]

## pH-range

## Temperature optimum (°C)

## Temperature range (°C)

## 3 ENZYME STRUCTURE

## Molecular weight

95000–110000 (overview [9], rat [1, 5], pig [11], Neurospora crassa [12], Pseudomonas marginalis [15]) [1, 5, 9, 11, 12, 15]

## Subunits

Dimer (2 × 50000–55000, rat [1, 5], overview [9]) [1, 5, 9]

## Glycoprotein/Lipoprotein

–

# 4 ISOLATION/PREPARATION

## Source organism
Rat [5, 19]; Pig [11, 18]; Neurospora crassa [8, 10, 12, 14, 16, 17]; Pseudomonas marginalis [3, 4, 15]; Rhizopus stolonifer [13]; Aspergillus niger (inducible enzyme) [13]; Penicillium roqueforti (inducible enzyme) [13]; Pseudomonas fluorescens (inducible enzyme) [13]; Pseudomonas acidovorans [2]; Streptomyces parvulus [6]; More (overview) [9]

## Source tissue
Liver [1, 5, 11, 18, 19]; Cell [17]

## Localisation in source
Cytoplasm [1]

## Purification
Pig [11]; Neurospora crassa [17]; Pseudomonas marginalis [15]; Rhizopus stolonifer [13]; Aspergillus niger [13]; Penicillium roqueforti [13]; Pseudomonas fluorescens [13]; Rat [5]

## Crystallization
[9, 12, 15]

## Cloned
–

## Renaturated
–

# 5 STABILITY

## pH
5.2 [17]; 5.8–8.0 [15]

## Temperature (°C)
70 (inactivated at) [1]

## Oxidation

## Organic solvent

## General stability information
Unstable without pyridoxal phosphate [5]

## Storage
Dialyzed against ammonium sulfate solution, 4°C, in the dark [5]

# 6 CROSSREFERENCES TO STRUCTURE DATABANKS

PIR/MIPS code

Brookhaven code

# 7 LITERATURE REFERENCES

[1] Stevens, J.L.: J. Biol. Chem., 260 (13) , 7945–7950 (1985)
[2] Palcic, M.M., Antoun, M., Tanizawa, K., Soda, K., Floss, H.G.: J. Biol. Chem., 260 (9) , 5248–5251 (1985)
[3] Bild, G.S., Morris, J.C.: Arch. Biochem. Biophys., 235 (1) , 41–47 (1984)
[4] Kishore, G.M.: J. Biol. Chem., 259 (17) , 10667–10674 (1984)
[5] Takechi, F., Otsuka, H., Shibata, Y.: J. Biochem., 88, 987–994 (1980)
[6] Troost, T., Hitchcock, M.J.M., Katz, E.: Biochim. Biophys. Acta, 612, 97–106 (1980)
[7] Tanizawa, K., Soda, K.: J. Biochem., 86, 1199–1209 (1979)
[8] Tanizawa, K., Soda, K.: J. Biochem., 86, 499–508 (1979)
[9] Soda, K., Tanizawa, K.: Adv. Enzymol. Relat. Areas Mol. Biol., 49, 1–40 (1979)
[10] Tanizawa, K., Soda, K.: J. Biochem., 85, 1367–1375 (1979)
[11] Tanizawa, K., Soda, K.: J. Biochem., 85, 901–906 (1979)
[12] Tanizawa, K., Yamamoto, T., Soda, K.: FEBS Lett., 70 (1) , 235–238 (1976)
[13] Shetty, A.S., Gaertner, F.H.: J. Bacteriol., 122 (1) , 235–244 (1975)
[14] Moriguchi, M., Soda, K.: Biochemistry, 12 (16) , 2975–2980 (1973)
[15] Moriguchi, M., Yamamoto, T., Soda, K.: Biochemistry, 12 (16) , 2969–2974 (1973)
[16] Jacoby, W.B., Bonner, D.M.: J. Biol. Chem., 205, 709–715 (1953)
[17] Jacoby, W.B., Bonner, D.M.: J. Biol. Chem., 205, 699–707 (1953)
[18] Wiss, O., Weber, F.: Hoppe-Seyler's Z. Physiol. Chem., 304, 232–240 (1956)
[19] Knox, W.E.: Biochemistry, 543, 379–385 (1953)

# 1 NOMENCLATURE

**EC number**
3.7.1.4

**Systematic name**
2, 4, 4, 6-Tetrahydroxydehydrochalcone 1, 3, 5-trihydroxybenzene hydrolase

**Recommended name**
Phloretin hydrolase

**Synonymes**
Hydrolase, phloretin
Lactase-phlorizin hydrolase (phloretin hydrolase and beta-glucosidase could not be seperated) [1, 2]

**CAS Reg. No.**
37289-38-6

# 2 REACTION AND SPECIFICITY

**Catalysed reaction**
Phloretin + $H_2O$ →
→ phloretate + phloroglucinol

**Reaction type**
Hydrolytic C-C bond cleavage
Ketone hydrolysis

**Natural substrates**
Phloretin + $H_2O$ [3, 4]

**Substrate spectrum**
1 Phloretin + $H_2O$ [3, 4]
2 3-Methylphloracetophenone + $H_2O$ [3]
3 2, 4, 4-Trihydroxydihydrochalcone + $H_2O$ [3]
4 Phlorizin + $H_2O$ [3]
5 2, 3, 4, 6-Tetrahydroxy-4-methoxy-dihydrochalcone + $H_2O$ [3]
6 2, 3, 4, 6-Tetrahydroxydihydrochalcone + $H_2O$ [3]
7 More (not phlorizin, Erwinia herbicola [4]) [3, 4]

## Product spectrum
1 Phloretate + phloroglucinol [3, 4]
2 Methylphloroglucinol [3]
3 Phloretate [3]
4 Phloretate [3]
5 Phloroglucinol [3]
6 Phloroglucinol [3]
7 More [3, 4]

## Inhibitor(s)
$Hg^{2+}$ [3, 4]; $Cu^{2+}$ [3, 4]; $Cd^{2+}$ [3]; $Zn^{2+}$ [3]; $Fe^{3+}$ [3]; $Al^{3+}$ [3]; p-Chloromercuribenzoate [3]

## Cofactor(s)/prostethic group(s)

## Metal compounds/salts

## Turnover number (min$^{-1}$)

## Specific activity (U/mg)
213 (Erwinia herbicola) [4]; 0.06 (Aspergillus niger) [3]; 0.45 (monkey, phlorizin) [1]

## $K_m$-value (mM)
0.038 (Erwinia herbicola) [4]; 0.3–0.4 (Aspergillus niger) [3]; 0.15 (3-methyl-phloracetophenone, Aspergillus niger) [3]; 0.4 (phlorizin) [1]

## pH-optimum
3.9–9.5 (Erwinia herbicola) [4]

## pH-range

## Temperature optimum (°C)

## Temperature range (°C)

## 3 ENZYME STRUCTURE

## Molecular weight

## Subunits
? (x × 132000, rat, SDS-PAGE, lactase-phlorizin hydrolase, 3–4 isoenzymes) [2]

## Glycoprotein/Lipoprotein
Glycoprotein (17% carbohydrate, lactase-phlorizin hydrolase) [2]

## 4 ISOLATION/PREPARATION

**Source organism**
Erwinia herbicola [4]; Fungi [3]; Mammalia (lactase-phlorizin-hydrolase) [1, 2]

**Source tissue**
Cells [3, 4]; Intestinal mucosa (lactase-phlorizin hydrolase) [1, 2]

**Localisation in source**
Membrane (bound, lactase-phlorizin-hydrolase) [2]

**Purification**
Erwinia herbicola [3]; Aspergillus niger (partial) [4]; Mammalia (lactase-phlorizin hydrolase) [2]

**Crystallization**
–

**Cloned**
–

**Renaturated**
–

## 5 STABILITY

**pH**

**Temperature (°C)**

**Oxidation**

**Organic solvent**

**General stability information**

**Storage**
For months at –20°C (lactase-phlorizin hydrolase) [2]

## 6 CROSSREFERENCES TO STRUCTURE DATABANKS

**PIR/MIPS code**

**Brookhaven code**

# 7 LITERATURE REFERENCES

[1] Ramaswamy, S., Radhakrishnan, A.N.: Biochim. Biophys. Acta, 403, 446–455 (1975)
[2] Birkenmeier, E., Alpers, D.H.: Biochim. Biophys. Acta, 350, 100–112 (1974)
[3] Minamikawa, T., Jaysankar, N.P., Bohm, B.A., Taylor, I.E.P., Towers, G.H.N.: Biochem. J., 116, 889–897 (1970)
[4] Chatterjee, A.K., Gibbins, L.N.: J. Bacteriol., 100, 594–600 (1969)

## 1 NOMENCLATURE

**EC number**
3.7.1.5

**Systematic name**
3-Acylpyruvate acylhydrolase

**Recommended name**
Acylpyruvate hydrolase

**Synonymes**
Hydrolase, acylpyruvate

**CAS Reg. No.**
54004-67-0

---

## 2 REACTION AND SPECIFICITY

**Catalysed reaction**
A 3-acylpyruvate $+ H_2O \rightarrow$
$\rightarrow$ a fatty acid anion $+$ pyruvate

**Reaction type**
Hydrolytic C-C bond cleavage
Ketone hydrolysis

**Natural substrates**
3-Formyl-pyruvate $+ H_2O$ [1]

**Substrate spectrum**
1 3-Formyl-pyruvate $+ H_2O$
2 2,4-Dioxovalerate $+ H_2O$
3 2,4-Dioxohexanoate $+ H_2O$
4 2,4-Dioxoheptanoate $+ H_2O$

**Product spectrum**
1 Pyruvate $+$ formate
2 ?
3 ?
4 ?

---

**Inhibitor(s)**
$Zn^{2+}$ [1]; $Ni^{2+}$ [1]; $Cu^{2+}$ [1]

---

**Cofactor(s)/prostethic group(s)**

**Metal compounds/salts**

---

**Turnover number (min⁻¹)**

Turnover number (min$^{-1}$)

**Specific activity (U/mg)**

**K$_m$-value (mM)**
0.002 (3-formylpyruvate) [1]; 0.011 (2, 4-dioxovalerate) [1]; 0.007 (2,
4-dioxohexanoate) [1]; 0.007 (2, 4-dioxoheptanoate) [1]

**pH-optimum**
8.0 [1]

**pH-range**
7–8.5 [1]

**Temperature optimum (°C)**

**Temperature range (°C)**

---

## 3 ENZYME STRUCTURE

**Molecular weight**
72000 (Agrobacterium, gel chromatography) [1]

**Subunits**

**Glycoprotein/Lipoprotein**
–

---

## 4 ISOLATION/PREPARATION

**Source organism**
Agrobacterium [1]

**Source tissue**

**Localisation in source**

**Purification**

**Crystallization**
–

**Cloned**
–

Renaturated
–

---

# 5 STABILITY

pH

**Temperature** (°C)
55 (50% activity lost after 3 minutes) [1]

Oxidation

Organic solvent

General stability information

Storage

---

# 6 CROSSREFERENCES TO STRUCTURE DATABANKS

PIR/MIPS code

Brookhaven code

---

# 7 LITERATURE REFERENCES

[1] Watson, G.K., Houghton, C., Cain, R.B.: Biochem. J., 140, 277–292 (1974)

---

## 1 NOMENCLATURE

**EC number**
3.7.1.6

**Systematic name**
2, 4-Dioxopentanoate acetylhydrolase

**Recommended name**
Acetylpyruvate hydrolase

**Synonymes**
Hydrolase, acetylpyruvate

**CAS Reg. No.**
56214-30-3

## 2 REACTION AND SPECIFICITY

**Catalysed reaction**
Acetylpyruvate + $H_2O$ →
→ acetate + pyruvate

**Reaction type**
Hydrolytic C-C bond cleavage

**Natural substrates**
Acetylpyruvate + $H_2O$ [1]

**Substrate spectrum**
1 Acetylpyruvate + $H_2O$ [1]
2 More (highly specific, does not act on pyruvate, oxaloacetate, maleyl-pyruvate, fumarylpyruvate or acetylacetone)

**Product spectrum**
1 Acetate + pyruvate + $H_2O$ [1]
2 ?

**Inhibitor(s)**
Oxaloacetate [1]; $Cu^{2+}$ (slightly) [1]; Pyruvate [1]; Oxalate [1]

**Cofactor(s)/prostethic group(s)**

**Metal compounds/salts**
$Mn^{2+}$ (divalent cations, especially $Mn^{2+}$ stimulate) [1]

**Turnover number** (min⁻¹)
   36 [1]

**Specific activity** (U/mg)
   0.8 [1]

**$K_m$-value** (mM)
   0.07–0.11 (acetylpyruvate) [1]

**pH-optimum**
   7.4 [1]

**pH-range**

**Temperature optimum** (°C)

**Temperature range** (°C)

---

## 3 ENZYME STRUCTURE

**Molecular weight**
   38000 (Pseudomonas putida, SDS-PAGE, gel filtration) [1]

**Subunits**
   Monomer (Pseudomonas putida) [1]

**Glycoprotein/Lipoprotein**
   –

---

## 4 ISOLATION/PREPARATION

**Source organism**
   Pseudomonas putida [1]

**Source tissue**
   Cells [1]

**Localisation in source**

**Purification**
   Pseudomonas putida [1]

**Crystallization**
   –

**Cloned**
   –

**Renaturated**
   –

## 5 STABILITY

pH

Temperature (°C)

Oxidation

Organic solvent

General stability information

Storage

---

## 6 CROSSREFERENCES TO STRUCTURE DATABANKS

PIR/MIPS code

Brookhaven code

---

## 7 LITERATURE REFERENCES

[1] Davey, J.F., Ribbons, D.W.: J. Biol. Chem., 250, 3826–3830 (1975)

---

## 1 NOMENCLATURE

**EC number**
  3.8.1.1

**Systematic name**
  Alkyl-halide halidohydrolase

**Recommended name**
  Alkylhalidase

**Synonymes**
  Halogenase
  Haloalkane halidohydrolase [1]
  Haloalkane dehalogenase [5]

**CAS Reg. No.**
  9025-22-3

## 2 REACTION AND SPECIFICITY

**Catalysed reaction**
  Bromochloromethane + $H_2O$ →
  → formaldehyde + bromide + chloride

**Reaction type**
  C-Halide hydrolysis

**Natural substrates**
  Haloalkanes + $H_2O$ (mid-chain, $C_1$-$C_8$, chloro-, bromo-, iodo-) [5]

**Substrate spectrum**
  1 Bromochloromethane + $H_2O$
  2 1-Chloroalkanes + $H_2O$
  3 1-Bromoalkanes + $H_2O$
  4 1-Iodoalkanes + $H_2O$
  5 Haloalkanes [1]
  6 Mono-chloralkane + $H_2O$ [2]
  7 Dichloralkane + $H_2O$ [2]
  8 Alkanes (alpha-omega-dichlorosubstituted) + $H_2O$ [1]
  9 Alkanes ($C_1$-$C_8$ chloro, $C_2$-$C_8$ monobromo-, $C_2$-$C_7$ monoiodo-) + $H_2O$

## Product spectrum
1 Formaldehyde + chloride + bromide
2 ?
3 ?
4 ?
5 ?
6 ?
7 ?
8 ?
9 ?

## Inhibitor(s)
p-Chloromercuribenzoate [5]; Iodoacetamide [5]; HgCl$_2$ [5];
N-Ethylmaleimide [5]

## Cofactor(s)/prostethic group(s)

## Metal compounds/salts

## Turnover number (min$^{-1}$)

## Specific activity (U/mg)
6.0 [5]

## K$_m$-value (mM)
1.1 (1, 2-dichlorethane) [5]

## pH-optimum
8.2 [5]

## pH-range

## Temperature optimum (°C)
37 [5]

## Temperature range (°C)

## 3 ENZYME STRUCTURE

## Molecular weight
36000 (Xanthobacter autotrophicus GJ 10, gel filtration) [5]

## Subunits
Monomer (1 × 36000, Xanthobacter autotrophicus GJ 10, SDS-PAGE) [5]
Dimer (2 × 36000, Xanthobacter autotrophicus GJ 10, SDS-PAGE) [5]

Glycoprotein/Lipoprotein

–

---

## 4 ISOLATION/PREPARATION

**Source organism**
Coryneforme bacteria [1]; Arthrobacter sp. [1]; Methylotrophic bacteria; Pseudomonas [2, 4]; Xanthobacter autotrophicus GJ 10 [3]

**Source tissue**
Cell [1, 2]

**Localisation in source**
Soluble [1]

**Purification**
Xanthobacter autotrophicus GJ 10 [5]

**Crystallization**
–

**Cloned**
–

**Renaturated**
–

---

## 5 STABILITY

**pH**
7.5 (around) [5]

**Temperature (°C)**
50 (instable at, stable up to [3]) [3, 5]

**Oxidation**

**Organic solvent**

**General stability information**

**Storage**
Several weeks at 4°C [5]

# 6 CROSSREFERENCES TO STRUCTURE DATABANKS

PIR/MIPS code

Brookhaven code

---

# 7 LITERATURE REFERENCES

[1] Scholtz, R., Schmuckle, A., Cock, A.M., Leisinger, T.: J. Gen. Microbiol., 133, 267–274 (1987)
[2] Yokota, T., Fuse, H., Omori, T., Minoda, Y.: Agric. Biol. Chem., 50 (2) , 453–460 (1986)
[3] Janssen, D.B., Scheper, A., Dijkluizen, L., Witholt, B.: Appl. Environ. Microbiol., 49 (3), 673–677 (1985)
[4] Omori, T., Alexander, M.: Appl. Environ. Microbiol., 35 (5) , 867–871 (1978)
[5] Keuning, S., Janssen, D.B., Witholt, B.: J. Bacteriol. , 163 (2) , 635–639 (1985)

## 1 NOMENCLATURE

**EC number**
3.8.1.2

**Systematic name**
2-Haloacid halidohydrolase

**Recommended name**
2-Haloacid dehalogenase

**Synonymes**
2-Haloalkanoid acid halidohydrolase
2-Haloalkanoic acid dehalogenase [2]
L-2-Haloacid dehalogenase:
DL-2-Haloacid dehalogenase [3]

**CAS Reg. No.**
37289-39-7

## 2 REACTION AND SPECIFICITY

**Catalysed reaction**
(S)-2-Haloacid + $H_2O$ →
→ (R)-2-hydroxyacid + halide (acts on acids of short chain lengths, $C_2$ to $C_4$, with inversion at C-2)

**Reaction type**
C-Halide hydrolysis

**Natural substrates**
Monohalo-fatty-acids ($C_2$ to $C_4$)

**Substrate spectrum**
1 (S)-2-Haloacid + $H_2O$
2 Fatty acids (monohaloacids, short chain length $C_2$ to $C_4$) + $H_2O$ [1, 4]

**Product spectrum**
1 (R)-2-Hydroxyacid + halide
2 ?

**Inhibitor(s)**
$HgCl_2$ (high concentrations) [1]; $Zn^{2+}$ [2]; $Pb^{2+}$ [2]; $Hg^{2+}$ [2]; $MnSO_4$ [3];
$Cu^{2+}$ [4]; $Ag^{2+}$ [4]

**Cofactor(s)/prostethic group(s)**

**Metal compounds/salts**

---

**Turnover number (min⁻¹)**

**Specific activity (U/mg)**
More [1, 4]; 120 [4]

**K_m-value (mM)**
4.5 (D-2-chloropropionate); 1.0 (L-2-chloropropionate) [2]; 3.2
(DL-2-chloropropionate); 5.0 (monochloroacetate) [3]

**pH-optimum**
9.5 [1, 2]; 10.5 [4]; 9.0–9.4 [5]

**pH-range**

**Temperature optimum (°C)**
45 [3, 4]

**Temperature range (°C)**

---

**3 ENZYME STRUCTURE**

**Molecular weight**
41000 (gel filtration, Pseudomonas sp.) [1]
68000 (sedimentation equilibrium, Pseudomonas sp.) [3]
34000 (Pseudomonas putida, gel filtration) [4]

**Subunits**
Dimer (2 × 28000, SDS-PAGE, Pseudomonas sp.) [1]
Dimer (2 × 35000, Pseudomonas sp., SDS-PAGE) [3]
Monomer (1 × 25000, Pseudomonas putida) [4]

**Glycoprotein/Lipoprotein**
–

---

**4 ISOLATION/PREPARATION**

**Source organism**
Pseudomonas sp. [1–3, 6]; Pseudomonas putida [4, 6]

**Source tissue**

**Localisation in source**

## Purification
Pseudomonas sp. [1, 3]; Pseudomonas putida [4]

## Crystallization
–

## Cloned
–

## Renaturated
–

---

## 5 STABILITY

**pH**
7–10 [3]; 6–11 [4]

**Temperature (°C)**

**Oxidation**

**Organic solvent**

**General stability information**

**Storage**
6 months at –18°C, pH 7.5 [1]; 1 year at –20°C [3]

---

## 6 CROSSREFERENCES TO STRUCTURE DATABANKS

**PIR/MIPS code**

**Brookhaven code**

---

## 7 LITERATURE REFERENCES

[1] Klages, M., Krauss, S., Lingens, F.: Hoppe-Seyler's Z. Physiol. Chem., 364, 529–535 (1983)
[2] Motosugi, K., Esahi, N., Soda, K.: Arch. Microbiol., 131, 179–183 (1982)
[3] Motosugi, K., Esahi, N., Soda, K.: J. Bacteriol., 150 (2) , 522–527 (1982)
[4] Motosugi, K., Esahi, N., Soda, K.: Agric. Biol. Chem., 46 (3) , 837–838 (1982)
[5] Goldman, P., Milne, G.W.A., Kleister, D.B.: J. Biol. Chem., 243 (2) , 428–434 (1968)
[6] Motosugi, K., Esahi, N., Soda, K.: Biotechnol. Bioeng. , 26, 805–806 (1984)

# 1 NOMENCLATURE

**EC number**
3.8.1.3

**Systematic name**
Haloacetate halidohydrolase

**Recommended name**
Haloacetate dehalogenase

**Synonymes**
Haloacetate halidohydrolase (H1 /inducible, H2 /constitutive)
Monohaloacetate dehalogenase [7]

**CAS Reg. No.**
37289-40-0

---

# 2 REACTION AND SPECIFICITY

**Catalysed reaction**
Haloacetate + $H_2O$ →
→ glycolate + halide

**Reaction type**
C-Halide hydrolysis

**Natural substrates**
Haloacetate (monohaloacetate)

**Substrate spectrum**
1 Haloacetate + $H_2O$ (ir) (e.g. monochloroacetate (H2, H1) [1],
monobromoacetate (H2) [1], monoiodoacetate (H2) [1],
monofluoroacetate (H1) [2, 4]) [1, 2, 4]

**Product spectrum**
1 Glycolate + halide

---

**Inhibitor(s)**
p-Chloromercuribenzoate [1]; p-Chloromercuriphenylsulfonate [1];
N-Ethylmaleimide [1]; $HgCl_2$ [1]; $AgCl_2$ [1]; Thiol reagents [1, 2]; Fatty acids
[6]; Aromatic acids (competitive) [6]

**Cofactor(s)/prostethic group(s)**

**Metal compounds/salts**

---

**Turnover number** (min⁻¹)

**Specific activity** (U/mg)
More [1, 2]; 120 [7]

**Kₘ-value** (mM)
2.5 (monochloroacetate); 0.5 (monobromoacetate); 1.1 (monoiodoacetate) [1]; 2.0 (monofluoroacetate) [2]; 20 (chloroacetate) [6]

**pH-optimum**
9.5 [1]; 9.0 [2]; 9.3 [6]; 9.5 [7]

**pH-range**
8–11 [1]; 7–10 [7]

**Temperature optimum** (°C)
50 [1, 2]

**Temperature range** (°C)
30–60 [1]

---

## 3 ENZYME STRUCTURE

**Molecular weight**
43000 (gel filtration, Moraxella sp.) [1]
42000 (gel filtration, Pseudomonas sp.) [2]
41000 (mouse) [4]

**Subunits**
Dimer (2 × 26000, SDS-PAGE, Moraxella sp.) [1]
Dimer (2 × 33000, SDS-PAGE, Pseudomonas sp.) [2]
Dimer (2 × 27000, SDS-PAGE, mouse) [4]

**Glycoprotein/Lipoprotein**
–

---

## 4 ISOLATION/PREPARATION

**Source organism**
Moraxella sp. [1, 3]; Pseudomonas sp. [1, 2, 5–7]; Mouse [4]

**Source tissue**
Cell [1, 3, 7]

**Localisation in source**
Cytoplasm [4]

**Purification**
Moraxella sp. (H2) [1]; Pseudomonas sp. (H1) [2]; Mouse (H1) [4]

**Crystallization**
–

**Cloned**
–

**Renaturated**
–

---

## 5 STABILITY

**pH**
5–10 [1]; 6.0–10.0 [2]; 7.0 [7]

**Temperature (°C)**
60 [1]

**Oxidation**

**Organic solvent**

**General stability information**

**Storage**
Several days, 0–5°C [6]; Several months, –196°C [6]

---

## 6 CROSSREFERENCES TO STRUCTURE DATABANKS

**PIR/MIPS code**

**Brookhaven code**

# 7 LITERATURE REFERENCES

[1] Kawasaki, H., Tone, N., Tonomura, K.: Agric. Biol. Chem., 45 (1) , 35–42 (1981)
[2] Kawasaki, H., Miyoshi, K., Tonomura, K.: Agric. Biol. Chem., 45 (1) , 543–544 (1981)
[3] Kawasaki, H., Tone, N., Tonomura, K.: Agric. Biol. Chem., 45 (1) , 29–34 (1981)
[4] Soiefer, A.I., Kostyniak, P.J.: J. Biol. Chem., 259 (17) , 10787–10792 (1984)
[5] Goldman, P., Milne, G.W.A.: J. Biol. Chem., 241 (23) , 5557–5559 (1966)
[6] Goldman, P.: J. Biol. Chem., 240 (8) , 3434–3438 (1965)
[7] Little, M., Williams, P.A.: Biochem. J., 114 (1) , 11–12 (1969)

## 1 NOMENCLATURE

**EC number**
3.8.1.4

**Systematic name**
L-Thyroxine iodohydrolase (reducing)

**Recommended name**
Thyroxine deiodinase

**Synonymes**
Thyroxine 5-deiodinase
Diiodothyronine 5'-deiodinase
Iodothyronine outer ring monodeiodinase
Iodothyronine 5'-deiodinase

**CAS Reg. No.**
70712-46-8

## 2 REACTION AND SPECIFICITY

**Catalysed reaction**
L-Thyroxine + $AH_2$ →
→ 3, 5, 3'-L-triiodo-L-thyronine + iodide + A + $H^+$ (A = cosubstrate, e.g. dithiols, $AH_2$ = reduced cosubstrate)

**Reaction type**
C-Halide hydrolysis

**Natural substrates**
L-Thyroxine [2]
3, 3', 5-Triiodothyronine

**Substrate spectrum**
1 L-Thyroxine + $AH_2$

**Product spectrum**
1 3, 5, 3'-L-Triiodo-L-thyronine + iodide + A + $H^+$

**Inhibitor(s)**
2-Thiouracil [6, 8]; Coumarin (anticoagulants) [1]; Dicoumarol [1]; Warfarin [1]; Salicylate [1]; Tetraiodothyroacetic acid [2]; More (plant extracts) [3]; Thiourea (derivatives) [2]

**Cofactor(s)/prostethic group(s)**
More (reduced dithiols, thiol cofactor [1], phospholipid required [4]) [1, 4]

**Metal compounds/salts**

---

**Turnover number** (min$^{-1}$)

**Specific activity** (U/mg)

**$K_m$-value** (mM)
0.003 (thyroxine) [7, 8]

**pH-optimum**
7.2 [5]; 8 [6]; 6.8 [7]; 6.5 [8]

**pH-range**
6.5–8 [6]

**Temperature optimum** (°C)
37 [4]

**Temperature range** (°C)

---

## 3 ENZYME STRUCTURE

**Molecular weight**
49900 (rat, gel filtration) [4]

**Subunits**

**Glycoprotein/Lipoprotein**
–

---

## 4 ISOLATION/PREPARATION

**Source organism**
Rat [1–5, 7, 8]

**Source tissue**
Liver [2, 3, 5, 7]; Kidney [1, 4, 5]; Thyroid [8]

**Localisation in source**
Membrane (bound) [4]; Lysosomes [5]; Microsomes [1–3, 7];
More (overview) [7]

**Purification**
Rat [4]

## Crystallization
[2]

## Cloned
–

## Renaturated
–

---

## 5 STABILITY

**pH**

**Temperature** (°C)
60 (inactivation after 30 minutes) [8]

**Oxidation**

**Organic solvent**

**General stability information**

**Storage**
–20°C, dithioerythritol [7]

---

## 6 CROSSREFERENCES TO STRUCTURE DATABANKS

**PIR/MIPS code**
A31118 (rat, fragment)

**Brookhaven code**

---

## 7 LITERATURE REFERENCES

[1] Goswami, A., Leonard, J.L., Rosenberg, I.N.: Biochem. Biophys. Res. Commun., 104 (4) , 1231–1238 (1982)
[2] Koehrle, J., Auf'mkolk, M., Rokos, H., Hesch, R.-D., Cody, V.: J. Biol. Chem., 261 (25) , 11613–11622 (1986)
[3] Auf'mkolk, M., Koehrle, J., Hesch, R.-D., Cody, V.: J. Biol. Chem., 261 (25) , 11623–11630 (1986)
[4] Leonard, J.L., Rosenberg, I.N.: Biochim. Biophys. Acta, 659, 205–218 (1981)
[5] Colquhoun, E.Q., Thomson, R.M.: FEBS Lett., 177 (2) , 221–226 (1984)
[6] Visser, T.J.: Trends Biochem. Sci., 5 (8) , 222–224 (1980)
[7] Auf Dem Brinke, D., Hesch, R.-D., Köhrle, J.: Biochem. J., 180, 273–279 (1979)
[8] Ericksen, V.J., Cavalieri, R.R., Rosenberg, L.L.: Endocrinology, Baltimore, 108 (4) , 1257–1264 (1981)

---

## 1 NOMENCLATURE

**EC number**
3.8.2.1

**Systematic name**
Diisopropyl-fluorophosphate fluorohydrolase

**Recommended name**
Diisopropyl-fluorophosphatase

**Synonymes**
Diisopropylfluorophosphate halogenase
Tabunase
DFPase
Di-isopropylphosphofluoridase
Dialkylphosphofluoridase [4]
Squid nerve IDFase [6]

**CAS Reg. No.**
9032-18-2

## 2 REACTION AND SPECIFICITY

**Catalysed reaction**
Diisopropyl fluorophosphate + $H_2O$ →
→ diisopropyl phosphate + fluoride

**Reaction type**
Phosphohalide hydrolysis

**Natural substrates**
? (metabolism of isothionate)

**Substrate spectrum**
1 Diisopropyl fluorophosphate + $H_2O$
2 More (phosphofluoridates, splits acid anhydride group: P-F linkage,
   other organophosphorus compounds and nerve gases) [1]

**Product spectrum**
1 Diisopropyl phosphate + fluoride
2 ?

**Inhibitor(s)**
Chelating agents [2]; p-Chloromercuribenzoic acid [3]; Heavy metal ions
[2]; More (dialysis) [4]

**Cofactor(s)/prostethic group(s)**
No cofactors required [6]

**Metal compounds/salts**
$Co^{2+}$ (activation) [2]; $Mn^{2+}$ (activation) [2]

---

**Turnover number** (min$^{-1}$)

**Specific activity** (U/mg)
More [3]

**$K_m$-value** (mM)
14 (diisopropylphosphofluoridate) [5]; 4.3 (diisopropyl fluorophosphate)
[6]

**pH-optimum**
7.4 [1, 4]; 7.5–8.0 [4]

**pH-range**
7.5–8.5 [6]

**Temperature optimum** (°C)
37 [3]

**Temperature range** (°C)

---

**3 ENZYME STRUCTURE**

**Molecular weight**

**Subunits**

**Glycoprotein/Lipoprotein**
–

---

**4 ISOLATION/PREPARATION**

**Source organism**
Rabbit [2, 3]; Hog [5]

**Source tissue**
Liver [2, 3]; Kidney [2, 3, 5]; Testis [2, 3]; Intestinal mucosa [2, 3]; Ganglion
[6]

**Localisation in source**
Soluble [6]; More (particulate) [1–5]

**Purification**
  Hog [5]

**Crystallization**
  —

**Cloned**
  —

**Renaturated**
  —

---

## 5 STABILITY

**pH**
  7.0–9.8 [4]

**Temperature (°C)**

**Oxidation**

**Organic solvent**
  Ethanol (stable against high concentrations)

**General stability information**

**Storage**
  At 0°C, 18 hours [4]; At –20°C, 6 months [6]

---

## 6 CROSSREFERENCES TO STRUCTURE DATABANKS

**PIR/MIPS code**

**Brookhaven code**

---

## 7 LITERATURE REFERENCES

[1] Losch, H., Losch, K., Haselmeyer, K.H., Chemnitius, J.-M., Zech, R.: Arzneim. Forsch.,
    32 (II) , 12, 1523–1529 (1982)
[2] Mounter, L.A. in "The Enzymes", 2nd Ed. (Boyer, P.D., Ed.) Vol.4, 541–550 (1960)
[3] Cohen, J.A., Warringa, M.G.P.J.: Biochim. Biophys. Acta, 26, 29–39 (1957)
[4] Mazur, A.: Methods Enzymol., 1, 651–656 (1955)
[5] Mounter, L.A., Floyd, C.S., Chanutin, A.: J. Biol. Chem., 204, 221–232 (1953)
[6] Hashin, F.C.G., Long, R.J.: Arch. Biochem. Biophys., 150, 548–555 (1972)

---

## 1 NOMENCLATURE

**EC number**
3.9.1.1

**Systematic name**
Phosphamide hydrolase

**Recommended name**
Phosphoamidase

**Synonymes**
Creatine phosphatase (possibly identical with E.C. 3.1.3.16 or E.C. 3.1.3.9)

**CAS Reg. No.**
9001-79-0

## 2 REACTION AND SPECIFICITY

**Catalysed reaction**
N-Phosphocreatine + $H_2O$ →
→ creatine + orthophosphate

**Reaction type**
Phosphoamide hydrolysis

**Natural substrates**
N-Phosphoamidates + $H_2O$ [1, 3]

**Substrate spectrum**
1 N-Phosphoamidates + $H_2O$ (phosphoamides, unspecific) (ir) [1, 2]

**Product spectrum**
1 Amines + orthophosphate

**Inhibitor(s)**
Molybdate [3]; Pyrophosphate (competitive) [3]; Iodosobenzene [1];
p-Chloromercuribenzoate [1]; $F^-$ [1]; Phosphitin [1]

**Cofactor(s)/prostethic group(s)**

**Metal compounds/salts**

**Turnover number (min⁻¹)**

Enzyme Handbook © Springer-Verlag Berlin Heidelberg 1991
Duplication, reproduction and storage in data banks are only
allowed with the prior permission of the publishers

**Specific activity** (U/mg)
  48 [1]

**$K_m$-value** (mM)
  2.8 (phosphoamidates) [3]

**pH-optimum**
  6 (sodium phosphoamidates) [1]

**pH-range**
  5–10.5 [3]

**Temperature optimum** (°C)
  38 [1]

**Temperature range** (°C)

## 3 ENZYME STRUCTURE

**Molecular weight**

**Subunits**

**Glycoprotein/Lipoprotein**
  –

## 4 ISOLATION/PREPARATION

**Source organism**
  Bovine [1]; Rat [3]; Rabbit [2]

**Source tissue**
  Spleen [1]; Liver [2, 3]; Kidney [3]

**Localisation in source**
  Microsomes [2, 3]

**Purification**
  Bovine [1]; Rabbit [2]; Rat [3]

**Crystallization**
  –

**Cloned**
  –

**Renaturated**
  –

## 5 STABILITY

**pH**
  4.5 (stable above) [2]

**Temperature (°C)**

**Oxidation**

**Organic solvent**

**General stability information**

**Storage**
  Lyophilized, −18°C [1]

---

## 6 CROSSREFERENCES TO STRUCTURE DATABANKS

**PIR/MIPS code**

**Brookhaven code**

---

## 7 LITERATURE REFERENCES

[1] Singer, M.F., Fruton, J.S.: J. Biol. Chem., 229, 111–119 (1957)
[2] Dudkin, S.M., Ledneva, R.K., Shabarova, Z.A., Prokofiev, M.A.: FEBS Lett., 16 (1) ,
    48–50 (1971)
[3] Parvin, R., Smith, R.A.: Biochemistry, 8 (4) , 1748–1755 (1969)

## 1 NOMENCLATURE

**EC number**
3.10.1.1

**Systematic name**
N-Sulfo-D-glucosamine sulfohydrolase

**Recommended name**
N-Sulfoglucosamine sulfohydrolase

**Synonymes**
Sulfoglucosamine sulfamidase
Heparin sulfamidase
2-Desoxy-D-glucoside-2-sulphamate sulphohydrolase (sulphamate sulphohydrolase) [5, 6]

**CAS Reg. No.**
37289-41-1

## 2 REACTION AND SPECIFICITY

**Catalysed reaction**
N-Sulfo-D-glucosamine + $H_2O$ →
→ D-glucosamine + sulfate

**Reaction type**
Hydrolysis

**Natural substrates**
Heparin (degradation products of) + $H_2O$ [1, 2]
Heparane sulfate + $H_2O$ [3–7]

**Substrate spectrum**
1 N-Sulfo-D-glucosamine + $H_2O$ (ir)
2 Glucosamine 2, 6-disulfate + $H_2O$ (ir) [1]
3 Heparan sulfate + $H_2O$ (ir) [3]

**Product spectrum**
1 D-Glucosamine + sulfate (ir)
2 D-Glucosamine (ir)
3 Heparin

**Inhibitor(s)**
$F^-$ [4]; p-Nitrophenol sulphate [4]; Nitrocatechol sulphate [4]; More (overview) [4]

## Cofactor(s)/prostethic group(s)

**Metal compounds/salts**
$Mg^{2+}$ (1 mM) [1]

**Turnover number** (min$^{-1}$)
18125 [1]

**Specific activity** (U/mg)
More [2]

**$K_m$-value** (mM)
0.009 (heparan sulfate) [4]; 0.13 (N-sulfo-D-glucosamine) [1]

**pH-optimum**
7 [1]; 5 [3–7]

**pH-range**
6–8 [1]; 4.5–6 [4]

**Temperature optimum** (°C)
25 [1]; 37 [4]

**Temperature range** (°C)
25–70 [4]

## 3 ENZYME STRUCTURE

**Molecular weight**
120000 (ultracentrifugation, human, placenta) [4, 5]
190000 (gel filtration, human) [6]

**Subunits**
Dimer [6]

**Glycoprotein/Lipoprotein**
Glycoprotein [5]

## 4 ISOLATION/PREPARATION

**Source organism**
Flavobacterium heparinum [1, 2, 8]; Mammalia (enzyme does not act on
sulfoglucosamine [5]) [3–7]; Human [3, 5–7]; Rat [4]

**Source tissue**
Cell [1, 2, 8]; Spleen [4]; Liver [6, 7]; Placenta [5]; Lymphoid tissue [3]

Localisation in source
   Cytoplasm [1, 2, 8]

Purification

Crystallization
   –

Cloned
   –

Renaturated
   –

---

## 5 STABILITY

pH
   5.0–7.4 (stable between) [2, 4]

Temperature (°C)
   45 (almost no more activity) [1, 3]; 0 (stable at) [4]

Oxidation

Organic solvent

General stability information

Storage

---

## 6 CROSSREFERENCES TO STRUCTURE DATABANKS

PIR/MIPS code

Brookhaven code

---

## 7 LITERATURE REFERENCES

[1] Dietrich, C.P.: Biochem. J., 111, 91–95 (1969)
[2] Dietrich, C.P., Silva, M.E., Michelcci, Y.M.: J. Biol. Chem., 248 (18) , 6408–6415 (1973)
[3] Kresse, H.: Biochem. Biophys. Res. Commun., 54 (3) , 1111 (1973)
[4] Friedman, Y., Arsenis, Ch.: Biochem. J., 139, 699–708 (1974)
[5] Paschke, E., Kresse, H.: Biochem. J., 181, 677–684 (1978)
[6] Mahuran, D., Clements, P., Hopwood, J.: Biochim. Biophys. Acta, 757, 359–365 (1983)
[7] Freeman, C., Hopwood, J.J.: Biochem. J., 234, 83–92 (1986)
[8] Bruce, J.S., McLean, M.W., Long, W.F., Williamsen, F. B.: Eur. J. Biochem., 165,
      633–638 (1987)

---

Enzyme Handbook © Springer-Verlag Berlin Heidelberg 1991
Duplication, reproduction and storage in data banks are only
allowed with the prior permission of the publishers

# 1 NOMENCLATURE

**EC number**
3.10.1.2

**Systematic name**
Cyclohexylsulfamate sulfohydrolase

**Recommended name**
Cyclamate sulfohydrolase

**Synonymes**
Cyclamate sulfamatase
Cyclamate sulfamidase
Amidase, cyclohexylsulfamate sulf-

**CAS Reg. No.**
52228-00-9

---

# 2 REACTION AND SPECIFICITY

**Catalysed reaction**
Cyclohexylsulfamate + $H_2O$ →
→ cyclohexylamine + sulfate

**Reaction type**
Hydrolysis

**Natural substrates**

**Substrate spectrum**
1 Cyclohexylsulfamate + $H_2O$ [1]
2 Sulfamates (aliphathic, 3–8 carbons) + $H_2O$ [1]

**Product spectrum**
1 Cyclohexylamine + sulfate
2 Amine + sulfate

---

**Inhibitor(s)**
Cyclohexylamine [1]; $ZnSO_4$ [1]; $NiSO_4$ [1]; $CoSO_4$ [1]; $CuSO_4$ [1]; $(NH_4)_2SO_4$ [1]; $NaCO_3$ [1]; $Na_2SO_3$ [1]; $Na_2B_4O_7$ [1]; $H_3BO_4$ [1]; K-Citrate [1]; EDTA (metal ions reverse inhibition) [1]; $Hg^{2+}$ [1]; $Cd^{2+}$ [1]; Iodoacetic acid [1]; $NaN_3$ [1]; p-Chloromercuribenzoate [1]; $BaCl_2$ [1]; $NH_2OH$ [1]; Phosphate [1]; $MgSO_4$ [1]; $MnCl_2$ [1]

---

**Cofactor(s)/prostethic group(s)**

**Metal compounds/salts**
$Ca^{2+}$ (stimulation) [1]; More (metal ions reverse inhibition by EDTA) [1]

**Turnover number** (min⁻¹)

**Specific activity** (U/mg)
160 [1]

**$K_m$-value** (mM)
5 (cyclamate) [1]

**pH-optimum**
6.5–6.7 [1]

**pH-range**
5.9–7.5 [1]

**Temperature optimum** (°C)
50 [1]

**Temperature range** (°C)

# 3 ENZYME STRUCTURE

**Molecular weight**

**Subunits**

**Glycoprotein/Lipoprotein**
–

# 4 ISOLATION/PREPARATION

**Source organism**
Pseudomonas sp.. [1]; Corynebacterium [1]; Enterococci [1]; Enterobacteria [1]; Clostridia [1]

**Source tissue**

**Localisation in source**

**Purification**
Pseudomonas sp. [1]

**Crystallization**
–

Cloned

–

Renaturated

–

# 5 STABILITY

pH

**Temperature (°C)**
50 (2 hours without loss of activity) [1]; 60 (40% loss of activity after 20
minutes) [1]; 70 (completely inactivated after 5 minutes) [1]

**Oxidation**

**Organic solvent**

**General stability information**
Gelatin is effective stabilizer [1]; In Tris buffer minor loss of activity than in
phosphate buffer [1]

**Storage**
36 hours, –20°C [1]

# 6 CROSSREFERENCES TO STRUCTURE DATABANKS

**PIR/MIPS code**

**Brookhaven code**

# 7 LITERATURE REFERENCES

[1] Niimura, T., Tokiedo, T., Yamaha, T.: J. Biochem., 75, 407–417 (1974)

## 1 NOMENCLATURE

**EC number**
3.11.1.1

**Systematic name**
2-Oxoethylphosphonate phosphonohydrolase

**Recommended name**
Phosphonoacetylaldehyde hydrolase

**Synonymes**
Hydrolase, phosphonoacetylaldehyde
Phosphonatase [3]
2 Phosphonoacetylaldehyde phosphonohydrolase [3]

**CAS Reg. No.**
37289-42-2

## 2 REACTION AND SPECIFICITY

**Catalysed reaction**
Phosphonoacetylaldehyde + $H_2O$ →
→ acetaldehyde + orthophosphate

**Reaction type**
Hydrolysis

**Natural substrates**
Phosphonoacetaldehyde + $H_2O$

**Substrate spectrum**
1 Phosphonoacetylaldehyde + $H_2O$ [1–3]
2 p-Nitrophenylphosphate + $H_2O$ [3]

**Product spectrum**
1 Acetaldehyde + orthophosphate [1–3]
2 p-Nitrophenol + orthophosphate [3]

**Inhibitor(s)**
$NaBH_4$ (complete loss of activity in presence of substrate or product) [1, 2];
Acetonylphosphonate (competitive) [1]; Orthophosphite [2, 3]; $S^{2-}$ [2, 3];
SO3-[2, 3]; Cyanide [2, 3]; EDTA [3]; $Ca^{2+}$ (causes inactivation in the absence of $Mg^{2+}$) [3]; $Zn^{2+}$ (causes inactivation in the absence of $Mg^{2+}$) [3];
NaCl [3]

**Cofactor(s)/prostethic group(s)**
$Mg^{2+}$ (stabilizes active homodimer form of the enzyme) [1, 3]

**Metal compounds/salts**
$Mg^{2+}$ [1, 3]

---

**Turnover number** (min$^{-1}$)
1000 [1, 3]

**Specific activity** (U/mg)
13.9 [3]

**$K_m$-value** (mM)
0.04 (phosphonoacetaldehyde) [1, 3]

**pH-optimum**
8–9 [3]

**pH-range**

**Temperature optimum** (°C)
45 [3]

**Temperature range** (°C)

---

## 3 ENZYME STRUCTURE

**Molecular weight**
83000 (gel filtration (Sephadex G-150), Bacillus cereus) [3]
68000 (gel filtration (polyacrylamide gels), Bacillus cereus) [3]

**Subunits**
Dimer (2 × 33000–37000, Bacillus cereus) [3]

**Glycoprotein/Lipoprotein**
–

---

## 4 ISOLATION/PREPARATION

**Source organism**
Bacillus cereus [1–3]; Bacteria [1]

**Source tissue**

**Localisation in source**

**Purification**
Bacillus cereus [1, 3]

Crystallization
–

Cloned
–

Renaturated
–

---

## 5 STABILITY

**pH**
7–8 (higher stability at pH 7–8 than at pH 9–10) [3]

**Temperature (°C)**
More (heat labile) [3]

**Oxidation**

**Organic solvent**

**General stability information**

**Storage**
–20°C, Tris-MgCl$_2$-dithiothreitol buffer, several months [3]

---

## 6 CROSSREFERENCES TO STRUCTURE DATABANKS

**PIR/MIPS code**

**Brookhaven code**

---

## 7 LITERATURE REFERENCES

[1] Olsen, D.B., Hepburn, T.W., Moos, M., Mariano, P.S., Dunaway-Mariano, D.:
    Biochemistry, 27, 2229–2234 (1988)
[2] La Nauze, J.M., Coggins, J.R., Dixon, H.B.F.: Biochem. J., 165, 409–411 (1977)
[3] La Nauze, J.M., Rosenberg, H., Shaw, D.C.: Biochim. Biophys. Acta, 212, 332–350
    (1970)